ENCYCLOPEDIA OF
FISHES

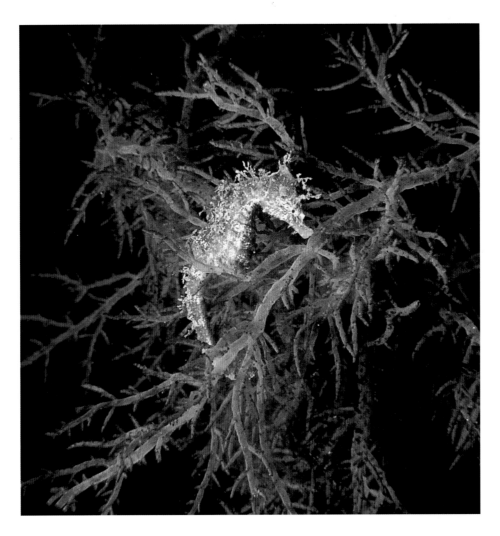

ENCYCLOPEDIA OF
FISHES

CONSULTANT EDITORS
Dr John R. Paxton
Australian Museum, Sydney, Australia
&
Dr William N. Eschmeyer
California Academy of Sciences, San Francisco, USA

ILLUSTRATIONS BY
Dr David Kirshner

Published in the United States by
Academic Press, A Division of Harcourt Brace & Company
525 B Street, Suite 1900, San Diego, CA 92101-4495, USA
Distributed worldwide exclusively by Academic Press
(except in Australia and New Zealand)

Senior Editor, Life Sciences: Charles R. Crumly Ph.D.

Conceived and produced by Weldon Owen Pty Limited
59 Victoria Street, McMahons Point, NSW 2060, Australia
A member of the Weldon Owen Group of Companies
Sydney • San Francisco

First published in 1994
Second edition 1998
Copyright © Weldon Owen Pty Limited 1998

Publisher: Sheena Coupe
Associate Publisher: Lynn Humphries
Project Editors: Jenni Bruce; Helen Cooney
Editorial Assistant: Veronica Hilton
Captions: David Kirshner; Terence Lindsey
Index: Garry Cousins
Designers: Denese Cunningham; Sue Rawkins
Picture Research: Esther Beaton; Annette Crueger
Production Manager: Caroline Webber
Production Assistant: Kylie Lawson
Vice President International Sales: Stuart Laurence

Co-ordination of scientific and editorial contributors by
Linda Gibson, Project Manager, Australian Museum Business Services

ISBN 0-12-547665-5

A catalog record for this book is available from
the Library of Congress, Washington, DC.

Printed by Kyodo Printing Co. (Singapore) Pte Ltd
Printed in Singapore

A WELDON OWEN PRODUCTION

Endpapers: A cloud of purple anthias *Pseudanthias tuka* mills about a coral reef in the
Bismarck Sea. Photo by David Doubilet
Page 1: Besides their bizarre shape, seahorses, such as *Hippocampus ramulosus*,
also differ from most fishes in their prehensile tails. Photo by Sophie Wilde/
AUSCAPE International
Pages 2–3: A scalloped hammerhead shark *Sphryna lewini* in the waters of the
Galapagos Islands. Photo by Doug Perrine/AUSCAPE International
Pages 4–5: *Rhinopias aphanes*, New Guinea. Most scorpionfishes have highly
venomous dorsal spines.
Page 7: Sport fisherman's delight, the rainbow trout *Oncorhynchus mykiss* has been
introduced to fresh waters worldwide.
Pages 10–11: Stingrays *Dasyatis* in the shallows. Named for their venomous tail
spines, stingrays have an almost worldwide distribution.
Pages 12–13: A pair of pink anemonefish *Amphiprion perideraion* peer from the
safety of their sea-anemone host. Photo by Kevin Deacon/Dive 2000
Pages 54–55: Ominous in silhouette, a gray reef shark (genus *Carcharhinus*) patrols
a coral reef. Photo by Marty Snyderman

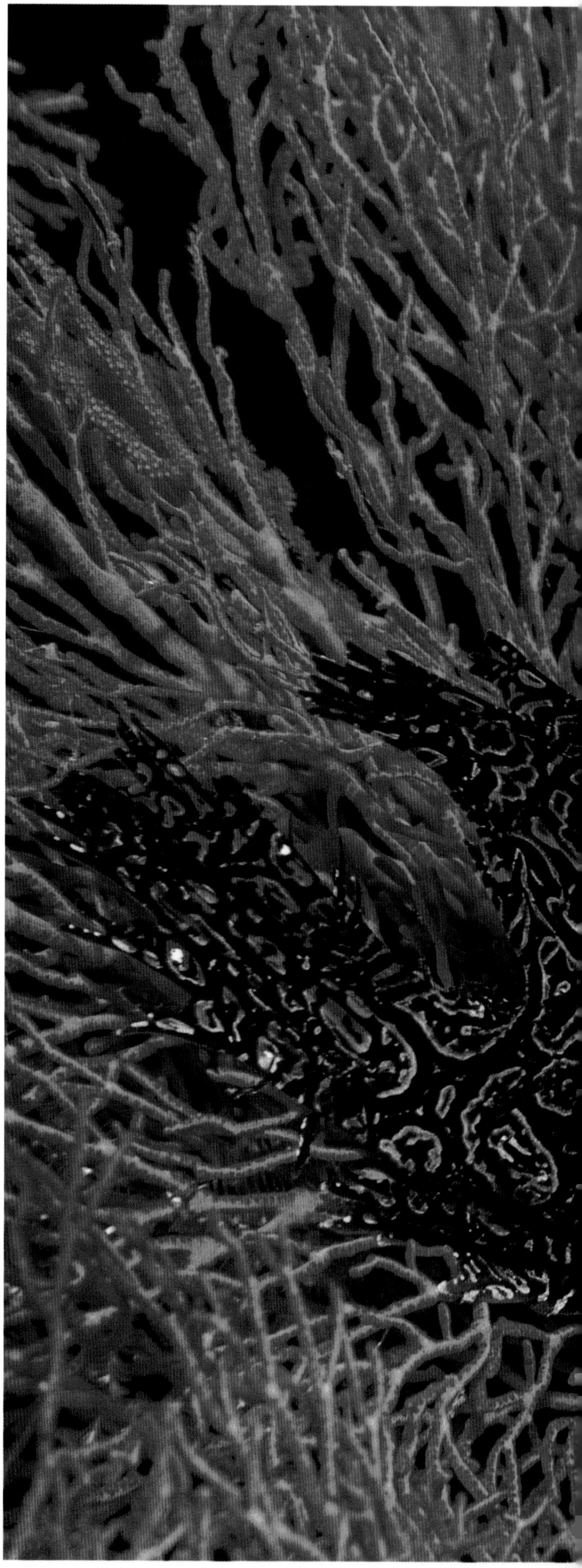

Carl Roessler

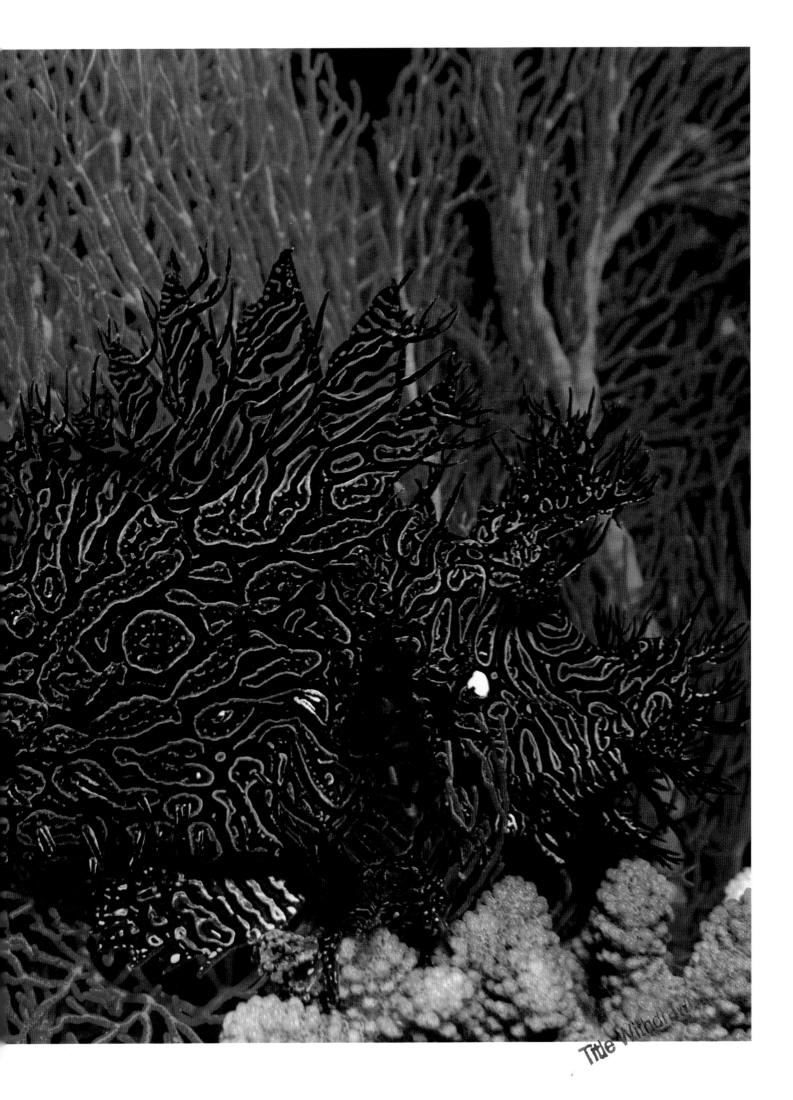

Title Withdrawn

CONSULTANT EDITORS & CONTRIBUTORS

DR JOHN R. PAXTON
Research Fellow,
Australian Museum, Sydney, Australia

DR WILLIAM N. ESCHMEYER
Senior Curator,
California Academy of Sciences, San Francisco, USA

CONTRIBUTORS

DR GERALD R. ALLEN
Senior Curator,
Western Australian Museum, Perth, Australia

DR PATRICIA C. ALMADA-VILLELA
Consultant in Conservation and
Aquatic Biology Co-chair,
IUCN SSC Coral Reef Fish Specialist Group,
Cambridge, UK

DR KUNIO AMAOKA
Professor of Marine Zoology,
Hokkaido University, Japan

DR KEITH E. BANISTER
Consultant and Visiting Research Fellow,
University of Kent, UK

DR DAVID R. BELLWOOD
Senior Lecturer, Marine Biology Department,
James Cook University, Australia

DR. E. BERTELSEN †
Zoological Museum,
University of Copenhagen, Denmark

PROFESSOR JOHN C. BRIGGS
Museum of Natural History,
University of Georgia, USA

PROFESSOR MICHAEL N. BRUTON
Director, Two Oceans Aquarium,
Cape Town, South Africa

DR FRANÇOIS CHAPLEAU
Associate Professor, Department of Biology,
University of Ottawa,
Ontario, Canada

DR J. HOWARD CHOAT
Professor, Marine Biology Department,
James Cook University, Australia

DR DANIEL M. COHEN
Deputy Director Emeritus,
Natural History Museum, Los Angeles, USA

DR BRUCE B. COLLETTE
Director, National Systematics Laboratory,
National Marine Fisheries Service,
Washington D.C., USA

DR CARL J. FERRARIS, JR
Research Associate,
California Academy of Sciences,
San Francisco, USA

DR WILLIAM L. FINK
Professor of Biology and Curator of Fishes,
University of Michigan, USA

DR ANTHONY C. GILL
Researcher, Fishes,
The Natural History Museum,
London, UK

DR LANCE GRANDE
Curator, Department of Geology,
Field Museum of Natural History,
Chicago, USA

DR P. HUMPHRY GREENWOOD †
JLB Smith Institute of Ichthyology,
Grahamstown, South Africa

DR DOUGLASS F. HOESE
Chief Scientist,
Australian Museum, Sydney, Australia

DR P. ALEXANDER HULLEY
Deputy Director, South African Museum,
Cape Town, South Africa

DR J. BARRY HUTCHINS
Head, Department of Aquatic Zoology,
Western Australian Museum, Perth, Australia

DR HANS-CHRISTIAN JOHN
Scientist,
"Taxonomische Arbeitsgruppe" der Biologischen
Anstalt Helgoland,
Hamburg, Germany

DR G. DAVID JOHNSON
Chairman, Department of Vertebrate Zoology,
National Museum of Natural History
Smithsonian Institution
Washington D.C., USA

DR ROBERT K. JOHNSON
Professor of Biology,
University of Charleston,
Charleston, USA

DR CHRISTINE KARRER
Technician, Institute für Seefischerei,
Hamburg, Germany

DR PETER R. LAST
Senior Research Scientist, Marine Laboratories,
Commonwealth Scientific and Industrial Research
Organization, Hobart, Australia

DR KAREL F. LIEM
Curator of Ichthyology, Professor of Biology,
Harvard University, USA

DR JOHN E. MCCOSKER
Chair of Aquatic Biology,
California Academy of Sciences,
San Francisco, USA

DR ROBERT M. MCDOWALL
Research Scientist,
National Institute of Water and
Atmospheric Research,
Christchurch, New Zealand

DR KEIICHI MATSUURA
Chief Curator,
National Science Museum,
Tokyo, Japan

DR GARETH NELSON
Senior Associate, School of Botany,
University of Melbourne, Australia

DR JØRGEN G. NIELSEN
Assistant Professor, Curator of Fishes,
Zoological Museum,
University of Copenhagen, Denmark

DR JOHN E. OLNEY
Assistant Professor,
Virginia Institute of Marine Sciences, USA

DR JAMES WILDER ORR
Research Zoologist,
Alaska Fisheries Science Center,
National Marine Fisheries Service,
Seattle, USA

DR LYNNE R. PARENTI
Curator,
National Museum of Natural History,
Smithsonian Institution,
Washington D.C., USA

DR NIKOLAY V. PARIN
Head of Ichthyological Laboratory,
P. P. Shirshov Institute of Oceanology,
Russian Academy of Sciences,
Moscow, Russia

DR JULIA K. PARRISH
Research Assistant Professor,
Zoology Department,
University of Washington, Seattle, USA

DR THEODORE W. PIETSCH
Professor and Curator of Fishes,
University of Washington, Seattle, USA

DR IAN C. POTTER
Professor of Animal Biology,
Murdoch University, Perth, Australia

DR JOHN E. RANDALL
Senior Ichthyologist Emeritus, Bishop Museum,
Hawaii, USA

DR VICTOR G. SPRINGER
Curator, National Museum of National History,
Smithsonian Institution,
Washington D.C., USA

DR JOHN STEVENS
Principal Research Scientist,
Marine Laboratories,
Commonwealth Scientific and
Industrial Research Organization,
Hobart, Australia

DR MELANIE L.J. STIASSNY
Curator and Chair,
American Museum of Natural History,
New York, USA

DR JAMES C. TYLER
Senior Scientist,
National Museum of Natural History,
Smithsonian Institution,
Washington D.C., USA

DR RICHARD P. VARI
Curator, National Museum of Natural History,
Smithsonian Institution,
Washington D.C., USA

DR STANLEY H. WEITZMAN
Curator, National Museum of Natural History
Smithsonian Institution,
Washington D.C., USA

DR EDWARD O. WILEY
Curator of Fishes,
Museum of Natural History, and
Professor of Systematics and Ecology,
University of Kansas, USA

DR MARK V. H. WILSON
Professor, Department of Biological Sciences,
University of Alberta, Canada

CONTENTS

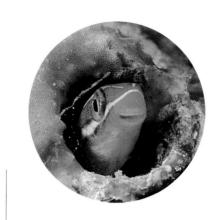

FOREWORD

Fishes form the earliest and the greatest group of all the backboned animals. They can be found throughout the vast ocean realm, from tropical to polar waters, and from the well-lit surface to the total dark and huge pressure of the deepest trenches, nearly 11 kilometers (7 miles) down. Evolving in the sea, they have invaded rivers and lakes many times during their evolutionary history, and many can live in both salt and fresh water.

Fishes are remarkably various in size, shape, color, physiology, and behavior. They are vertebrates like ourselves, yet we can find species that can change sex in a matter of days. There are species where all individuals are females and males are unnecessary, for young are produced without the eggs being fertilized. In one species of shark, cannibalism inside the mother among her voracious babies usually results in only one large individual being born. There are fishes that mimic the background, a floating seagrass leaf, or a poisonous fish. There is parental care of many different kinds—including the delightful little seahorse male brooding his young in a neat pouch. There are fishes that can walk on land and fishes that can glide for hundreds of meters on gossamer wings.

Fishes rival the birds and butterflies of terrestrial habitats: on a coral reef, two hundred species may live together within a very small area, each playing out its separate role in complex, intermeshing systems. At the brutal end of the fish scale are massive sharks, capable of severing a human leg with one powerful bite or smashing the solid head of a thigh bone as though it had been struck with a heavyweight hammer. But, as is often so in life, the real giants are gentle: sharks that feed on plankton and very small fishes, harmless to humans, yet reaching lengths of over 16 meters (50 feet) and many tonnes in weight. At the other extreme is the smallest of all living backboned animals, a delicate, transparent goby, only as long as the last joint of your little finger, weighing perhaps one-millionth the weight of the largest sharks. The variety of fishy form and living pattern seems endless. My own enthusiasm for fishes came from living beside the sea as a child and peering into tidal pools, watching the inhabitants going about their different and intriguing ways. May this book aid the enjoyment and understanding of the wonderful world of fishes for you, your children, and your grandchildren, for years to come.

FRANK H. TALBOT

Adjunct Professor,
Graduate School of the Environment,
Macquarie University, Sydney, Australia

D. Doubilet

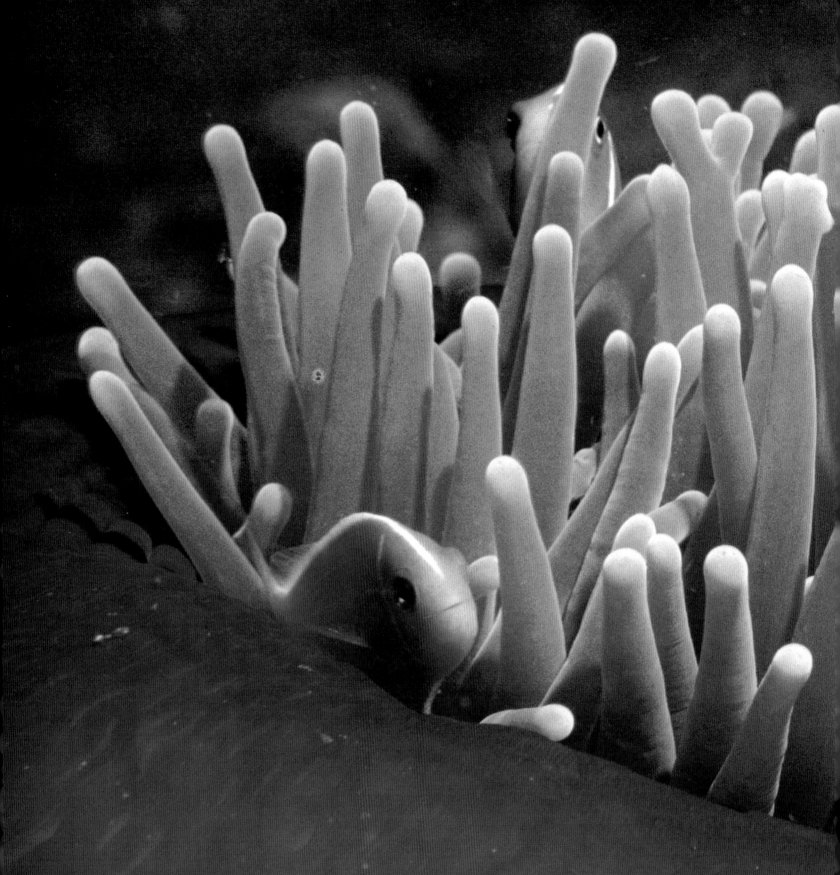

PART ONE
THE WORLD

OF FISHES

INTRODUCING FISHES

S cientists divide the animal kingdom into several major groups for classification purposes. By far the largest group is the invertebrates: it contains about 95 percent of the millions of known species of animals, including sponges, mollusks, crustaceans, and insects. With some 24,000 species, fishes are overwhelmingly the predominant vertebrates. They have succeeded in occupying and exploiting every kind of habitat in the hydrosphere. A few species of tiny catfishes, called candiru, have even found a niche inside the gill cavity of larger fishes where they feed on the blood of the gill filaments. In size, fishes range from diminutive gobies of 1 centimeter (²/₅ inch) to giant whale sharks and basking sharks, which grow to at least 12 meters (40 feet). The widths of the Pacific and Atlantic oceans are no barriers to tunas and salmon; their streamlined bodies and large muscle masses are perfectly adapted for covering large distances. Flexible and finely controllable fins enable fishes to execute a wide spectrum of precise swimming movements, including hovering and swimming backwards. Some fins are used for gliding distances up to several hundred meters, while other fins are used as limbs under water as well as on land. The possession of fins immediately differentiates fishes from other vertebrates. Only a few fishes have lost their fins.

FINS, MUSCLES, AND SWIMMING

Fishes swim in water, a medium that is 800 times denser than air. Typically they oscillate the tail fin and body surfaces across the line of motion, exerting backward thrusts that propel the fish forward. Most fishes have streamlined bodies rounded at the center and tapering toward the head and tail. They depend on rapid oscillations of the tail fin for movement through the dense medium of water. Because swimming movements demand large muscles, swimming muscles make up 40 to 65 percent of the body weight of a fish.

The oscillations of the fish's body are produced by the contractions of a series of muscles, the myotomes, that lie on either side of the backbone. These contractions result in a gentle bending of the body in a concave wave that deepens as it moves from head to tail. Swimming is a coordinated switching of waves of contraction and relaxation from one side of the body muscles to the other. Undulatory movements of the body reach an extreme level in eels, whose tail fins are rather reduced so that all the thrust comes from the exaggerated waves that travel down the body.

In sharp contrast, in tunny-like fishes such as tuna, no undulatory waves travel down the body. The tunny's immense body muscles pull on the stiff forked, sickle-shaped tail fin which, because of its dynamic structure, does not produce turbulence and drag when oscillating in the water. Consequently all the power generated by the body muscles results in thrust, with minimal loss as drag. All fast-swimming fishes have sickle-shaped tail fins. Most fishes are spindle-shaped and fall in the middle part of the spectrum between the two extremes of the eels and the tunnies.

The development of versatile, finely controllable fins has been a significant feature in fish evolution, and many fishes swim by means other than body and tail for oscillations. Wrasses, parrotfishes, and surgeonfishes use their pectoral fins as oars, with the tail fin acting simply as a rudder or stabilizer. Undulations down the long dorsal and anal fins can provide forward and backward thrust, as in triggerfishes, seahorses, and pipefishes. Side to side sculling actions of the dorsal and anal fins, in conjunction with paddling strokes of the pectoral fins, are the preferred mode of swimming of trunkfishes. Sharks use their pectoral fins as hydroplanes. The cambered pectoral fins function like the wings of an aircraft, giving the shark a lift

David Fleetham/Oxford Scientific Films

▲ *A blue-spotted stingray Taeniura lymma, Red Sea. The graceful, flapping, almost birdlike flight of rays through the water is quite different from the swimming actions of other fishes.*

▶ *A generalised bony fish showing major morphological features. Fishes vary widely in size, position, and arrangement of fins and other appendages, and no single species has all the features shown.*

▶ *Sharks differ most obviously from bony fishes in their multiple gill slits and in the outstretched set of their pectoral fins, used as hydroplanes rather than oars.*

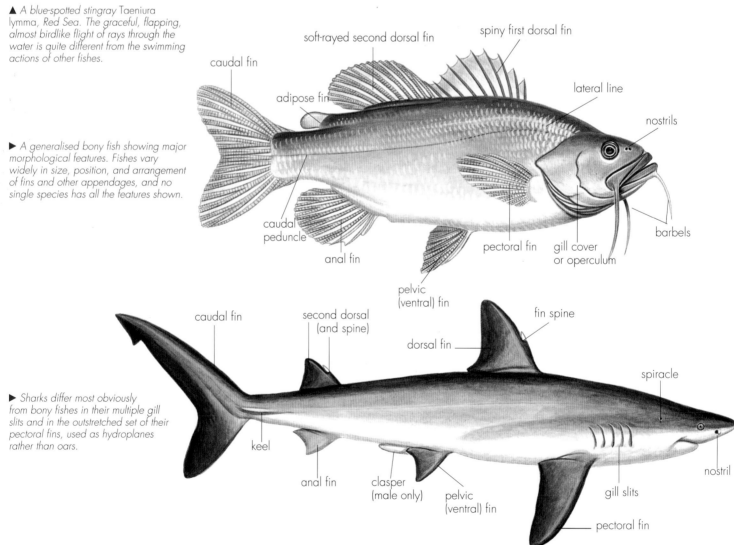

as it swims forward; this offsets its tendency to sink, as the shark's body is heavier than sea water. For large sharks, the ability to maintain height in the water is particularly important since they feed in the ocean's surface layers.

NEUTRAL BUOYANCY

Unlike sharks, many bony fishes (the teleosts) are endowed with weightlessness in water, or neutral buoyancy. Once neutral buoyancy is achieved, the fish gains greater freedom to maneuver and the pectoral fins no longer have to function as fixed hydroplanes. Many teleost fishes have a large gas-filled bladder (often called a swim bladder) in their body cavity, which eliminates the weight of the fish in water. Neutrally buoyant fishes can hover in the water and swim with much less energy. However, there is only one depth at which a fish has neutral buoyancy, so it is important for the fish to be able to regulate the amount of gas in the swim bladder to maintain neutral buoyancy at different depths. When a fish dives, the swim bladder is compressed by the increase in the surrounding pressure and the fish becomes heavier than water. To regain neutral buoyancy it must inflate the bladder to the proper volume. When the fish surfaces, the swim bladder will expand as the surrounding pressure decreases, and the fish must deflate the swim bladder to compensate for the reduced pressure. Many fishes with swim bladders have a gas-producing mechanism that keeps the swim bladder at just the right buoyancy volume. The ability of fishes to maintain neutral buoyancy is a decisive factor in their exploitation of all parts of the hydrosphere. However, buoyancy would be a real hindrance for bottom dwelling fishes as they exploit living spaces on rocks, sand, and mud, and in crevices. Most bottom dwelling fishes lack swim bladders.

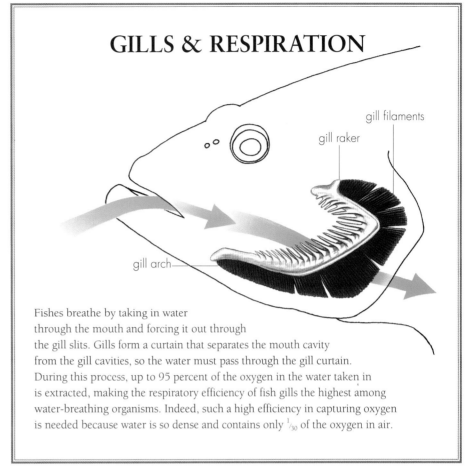

GILLS & RESPIRATION

gill filaments

gill raker

gill arch

Fishes breathe by taking in water through the mouth and forcing it out through the gill slits. Gills form a curtain that separates the mouth cavity from the gill cavities, so the water must pass through the gill curtain. During this process, up to 95 percent of the oxygen in the water taken in is extracted, making the respiratory efficiency of fish gills the highest among water-breathing organisms. Indeed, such a high efficiency in capturing oxygen is needed because water is so dense and contains only $\frac{1}{30}$ of the oxygen in air.

FEEDING

Feeding on prey and other food items in the dense medium of water poses special challenges. Many fish species have a diet of algae, insect larvae, worms, snails, clams, mussels, shrimps, crabs, plankton, fishes, amphibians, feces, and swarms of insects that have been blown into the water. Most fishes have a conical mouth cavity with a relatively

Georgette Dowma/Planet Earth Pictures

▲ *Almost every conceivable form of foraging behavior is found among fishes. Here a striped butterflyfish* Chaetodon fasciatus *uses its highly modified tube-like jaws to peck at the legs of a starfish trapped in a Triton's trumpet shell* Charonia tritonis.

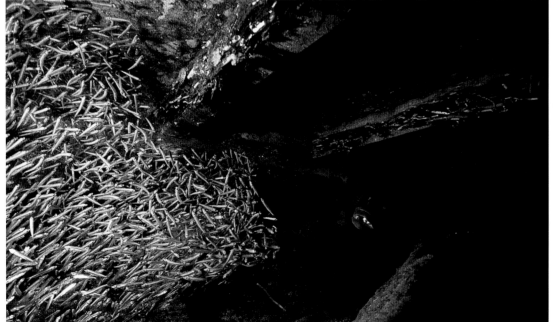

Marty Snyderman

◀ *A hunter in action: a black grouper* Mycteroperca bonaci *(family Serranidae) stalks silversides (family Atherinidae) from the shelter of a jetty in the Caribbean.*

SCALES

placoid

ganoid

ctenoid

cycloid

▨ = embedded part of scale

▲ The four kinds of fish scales. Tooth-like, or placoid, scales are characteristic of sharks, while sturgeons and some other armored fishes have hard, ganoid scales. Most bony fishes have tough scales interleaved like tiles on a roof; these may be either serrated (ctenoid) or smooth (cycloid) along the rear margins.

small mouth opening in front and an expandable rear compartment. As in a pipette, the sudden expansion of the rear compartment draws in water by suction. Prey and food suspended in the water are drawn in with the in-rushing stream. Thus, in suction feeding, the fish becomes stationary just before the feeding act, and the prey is sucked into the mouth by the sudden expansion of the rear compartment of the fish's mouth cavity. Fishes can generate enormous suction at an amazingly high speed. The ability to exploit a broad spectrum of food and prey by suction feeding is a hallmark of fishes.

The limiting factor for suction feeding is the size of the prey. When the prey is too large to be sucked in, it can be captured by those fishes that have very large mouth openings and cylindrically shaped mouth cavities. Fishes with such features cannot feed by suction. Instead they approach the prey at high speed, open their large mouths really wide and simply overtake the prey. This strategy is known as ram feeding. Ram feeders usually have specially designed body shapes and fins that enable them to accelerate very rapidly.

Some fishes, among them biters and algae scrapers, make use of their teeth in collecting the food. Some swallow food whole, while others, such as the parrotfishes, masticate extensively with molar teeth on jaws located in their throat.

SENSE ORGANS

As fishes exploit fresh waters that range from torrents to still or sluggish waters, and sea waters from surf-swept shores to the pitch dark abyss, they have evolved a great diversity of sense organs to analyze their contrasting environments. These sense organs take advantage of the physical and chemical properties of water.

Lateral Line

The lateral line organs are receptors for mechanical disturbances in water. In their simplest form these sense organs consist of a lump of jelly, the cupula, which rests on sensory cells or hair cells and projects from the surface of the fish. Such sense organs are called neuromasts. When a current of water bends the cupula, which contains tiny hairlike structures, it is perceived by the hair cells. However, in most cases these neuromasts are enclosed in canals that lie just below the surface of the body. Lateral line canals run along each side of the body, with a system of branching canals strategically positioned on the head. Each canal has a series of openings to the surface and contains a neuromast between each opening and the next. The cupulae project in the cavity of the canal. With their lateral line organs, fishes can perceive even the smallest disturbances in the water due to moving objects. They use the lateral line as "distance-touch sense" in locating prey and finding their way among obstacles.

Ears and amplification devices

Sound waves are pressure changes that can be perceived by a fish's inner ear. Although most fishes have limited perception of sound, fishes with connections between the ear and the swim bladder can hear extremely well. Water in contact with a gas bubble vibrates with greater amplitude than it would if there were no bubble. Since a fish's swim bladder is filled with gas, it acts as a sound amplifier and improves the hearing of the fish. Characins, carps, and minnows hear extremely well. A series of small bones, known as the Weberian apparatus, link the swim bladder with the ear. Fishes with a Weberian apparatus hear a wider frequency of sounds, have much sharper hearing, and can discriminate pitch more effectively. They are able to detect enemies and potential prey with this acute hearing, and can communicate with each other by producing various sounds.

Electric organs and receptors

Water conducts electric currents very well and many fish have evolved sense organs that can detect even the weakest electric fields. These electroreceptors are modified lateral line organs. Most conspicuous among them are those that occur in the head region of the sharks. They appear as pores filled with a jelly which has an electrical conductivity twice as high as that of the body fluids. This jelly is therefore capable of concentrating the electrical current. Sharks can detect the electrical potentials generated by swimming muscles and the heartbeat of fishes and other animal prey. They can even perceive the electromagnetic field of the earth. They use this sensory ability in homing and migratory movements. Many teleost fishes also possess jelly-filled organs that can detect electric currents and fields.

Electric-generating organs in fishes are modified muscle cells that have lost their ability to contract and serve only to generate electrical potentials. In most cases they are modified swimming muscles. Most electric fishes have only weak electric organs but the electric eel, electric ray, and electric catfish can produce several hundred volts of electricity in brief bursts. Fishes that generate these powerful electric bursts use them to stun prey and deter enemies. Weak electric fishes use their electric organs to communicate with their own kind and in contests with other species, much as birds use song.

REPRODUCTION

Fishes reproduce in more diverse ways than any other group of organisms. In most species, males and females are born as such. Some—salmon, for example—reproduce in the "big-bang" fashion, spawning only once, but in a spectacular way, after which they die. Others are repeat seasonal

spawners, and among these we can find a spectrum ranging from those that take care of their relatively small brood, to those that engage in group broadcast spawning in the surface waters, abandoning their numerous offspring to the mercy of water currents and countless predators. Some species lay millions of eggs, the ocean sunfish up to 28 million per season, while others lay as few as a dozen eggs. Most fishes are oviparous—the eggs are fertilized externally and hatch outside the mother. However, several families, including many sharks and the surfperches (family Embiotocidae), are viviparous. Fertilization is internal and the embryos undergo prolonged gestation periods of up to 10 months within the uterus or modified ovary of the mother.

In many fish families (such as wrasses, parrotfishes, and swamp eels) every individual is born as a female and males are produced by sex reversal from female to male. This phenomenon is known as protogyny. In yet other families (for example, clownfishes) each individual is born as a male and females appear by the sex reversal of males, a process called protandry. In the protogynous bass-like *Anthias*, as well as in the protandrous clownfishes, sex reversal is triggered by the nature of social interactions within the community. One of the killifish species, the molly *Poecilia formosa*, is all female. Reproduction proceeds by gynogenesis. The female molly

seduces an unsuspecting male from another species. His sperm is used merely to trigger the development of the molly's eggs, after which it is disposed of. The male of the other species does not contribute anything genetically. The molly simply produces clones, exact replicas of herself.

KAREL F. LIEM

▲ *A prickly anglerfish* Echinophryne crassispina *(family Antennariidae) guards its eggs. Parental care in fishes varies widely, from entirely lacking in many species to extremely elaborate in others.*

◄ *Fish reproductive strategies include some of the most extraordinary to be found among vertebrates. This remarkable photograph reveals the moment when a female seahorse (family Syngnathidae) uses her ovipositor to deposit her eggs into the male's pouch, where they will be fertilized, incubated, hatched, and in due course emerge as miniature adults.*

CLASSIFYING FISHES

There are 24,300 known kinds of living fishes placed in some 4,300 genera estimated by the authors of this volume. There are species living in almost all aquatic habitats from the deep oceans to tropical forests and mountain streams. In this book we classify fishes into 59 orders and 490 families. Although these figures indicate there are many kinds of fishes, they say little about the great degree of anatomical, functional, and behavioral diversity among them. Fishes are an extremely diverse group of animals and the difficult task of placing them in a classification that has evolutionary meaning is far from finished. Although the present classification is transitory, it should help make the great diversity of fishes evident.

FISH CLASSIFICATION

This book is organized mainly around a traditional classification, Nelson's *Fishes of the World,* which is based only partly on evolutionary history. This classification will change in the future as revisions based on hypothetical genealogies become more common. The term "fishes", as used here, is a common name for a group of aquatic animals arranged into four classes: Myxini (hagfishes), Cephalaspidomorphi (lampreys), Chondrichthyes (sharks, rays, and relatives), and Osteichthyes (bony fishes).

The species of each class are structurally quite different from those of the other classes and the evolutionary interrelationships of the classes are poorly understood. The concept "fishes", to which most of us are accustomed, lacks evolutionary meaning because it does not encompass a natural group, that is, one including all its descendants from a common ancestor. For example, in the conventional fish classification adopted here one of the major classes, Osteichthyes or bony fishes, excludes the vertebrate descendants of fishes (amphibians, reptiles, birds, and mammals) known as tetrapods. This kind of exclusion, still found in most classifications, is an artificial concession to traditional conventions of human thought that obscures certain aspects of evolutionary history.

Even though the term "fishes" does not represent a natural group, some of the included subgroups, for example, certain orders, families, and genera, are natural groups. Such groups have evolutionary meaning and are historically definable within a classification system.

Unfortunately whether many of the orders, families and literally hundreds of the genera are natural groups remains to be investigated by diligent, detailed comparative studies of their anatomical, biochemical, molecular and genetic structures. This is a major remaining task for systematic ichthyologists.

WHAT ARE FISH SPECIES?

A popular definition of a species among biologists, the biological species concept, defines species as groups of interbreeding natural populations of organisms which are actually or potentially reproductively isolated from other such populations. Another definition, the more recently conceived evolutionary species concept, states that a species is a single lineage of ancestor descendant populations that maintains its identity from other such lineages and has its own evolutionary

▼ *A blue and yellow ribbon-eel Rhinomuraena quaesita (family Muraenidae). Eels differ substantially from other fishes in a number of behavioral and anatomical features, including long sinuous body shape, lack of pelvic fins, and a larval stage; they are placed in a distinct order, Anguilliformes.*

Max Gibbs/Oxford Scientific Films

tendencies and history. These definitions are brief general predictions about what biologists think species may be like in nature as living entities in time. Unfortunately these definitions require some information unavailable from the population samples on which systematic biologists must base their species descriptions. For example, the first definition requires information about potential reproductive isolation. This is almost always unavailable. The second definition does not require potential or actual intrinsic genetic isolation but does not deny its existence. The second definition accepts, for example, simple geographical isolation as sufficient for isolating populations of different species. Most formal species descriptions in effect predict that the preserved population samples on which species descriptions are based represent populations in nature with the characteristics covered by one of these species definitions. In fact, population structure of organisms in nature often may be more complex than these definitions would seem to indicate.

Two italicized Greek or Latin names constitute the scientific name of a species. The first name, always with its first letter capitalized, is a generic name. When applied to more than one species it indicates a hypothetical evolutionary or phylogenetic relationship among the species bearing it. The second name, always with its first letter in lower case, is used to designate a particular species and must be different for each species placed in the same genus. Ideally scientific names such as *Cyprinus carpio*, the carp, are universally used throughout the world by systematists so that there can be no confusion about the species being discussed.

Common or vernacular names for a particular species often vary from country to country or locality to locality. Often the same common name is used for more than one species, for example the name "carp" is applied to several large species, some related to *Cyprinus carpio*, and some not.

Sometimes geographically isolated, or partly isolated, and somewhat different, but apparently closely related, populations of one species are identified as subspecies. Subspecies have a third italicized Greek or Latin word added to the scientific name. In fishes at least there has been a tendency in recent years not to recognize subspecies and to only recognize all truly distinct populations as separate species. Barely or mildly distinct populations are often discussed but not given scientific names.

▼ *White-collar butterflyfish Chaetodon collare, Maldives. This colorful species belongs to the order Perciformes, the largest group of fishes. The group contains about 9,000 species, nearly one-half of the world's total. Most live in the sea rather than freshwater, but otherwise the perciform fishes represent a huge diversity of color, pattern, shape and size.*

D. Parer & E.Parer-Cook / AUSCAPE International

▲ *A male weedy seadragon*
Phyllopteryx taeniolatus with eggs (left)
and a leafy seadragon Phycodorus
eques in kelp (right). These two
Australian species have features that
indicate each is more closely related to
a different genus of pipefishes. Their
similarities are the result of parallel
evolution in similar environments.

THE NATURE OF CLASSIFICATION

Classifications serve two major purposes. The first
is to describe the world's species (living and
extinct) so that we can know the diversity of the
world's organisms and can have names for species
of special interest such as food fishes. The second
is to provide hierarchical summaries of evolutionary
knowledge of the world's species that allows
humans to understand biological history and
diversity. It provides a comparative historical basis
for the study of all biological disciplines that are
best interpreted in a comparative context. In
addition to a basic human aesthetic enjoyment in
knowledge of biological diversity, classifications
are fundamental for humans to act wisely in the
use and treatment of life on this planet. Without
classifications, biological knowledge would be a
series of disconnected facts obscuring not only our
own evolutionary history but much of our ability
to use biological knowledge for conservation,
medicine, agriculture, and industry.

The nature of classification and especially the
methods for studying evolutionary relationships
among organisms has undergone theoretical
improvement in recent years. This improvement,
called cladistics or phylogenetics, recognizes
successive levels of a classification, such as a
genus, family, order, or class, as indicating
hierarchical degrees of evolutionary relationships
based on all species sharing descent from a
common ancestor. Another, older classification
system, called phenetic, is based simply on
similarity among taxa and ignores the concept of
lineages. Thus phenetic classifications are
sometimes historically misleading when there has
been evolutionary convergence leading to
similarity of anatomical or other biological
attributes. Another "system" of classification,

called evolutionary classification, attempts to
incorporate both the phenetic and the cladistic
information in one classification but when these
two approaches to classification are in conflict, a
common occurrence, the evolutionary
classification becomes internally inconsistent and
fails to indicate the genealogical aspect of
evolution and thus does not live up to its name.
Most classifications, like the one adopted here, are
combinations of cladistic and "evolutionary"
because the task of converting to cladistic
classifications is far from complete.

HOW ARE FISH CLASSIFICATIONS CONSTRUCTED ?

Historically the methods employed in devising
classifications were in principal quite simple.
Similar looking fishes were assumed to be related
and placed together in a genus, family or other
group and, in times before Darwin's ideas about
evolution, the groups had no evolutionary
meaning. Once it was understood that a
classification could have evolutionary significance,
attempts were made to interpret pre-evolutionary
classifications in terms of evolution.

Accompanying evolutionary interpretations of
classifications came a more critical evaluation of
the meaning of similarity. It was perceived that
only those similarities that represented homologies
were indicators of evolutionary relationship. In its
simplest form homology refers to a feature (for
example an anatomical, biochemical or genetic
character) shared by two or more species or groups
of species. Such a feature, if homologous, must be
inherited from an immediate common ancestor of
those species. Relationships based on homologies
are called phylogenetic. Homologous characters
are interpreted as indicative of common ancestry.

Howard Hall/Oxford Scientific Films

▲ *Like many other sharks, the sand tiger shark* Carcharius taurus *is built for speed, with a sleek, streamlined silhouette—an adaptation to a cruising, long-range, open-water lifestyle.*

▼ *Vastly different in lineage but similar in life style, the tunas (genus* Thunnus) *resemble sharks in being high-speed predators of the open ocean, and show correspondingly similar features of shape and structure—an example of convergent evolution.*

Constructing phylogenies and then classifications based on phylogenies is a search for many homologies shared by taxa that will allow the construction of hierarchies among species. We can only hypothesize that characters are homologous. There always is the possibility that some similar characters may be convergent, that is similarities that have evolved independently, without having an immediate common ancestor (a phenomenon called convergent evolution). Possible homologous characters have been examined for their potential use in phylogenetic reconstruction in at least two ways. Are the similarities superficial (for example in the comparison of bird wings and butterfly wings) indicating convergence or are they similar in many details (for example the hands of humans and monkeys) indicating evolutionary relationship? The process of cladistic analysis is considerably more complex than explained here but its results are a phylogenetic diagram or tree representing hypothetical evolutionary relationships based on lineages sharing common ancestors. Once a phylogenetic tree is found it can then be converted into a linear or sequential classification designed to convey the information on phylogeny with as little obfuscation as possible.

SCIENTIFIC NAMES

We have provided a few comments on the zoological aspects of classification and the use of scientific names. There is also a series of rules and guides for the application of zoological names called The International Code of Zoological Nomenclature which deals with problems of coining and publishing new scientific names and establishes rules regarding the priority of scientific names in cases where two or more such names have been applied to one species. The ultimate aim of these rules and guidelines is to provide stable scientific nomenclature for the world's fauna and a greater understanding of the world's organisms.

STANLEY H. WEITZMAN

Yves Gladu/AUSCAPE International

CLASSES, ORDERS, AND FAMILIES OF LIVING (OR RECENT) FISHES

The following list is mostly based on *Fishes of the World* (3rd edition) by Joseph S. Nelson, 1994, a Wiley-Interscience publication, as well as some more recent publications, with changes by the various authors of each chapter. Alternative common names are given in parentheses.

SUPERCLASS AGNATHA	*JAWLESS FISHES*	ORDER	
CLASS MYXINI	**HAGFISHES**	RAJIFORMES	SKATES
		Rajidae	Skates
ORDER			
MYXINIFORMES	HAGFISHES	ORDER	
MYXINIDAE	Hagfishes	PRISTIFORMES	SAWFISHES
		Pristidae	Sawfishes
CLASS			
CEPHALASPIDOMORPHI	**LAMPREYS**	ORDER	
	& ALLIES	TORPEDINIFORMES	ELECTRIC RAYS
		Torpedinidae	Electric rays
ORDER		Hypnidae	Coffinrays
PETROMYZONTIFORMES	LAMPREYS	Narcinidae	Narcinids
PETROMYZONTIDAE	Northern lampreys	Narkidae	Narkids
GEOTRIIDAE	Pouched lampreys		
MORDACIIDAE	Shorthead lampreys	ORDER	
		MYLIOBATIFORMES	STINGRAYS
SUPERCLASS GNATHOSTOMATA	*JAWED FISHES*		**& ALLIES**
CLASS CHONDRICHTHYES	**CARTILAGINOUS**	Potamotrygonidae	River rays
	FISHES	Dasyatididae	Stingrays
		Urolophidae	Round stingrays
SUBCLASS ELASMOBRANCHII	SHARKS & RAYS	Gymnuridae	Butterfly rays
		Hexatrygonidae	Sixgill rays
ORDER		Myliobatididae	Eagle rays
HEXANCHIFORMES	SIX- & SEVENGILL	Rhinopteridae	Cownose rays
	SHARKS	Mobulidae	Mantas, devil rays
Chlamydoselachidae	Frill shark		
Hexanchidae	Sixgill, sevengill	SUBCLASS HOLOCEPHALI	CHIMAERAS
	(cow) sharks		
		ORDER	
ORDER		CHIMAERIFORMES	CHIMAERAS
SQUALIFORMES	DOGFISH SHARKS	Callorhynchidae	Plownose chimaeras
	& ALLIES		(elephant fishes)
Echinorhinidae	Bramble sharks	Chimaeridae	Shortnose chimaeras
Squalidae	Dogfish sharks		(ghost sharks, ratfishes)
Oxynotidae	Roughsharks	Rhinochimaeridae	Longnose chimaeras
			(spookfishes)
ORDER			
PRISTIOPHORIFORMES	SAW SHARKS	**CLASS OSTEICHTHYES**	**BONY FISHES**
Pristiophoridae	Saw sharks		
		SUBCLASS SARCOPTERYGII	LUNGFISHES AND COELACANTH
ORDER			(FLESHY-FINNED FISHES)
HETERODONTIFORMES	HORN (BULLHEAD)		
	SHARKS	ORDER	
Heterodontidae	Horn (bullhead) sharks	CERATODONTIFORMES	LUNGFISHES
		Ceratodontidae	Australian lungfish
ORDER			
ORECTOLOBIFORMES	CARPET SHARKS	ORDER	
	& ALLIES	LEPIDOSIRENIFORMES	LUNGFISHES
Parascylliidae	Collared carpet sharks	Lepidosirenidae	South American lungfish
Brachaeluridae	Blind sharks	Protopteridae	African lungfishes
Orectolobidae	Wobbegongs		
Hemiscylliidae	Longtail carpet sharks	ORDER	
Stegostomatidae	Zebra shark	COELACANTHIFORMES	COELACANTH
Ginglymostomatidae	Nurse sharks	Latimeriidae	Coelacanth
Rhincodontidae	Whale shark		(gombessa)
ORDER		SUBCLASS ACTINOPTERYGII	RAY-FINNED FISHES
LAMNIFORMES	MACKEREL SHARKS		
	& ALLIES	ORDER	
Odontaspididae	Sand tigers	POLYPTERIFORMES	BICHIRS
	(grey nurse sharks)	Polypteridae	Bichirs
Mitsukurinidae	Goblin shark		
Pseudocarchariidae	Crocodile shark	ORDER	
Megachasmidae	Megamouth shark	ACIPENSERIFORMES	STURGEONS
Alopiidae	Thresher sharks		& ALLIES
Cetorhinidae	Basking shark	Acipenseridae	Sturgeons
Lamnidae	Mackerel sharks	Polyodontidae	Paddlefishes
	(makos)		
		ORDER	
ORDER		LEPISOSTEIFORMES	GARS
CARCHARHINIFORMES	GROUND SHARKS	Lepisosteidae	Gars
Scyliorhinidae	Catsharks		
Proscylliidae	Finback catsharks	ORDER	
Pseudotriakidae	False catshark	AMIIFORMES	BOWFIN
Leptochariidae	Barbled houndshark	Amiidae	Bowfin
Triakidae	Houndsharks		
Hemigaleidae	Weasel sharks	GROUP TELEOSTEI	TELEOSTS
Carcharhinidae	Requiem (whaler) sharks		
Sphyrnidae	Hammerhead sharks	ORDER	
		HIODONTIFORMES	MOONEYES
ORDER		Hiodontidae	Mooneyes
SQUATINIFORMES	ANGEL SHARKS		
Squatinidae	Angel sharks	ORDER	
		OSTEOGLOSSIFORMES	BONYTONGUES,
ORDER			ELEPHANTFISHES
RHINOBATIFORMES	SHOVELNOSE RAYS		& ALLIES
Platyrhinidae	Platyrhinids	Osteoglossidae	Bonytongues
Rhinobatidae	Guitarfishes	Pantodontidae	Freshwater butterflyfish
Rhynchobatidae	Sharkfin guitarfishes	Notopteridae	Featherbacks

Mormyridae	Elephantfishes		
Gymnarchidae	Aba		
ORDER			
ELOPIFORMES	TARPONS		
	& ALLIES		
Elopidae	Tenpounders		
	(ladyfishes)		
Megalopidae	Tarpons		
ORDER			
ALBULIFORMES	BONEFISHES		
	& ALLIES		
Albulidae	Bonefishes		
ORDER			
NOTACANTHIFORMES	SPINY EELS & ALLIES		
Halosauridae	Halosaurs		
Notacanthidae	Spiny eels		
ORDER			
ANGUILLIFORMES	EELS		
Anguillidae	Freshwater eels		
Heterenchelyidae	Shortfaced eels		
Moringuidae	Spaghetti eels		
Chlopsidae	False moray eels		
Myrocongridae	Myrocongrid		
Muraenidae	Moray eels		
Nemichthyidae	Snipe eels		
Muraenesocidae	Pike eels		
Synaphobranchidae	Cutthroat eels		
Ophichthidae	Snake eels,		
	worm eels		
Nettastomatidae	Duckbill eels		
Colocongridae	Colocongrids		
Congridae	Conger eels,		
	garden eels		
Derichthyidae	Narrowneck eels,		
	spoonbill eels		
Serrivomeridae	Sawtooth eels		
Cyematidae	Bobtail snipe eels		
Saccopharyngidae	Swallowers		
Eurypharyngidae	Gulper eels		
Monognathidae	Singlejaw eels		
ORDER			
CLUPEIFORMES	SARDINES,		
	HERRINGS,		
	ANCHOVIES		
	& ALLIES		
Denticipitidae	Denticle herring		
Clupeidae	Sardines, herrings,		
	pilchards		
Engraulididae	Anchovies		
Chirocentridae	Wolf herrings		
ORDER			
GONORYNCHIFORMES	MILKFISH,		
	BEAKED SALMONS		
	& ALLIES		
Chanidae	Milkfish		
Gonorynchidae	Beaked salmons		
Kneriidae	Shellears		
Phractolaemidae	Hingemouth		
ORDER			
CYPRINIFORMES	CARPS, MINNOWS		
	& ALLIES		
Cyprinidae	Carps, minnows		
Psilorhynchidae	Psilorhynchids		
Balitoridae	Hillstream loaches		
Cobitidae	Loaches		
Gyrinocheilidae	Algae eaters		
Catostomidae	Suckers		
ORDER			
CHARACIFORMES	CHARACINS		
	& ALLIES		
Citharinidae	Citharinids		
Distichodontidae	Distichodontins,		
	African fin eaters		
Hepsetidae	African pike characin		
Erythrinidae	Trahiras and allies		
Ctenoluciidae	American pike characins		
Lebiasinidae	Voladoras, pencil fishes		
Characidae	African tetras (characins),		
	characins, tetras, brycons,		
	piranhas & allies		
Curimatidae	Toothless characins		
Prochilodontidae	Flannelmouth characins		

Anostomidae	Anostomins, leporinins
Hemiodontidae	Hemiodontids
Chilodontidae	Headstanders
Gasteropelecidae	Freshwater hatchetfishes

ORDER
SILURIFORMES — **CATFISHES & KNIFEFISHES**

Suborder Siluroidei

Diplomystidae	Velvet catfishes
Nematogenyidae	Mountain catfish
Trichomycteridae	Spinyhead catfishes, candirus
Callichthyidae	Plated catfishes
Scoloplacidae	Spinynose catfishes
Astroblepidae	Andes catfishes
Loricariidae	Armored suckermouth catfishes
Ictaluridae	Bullhead catfishes, madtoms
Bagridae	Bagrid catfishes
Cranoglanididae	Armorhead catfishes
Siluridae	Sheatfishes, wels, glass catfishes
Schilbidae	Schilbid catfishes
Pangasiidae	Pangasiid catfishes
Amblycipitidae	Torrent catfishes
Amphiliidae	Loach catfishes
Akysidae	Asian banjo catfishes
Sisoridae	Hillstream catfishes
Clariidae	Labyrinth (airbreathing) catfishes
Heteropneustidae	Airsac catfishes
Chacidae	Square head (angler) catfishes
Olyridae	Olyrid catfishes
Malapteruridae	Electric catfishes
Ariidae	Sea catfishes
Plotosidae	Eeltail catfishes
Mochokidae	Upsidedown catfishes (squeakers)
Doradidae	Thorny catfishes
Auchenipteridae	Wood catfishes
Pimelodidae	Longwhisker, shovelnose catfishes
Ageneiosidae	Bottlenose (barbelless) catfishes
Helogenidae	Helogene catfishes
Cetopsidae	Carneros, shark (whale) catfishes
Hypophthalmidae	Loweye catfishes
Aspredinidae	Banjo catfishes

Suborder Gymnotoidei

Sternopygidae	Longtail knifefishes
Rhamphichthyidae	Longsnout knifefishes
Hypopomidae	Bluntnose knifefishes
Apternotidae	Black knifefishes
Gymnotidae	Banded knifefishes
Electrophoridae	Electric eel (electric knifefish)

ORDER
ARGENTINIFORMES — **HERRING SMELTS, BARRELEYES & ALLIES**

Argentinidae	Herring smelts
Microstomatidae	Deepsea smelts
Opisthoproctidae	Barreleyes (spookfishes)
Alepocephalidae	Slickheads
Bathylaconidae	Bathylaconids
Platytroctidae	Tubeshoulders

ORDER
SALMONIFORMES — **SALMONS, SMELTS & ALLIES**

Osmeridae	Northern smelts, ayu
Salangidae	Icefishes (noodlefishes)
Retropinnidae	Southern smelts
Galaxiidae	Galaxiids
Lepidogalaxiidae	Salamanderfish
Salmonidae	Trouts, salmons, chars, whitefishes

ORDER
ESOCIFORMES — **PIKES, PICKERELS & ALLIES**

Esocidae	Pikes
Umbridae	Mudminnows

ORDER
ATELEOPODIFORMES — **JELLYNOSE FISHES**

Ateleopodidae	Jellynose fishes

ORDER
STOMIIFORMES — **DRAGONFISHES, LIGHTFISHES & ALLIES**

Gonostomatidae	Lightfishes, bristlemouths
Sternoptychidae	Hatchetfishes and allies
Phosichthyidae	Lightfishes
Stomiidae	Viperfishes, dragonfishes, snaggletooths, loosejaws

ORDER
AULOPIFORMES — **LIZARDFISHES & ALLIES**

Aulopidae	Aulopids
Bathysauridae	Bathysaurids
Chlorophthalmidae	Greeneyes
Ipnopidae	Ipnopids, tripod fishes
Synodontidae	Lizardfishes, Bombay duck
Scopelarchidae	Pearleyes
Notosudidae	Waryfishes
Giganturidae	Telescope fishes
Paralepididae	Barracudinas
Anotopteridae	Daggertooth
Evermannellidae	Sabertooth fishes
Omosudidae	Omosudid
Alepisauridae	Lancetfishes
Pseudotrichonotidae	Pseudotrichonotids

ORDER
MYCTOPHIFORMES — **LANTERNFISHES & ALLIES**

Neoscopelidae	Neoscopelids
Myctophidae	Lanternfishes

ORDER
LAMPRIDIFORMES — **OARFISHES & ALLIES**

Lamprididae	Opahs
Veliferidae	Velifers
Lophotidae	Crestfishes
Radiicephalidae	Inkfishes
Trachipteridae	Ribbonfishes
Regalecidae	Oarfishes
Stylephoridae	Tube eye

ORDER
POLYMIXIIFORMES — **BEARDFISHES**

Polymixiidae	Beardfishes

ORDER
PERCOPSIFORMES — **TROUTPERCHES & ALLIES**

Percopsidae	Troutperches
Aphredoderidae	Pirate perch
Amblyopsidae	Cavefishes

ORDER
GADIFORMES — **CODS, HAKES & ALLIES**

Muraenolepididae	Muraenolepidids
Moridae	Morid cods (moras)
Melanonidae	Melanonids
Euclichthyidae	Euclichthyid
Bregmacerotidae	Codlets
Gadidae	Codfishes, haddocks & allies
Merlucciidae	Hakes
Steindachneriidae	Steindachneriid
Macrouridae	Rattails (grenadiers)

ORDER
OPHIDIIFORMES — **CUSKEELS, PEARLFISHES & ALLIES**

Ophidiidae	Cuskeels, brotulas & allies
Carapidae	Pearlfishes
Bythitidae	Livebearing brotulas
Aphyonidae	Aphyonids

ORDER
BATRACHOIDIFORMES — **TOADFISHES & MIDSHIPMEN**

Batrachoididae	Toadfishes & midshipmen

ORDER
LOPHIIFORMES — **ANGLERFISHES, GOOSEFISHES & FROGFISHES**

Lophiidae	Goosefishes (monkfishes)
Antennariidae	Frogfishes (shallow anglerfishes)
Tetrabrachiidae	Humpback angler
Lophichthyidae	Lophichthyid
Brachionichthyidae	Warty anglers (hand fishes)
Chaunacidae	Seatoads (gapers, coffinfishes)
Ogcocephalidae	Batfishes
Caulophrynidae	Fanfin anglers
Ceratiidae	Seadevils
Gigantactinidae	Whipnose anglers
Neoceratiidae	Needlebeard angler
Linophrynidae	Netdevils
Oneirodidae	Dreamers
Thaumatichthyidae	Wolftrap angler
Centrophrynidae	Hollowchin anglers
Diceratiidae	Double anglers
Himantolophidae	Footballfishes
Melanocetidae	Blackdevils

ORDER
GOBIESOCIFORMES — **CLINGFISHES & ALLIES**

Gobiesocidae	Clingfishes
Callionymidae	Dragonets
Draconettidae	Draconettids

ORDER
CYPRINODONTIFORMES — **KILLIFISHES & ALLIES**

Rivulidae	South American annuals
Aplocheilidae	African annuals
Profundulidae	Profundulids
Fundulidae	Killifishes
Valenciidae	Valenciids
Anablepidae	Foureyed fishes (cuatro ojos)
Poeciliidae	Livebearers (guppies, etc)
Goodeidae	Goodeids
Cyprinodontidae	Pupfishes

ORDER
BELONIFORMES — **RICEFISHES, FLYINGFISHES & ALLIES**

Adrianichthyidae	Ricefishes
Exocoetidae	Flyingfishes
Hemiramphidae	Halfbeaks
Belonidae	Needlefishes
Scomberesocidae	Sauries

ORDER
ATHERINIFORMES — **SILVERSIDES, RAINBOWFISHES & ALLIES**

Atherinidae	Silversides, topsmelts, grunions
Dentatherinidae	Dentatherinid
Notocheiridae	Surf silversides
Melanotaeniidae	Rainbowfishes
Pseudomugilidae	Blue eyes
Telmatherinidae	Sailfin silversides
Phallostethidae	Priapium fishes

ORDER
STEPHANOBERYCIFORMES — **PRICKLEFISHES, WHALEFISHES & ALLIES**

Stephanoberycidae	Pricklefishes
Melamphaidae	Bigscale fishes (ridgeheads)
Gibberichthyidae	Gibberfishes
Hispidoberycidae	Bristlyskin
Rondeletiidae	Orangemouth whalefishes
Barbourisiidae	Redvelvet whalefish
Cetomimidae	Flabby whalefishes
Mirapinnidae	Hairyfish, tapetails
Megalomycteridae	Mosaicscale (bignose) fishes

ORDER
BERYCIFORMES — **SQUIRRELFISHES & ALLIES**

Holocentridae	Squirrelfishes (soldier fishes)
Berycidae	Alfoncinos
Anoplogasteridae	Fangtooth fishes
Monocentrididae	Pineapple (pinecone) fishes
Anomalopidae	Flashlight (lanterneye) fishes
Trachichthyidae	Roughies (slimeheads)
Diretmidae	Spinyfins

ORDER ZEIFORMES — DORIES & ALLIES

Parazenidae	Parazen
Macrurocyttidae	Macrurocyttids
Zeidae	Dories
Oreosomatidae	Oreos
Grammicolepididae	Tinselfishes
Caproidae	Boarfishes

ORDER GASTEROSTEIFORMES — STICKLEBACKS, PIPEFISHES & ALLIES

Hypoptychidae	Sand eel
Aulorhynchidae	Tubesnouts
Gasterosteidae	Sticklebacks
Indostomidae	Paradox fish
Pegasidae	Seamoths
Aulostomidae	Trumpetfishes
Fistulariidae	Cornetfishes
Macroramphosidae	Snipefishes
Centriscidae	Shrimpfishes (razorfishes)
Solenostomidae	Ghost pipefishes
Syngnathidae	Pipefishes, seahorses

ORDER SYNBRANCHIFORMES — SWAMPEELS & ALLIES

Synbranchidae	Swampeels
Mastacembelidae	Spiny eels
Chaudhuriidae	Chaudhuriids

ORDER SCORPAENIFORMES — SCORPIONFISHES & ALLIES

Scorpaenidae	Scorpionfishes (rockfishes), stonefishes
Caracanthidae	Coral crouchers
Aploactinidae	Velvetfishes
Pataecidae	Prowfishes
Gnathanacanthidae	Red velvetfish
Congiopodidae	Pigfishes (horsefishes)
Triglidae	Sea robins (gurnards)
Dactylopteridae	Helmet (flying) gurnards
Platycephalidae	Flatheads
Bembridae	Deepwater flatheads
Hoplichthyidae	Spiny (ghost) flatheads
Anoplopomatidae	Sablefish, skilfish
Hexagrammidae	Greenlings
Zaniolepididae	Combfishes
Normanichthyidae	Southern sculpin
Rhamphocottidae	Grunt sculpin
Ereuniidae	Deepwater sculpins
Cottidae	Sculpins
Comephoridae	Baikal oilfishes
Abyssocottidae	Abyssocottids
Hemitripteridae	Hemitripterids
Psychrolutidae	Fatheads (blob fishes)
Bathylutichthyidae	Bathylutichthyids
Agonidae	Poachers
Cyclopteridae	Lumpfishes
Liparidae	Snailfishes, lumpsuckers

ORDER PERCIFORMES — PERCHES & ALLIES

Suborder Percoidei

Ambassidae	Glassfishes
Acropomatidae	Acropomatids
Epigonidae	Epigonids
Scombropidae	Gnomefishes
Symphysanodontidae	Symphysanodontids
Caesioscorpididae	Caesioscorpidids
Polyprionidae	Wreckfishes
Dinopercidae	Cavebasses
Centropomidae	Snooks
Latidae	Giant perches
Percichthyidae	South temperate basses
Moronidae	North temperate basses
Serranidae	Sea basses, groupers
Callanthiidae	Rosey perches
Pseudochromidae	Dottybacks
Grammatidae	Basslets
Opistognathidae	Jawfishes
Plesiopidae	Roundheads
Notograptidae	Bearded snakeblennies
Pholidichthyidae	Convict blennies
Cepolidae	Bandfishes
Glaucosomatidae	Pearl perches
Terapontidae	Grunters
Banjosidae	Banjosids
Kuhliidae	Aholeholes
Centrarchidae	Sunfishes
Percidae	Perches, darters
Priacanthidae	Bigeyes (catalufas)
Apogonidae	Cardinalfishes
Dinolestidae	Dinolestid
Sillaginidae	Smelt whitings (whitings)
Malacanthidae	Tilefishes
Labracoglossidae	Labracoglossids
Lactariidae	False trevallies
Pomatomidae	Bluefish (tailor)
Menidae	Moonfish
Leiognathidae	Ponyfishes (slipmouths)
Bramidae	Pomfrets
Caristiidae	Manefishes
Arripidae	Australian salmons
Emmelichthyidae	Rovers
Lutjanidae	Snappers (emperors)
Lobotidae	Tripletails
Gerreidae	Mojarras (silver biddies)
Haemulidae	Grunts
Inermiidae	Bonnetmouths
Sparidae	Porgies (breams)
Centracanthidae	Centracanthids
Lethrinidae	Scavengers (emperors)
Nemipteridae	Threadfin, monocle breams
Sciaenidae	Drums, croakers
Mullidae	Goatfishes
Monodactylidae	Moonfishes (fingerfishes)
Pempherididae	Sweepers
Leptobramidae	Beachsalmon
Bathyclupeidae	Bathyclupeids
Toxotidae	Archerfishes
Coracinidae	Galjoenfishes
Kyphosidae	Seachubs
Girellidae	Nibblers (blackfishes)
Scorpididae	Halfmoons, mado (sweeps)
Microcanthidae	Stripey
Ephippidae	Spadefishes
Scatophagidae	Scats
Chaetodontidae	Butterflyfishes
Pomacanthidae	Angelfishes
Enoplosidae	Oldwife
Pentacerotidae	Armorheads
Oplegnathidae	Knifejaws
Icosteidae	Ragfish
Kurtidae	Nurseryfishes
Scombrolabracidae	Scombrolabracid

Suborder Carangoidei

Nematistiidae	Roosterfish
Carangidae	Jacks (trevallies)
Echeneidae	Remoras
Rachycentridae	Cobia (black kingfish)
Coryphaenidae	Dolphins

Suborder Cirrhitoidei

Cirrhitidae	Hawkfishes
Chironemidae	Kelpfishes
Aplodactylidae	Aplodactylids
Cheilodactylidae	Morwongs
Latrididae	Trumpeters

Suborder Mugiloidei

Mugilidae	Mullets

Suborder Polynemoidei

Polynemidae	Threadfins

Suborder Labroidei

Cichlidae	Cichlids
Embiotocidae	Surfperches
Pomacentridae	Damselfishes, anemonefishes
Labridae	Wrasses
Odacidae	Rock whitings
Scaridae	Parrotfishes

Suborder Zoarcoidei

Bathymasteridae	Ronquils
Zoarcidae	Eelpouts
Stichaeidae	Pricklebacks
Cryptacanthodidae	Wrymouths
Pholididae	Gunnels
Anarhichadidae	Wolffishes
Ptilichthyidae	Quillfishes
Zaproridae	Prowfish
Scytalinidae	Graveldiver

Suborder Notothenioidei

Bovichtidae	Thornfishes
Nototheniidae	Nototheniids
Harpagiferidae	Plunderfishes
Bathydraconidae	Antarctic dragonfishes
Channichthyidae	Icefishes

Suborder Trachinoidei

Chiasmodontidae	Swallowers
Champsodontidae	Champsodontids
Trichodontidae	Sandfishes
Trachinidae	Weeverfishes
Uranoscopidae	Stargazers
Trichonotidae	Sanddivers
Creediidae	Sandburrowers
Leptoscopidae	Leptoscopids
Percophidae	Duckbills
Pinguipedidae	Sandperches
Cheimarrhichthyidae	Torrentfish
Ammodytidae	Sandlances

Suborder Blennioidei

Tripterygiidae	Triplefins
Dactyloscopidae	Sand stargazers
Labrisomidae	Labrisomids
Clinidae	Kelp blennies (kelpfishes)
Chaenopsidae	Tube blennies
Blenniidae	Combtooth blennies

Suborder Gobiodei

Rhyacichthyidae	Loach goby
Odontobutidae	Freshwater Asian gobies
Gobiidae	Gobies, sleepers (gudgeons)

Suborder Acanthuroidei

Siganidae	Rabbitfishes
Luvaridae	Louvar
Zanclidae	Moorish idol
Acanthuridae	Surgeonfishes (tangs)

Suborder Scombroidei

Sphyraenidae	Barracudas
Gempylidae	Snake mackerels, gemfishes
Trichiuridae	Cutlassfishes
Scombridae	Mackerels, tunas, bonitos
Xiphiidae	Swordfish
Istiophoridae	Billfishes

Suborder Stromateoidei

Amarsipidae	Amarsipids
Centrolophidae	Medusafishes
Nomeidae	Driftfishes
Ariommatidae	Ariommatids
Tetragonuridae	Squaretails
Stromateidae	Butterfishes

Suborder Anabantoidei

Badidae	Chameleonfish
Nandidae	Leaffishes
Pristolepidae	False leaffishes
Channidae	Snakeheads
Anabantidae	Climbing gouramies
Belontiidae	Gouramies
Helostomatidae	Kissing gourami
Osphronemidae	Giant gourami
Luciocephalidae	Pikehead

ORDER PLEURONECTIFORMES — FLATFISHES

Psettodidae	Spiny flatfishes
Citharidae	Citharids
Bothidae	Lefteye flounders
Pleuronectidae	Righteye flounders
Cynoglossidae	Tonguefishes (tongue soles)
Soleidae	Soles
Achiridae	American soles

ORDER TETRAODONTIFORMES — TRIGGERFISHES & ALLIES

Triacanthodidae	Spikefishes
Triacanthidae	Triplespines
Balistidae	Triggerfishes
Monacanthidae	Filefishes (leatherjackets)
Aracanidae	Robust boxfishes
Ostraciidae	Boxfishes, cowfishes, trunkfishes
Triodontidae	Pursefish (three toothed puffer)
Tetraodontidae	Pufferfishes (toados)
Diodontidae	Porcupinefishes
Molidae	Ocean sunfishes (molas)

FISHES THROUGH THE AGES

More than half of all living vertebrates are fishes, and it is probably a safe assumption that well over half of all extinct vertebrates were fishes. The earliest known fishes predate the earliest known animals with limbs (tetrapods) by more than 100 million years. The fossil record of fishes is generally better than that of non-aquatic vertebrates. Nearly all fossils are preserved within sedimentary rocks that originated in aquatic environments. Thus, it is generally more likely that a dead fish will end up being fossilized in reasonably good condition than a dead bird or land animal. The earliest fossils that may be the remains of fishes date back to the early Cambrian, or just over 500 million years ago. However, these early fossils are only scrappy fragments and some scientists have challenged whether they are vertebrate remains at all.

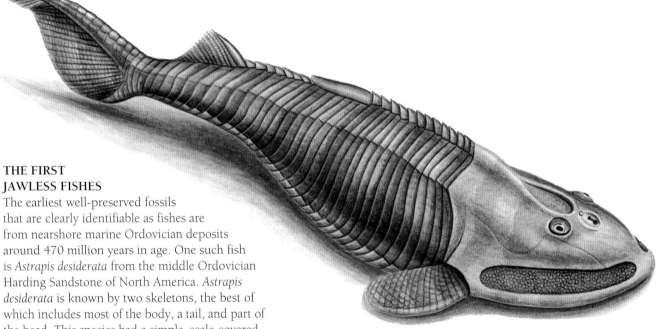

THE FIRST JAWLESS FISHES

The earliest well-preserved fossils that are clearly identifiable as fishes are from nearshore marine Ordovician deposits around 470 million years in age. One such fish is *Astrapis desiderata* from the middle Ordovician Harding Sandstone of North America. *Astrapis desiderata* is known by two skeletons, the best of which includes most of the body, a tail, and part of the head. This species had a simple, scale-covered body with a series of eight branchial openings in the head following a small opening for the eye.

These early known forms were all jawless with extensive body armor consisting of a solid exterior of bony plates and scales. We know very little about their internal skeletons; they possessed a cranium, but we do not know whether they all had vertebral elements. Most appear to have become extinct by the end of the Devonian (360 million years ago). Today, only two groups of jawless fishes survive: the hagfishes and the lampreys. Both these groups have a poor fossil record, probably because they had very few bony skeletal elements likely to be preserved as fossils. There are only two known fossil lampreys (both from the Carboniferous period, 360 to 285 million years ago) and no described fossil hagfish (although the first fossil species will be described shortly, from the early Carboniferous, about 360 million years ago).

During the Paleozoic (570 to 245 million years ago) there were several major groups of jawless fishes that are now extinct. One of these was the order Pteraspidiformes (Heterostraci). This order includes the earliest vertebrates and many other species in about 150 genera. This group is known only until the late Devonian. The earliest members were thought to be marine but some of the later species are preserved also in freshwater deposits. Most of the jawless fishes of the Devonian were small but pterapsids were an exception, sometimes reaching 1.5 meters (5 feet) in length. A number of other extinct orders of jawless fishes are also known from the Silurian (435 to 408 million years ago) and Devonian periods, including the orders Galeaspidiformes, Anaspidiformes, and Cephalaspidiformes. Galeaspidiforms are known

▲ *The Cephalaspidiformes were distant relatives of the living hagfishes and lampreys. Hemicyclaspis was a bottom-dweller with a solid bony shield enclosing its head. These fishes were jawless, had a single nostril, and possessed numerous gill pouches.*

by about 30 genera from the Devonian period of China. The predominantly freshwater anaspidiforms are from late Silurian to early Devonian deposits, although the greatest diversity of this group is from the Upper Silurian. Anaspidiforms, known by about 10 genera, have been reported from Europe and North America. Cephalaspidiforms, known from late Silurian to late Devonian deposits, are another group thought to have been predominantly freshwater, with 7 known families and about 50 described genera. Cephalaspidiforms and anaspidiforms are thought to be more closely related to lampreys than to the other groups of jawless fishes.

There are several other long-extinct groups of fishes that are somewhat obscure, such as the order Thelodontiformes and the genus *Paleospondylus*, and we are not even certain that they had no jaws. Thelodontiforms included about 5 families and 20 described genera and are well known only from the late Silurian and early Devonian. Isolated scales thought to be from thelodontiforms are also reported from Lower Silurian deposits. Fossils of this group have been found in Europe, Central America, North America, Spitsbergen (Arctic), Australia, and Antarctica. *Paleospondylus* is known from the Middle Devonian of Scotland.

THE FIRST JAWED FISHES

Gnathostomes, or jawed fishes, are first known from deposits of early Silurian age, about 100 million years after the first jawless fishes. The evolutionary relationships of the early groups of gnathostomes are not well understood.

The placoderms were one of the largest early groups. The head and shoulder girdles of these fishes were heavily armored with bony plates, and most species are known only by these plates; complete skeletons are rare. First known from the early Devonian, placoderms became quite diverse and geographically widespread throughout much of the Devonian, with over 35 families and 270 genera described. Some genera, such as *Dunkleosteus*, reached lengths of more than 2 meters (6½ feet), and were major predators of the Devonian seas. They appear to have become virtually extinct at the end of the Devonian period, with only one or two genera surviving into the early Carboniferous (360 million years ago). No placoderms are known past the Early Carboniferous.

CARTILAGINOUS FISHES

Another early group of jawed fishes are the cartilaginous fishes, class Chondrichthyes. This group survives today, represented by about 165 living genera and 960 living species of sharks, chimaeras, skates, and rays. Some scientists believe placoderms are more closely related to cartilaginous fishes than to bony fishes (class Osteichthyes), but evidence for this is weak. The

▼▶ Eastmanosteus calliaspis, a 370-million-year-old Devonian fossil placoderm. The head and trunk shield plates and even the jaws are present in this specimen, but the posterior two-thirds of the body, the fins and often the vertebrae are not fossilized. These striking fossils from the Gogo Formation of northwest Australia are preserved in the round in the football-sized nodules. The photo below shows a nodule partially broken open with the bones and plates in situ. The fossil formation consists of hills (fossil coral reefs) and the valleys (the reef lagoons) where the fish-containing nodules are lying on the ground.

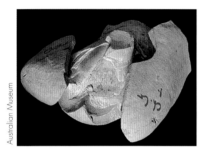

Australian Museum

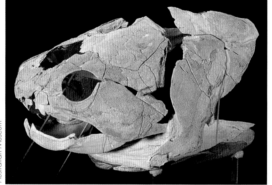

Australian Museum

▼ Placoderms were an early group of jawed fishes characterized by heavy armor on the head and shoulder girdle. One group (of Placoderms), the Arthrodires, consisted of large predatory species such as the monstrous Dunkleosteus, which attained a length of more than 2 meters.

earliest known chondrichthyan fossils are scale fragments from the late Silurian, but it was not until the Devonian that a major diversification of cartilaginous fishes occurred. Most of these groups appear to have become extinct by the end of the Paleozoic (245 million years ago). The diversification of modern chondrichthyan groups appears to have occurred in the Jurassic (200 to 145 million years ago) and Cretaceous (145 to 65 million years ago). Chondrichthyans appear always to have been predaceous marine fishes, as they are today.

Most fossil species of chondrichthyans are known only from fragmentary specimens, usually teeth and toothplates. The most common fossils are isolated teeth, and many taxonomic groups are named on the basis of teeth or toothplates only. This sometimes leads to taxonomic problems when more complete material is discovered. Several major fossil groups are also represented by at least a few well-preserved nearly complete skeletons, and much work is needed to determine more precisely their evolutionary relationships.

ACANTHODIANS

The name Teleostomi is sometimes used for a group containing Acanthodii and Osteichthyes. Acanthodians are another extinct early group of fishes. Well-preserved, nearly complete skeletons of acanthodians have been found in early Devonian to early Permian deposits, but their characteristic fin spines are known as far back as Lower Silurian (possibly even late Ordovician). The common name "spiny sharks" has often been used for these fishes because of the stout spines that strengthen the fins and the upwardly tilted tail. Although they superficially resemble sharks, these fishes have a few bony elements in their internal skeleton and are today usually grouped with the bony fishes (Osteichthyes). Most Silurian acanthodians are from preserved marine deposits, and most Devonian forms are from freshwater deposits. The group is geographically widespread (found even in Antarctica) and contains over 70 genera and 5 families. Acanthodians are the earliest known true jawed fishes, and in the past have been placed as the most primitive group of jawed fishes, or placed together with placoderms, or with chondrichthyans. Most recently they have been placed as the sistergroup to Osteichthyes.

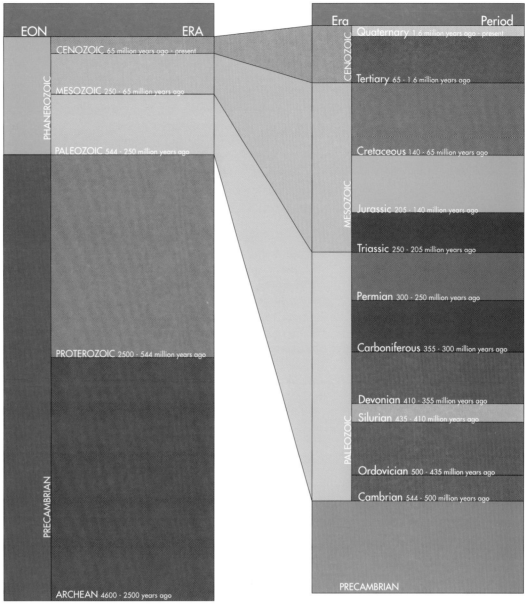

BONY FISHES

Osteichthyes, or the true bony fishes, consist of two major groups: the fleshy-finned fishes (Sarcopterygii) and the ray-finned fishes (Actinopterygii). Both groups are enormous, and contain nearly all vertebrate species. In addition to certain fishes, Sarcopterygii technically contains all tetrapods, giving it well over 20,000 living species. Actinopterygii contains the majority of living "fishes", and also comprises more than 20,000 living species. The fossil record of these groups is extremely large.

Fleshy-finned fishes

If we eliminate the tetrapods from Sarcopterygii and look only at the fishes, the group is relatively small, containing primarily lungfishes (subclass Dipneusti), rhipidistians (superorder Osteolepimorpha), and coelacanths (superorder Coelacanthimorpha). Some scientists do not consider coelacanths to be sarcopterygians but

▲ The geologic time scale. In North America the Carboniferous is usually divided into Mississippian (earlier) and Pennsylvanian (later). The Tertiary can be subdivided into five epochs (Paleocene, Eocene, Oligocene, Miocene, Pliocene) and the Quaternary into two (Pleistocene and Recent).

rather the sister-group to a group containing both the lobe-fins and the ray-finned fishes. The relationships of various other sarcopterygian groups are still quite controversial.

Lungfishes Today there are only six species of lungfishes in three genera, but there are many more fossil forms, including well over 60 described genera. Lungfishes were once widespread geographically, and were apparently at their greatest diversity during the Devonian. The earliest known lungfish fossils are from Lower Devonian deposits of China. Although living lungfishes are all freshwater fishes, fossil lungfishes have been found in both marine and freshwater deposits. Lungfish toothplates are fairly widespread in Devonian deposits. The evolutionary relationships of fossil lungfishes are not well understood. The ability to aestivate (spend the summer in a torpid condition) shown by living lungfishes is indicated also by the early Permian genus *Gnathorhiza*, frequently found preserved in burrows. Empty burrows which may also have been left by extinct species of lungfishes are known from the Carboniferous and possibly the Devonian.

and the Osteolepiformes. Osteolepiformes are typified by two of the best-known Devonian fossils, *Eusthenopteron* and *Osteolepis*. These Upper Devonian genera are known by numerous nearly complete skeletons, with many extremely well-preserved individuals. They were slender-bodied fishes with a pineal opening in the skull, and large fangs in the mouth.

Coelacanths The coelacanths (Actinistia or Coelacanthimorpha) are another major fleshy-finned group. They are represented today by only a single species, *Latimeria chalumnae,* but were once much more diverse. *Latimeria* is the classic example of a "living fossil"; it was not discovered until 1938, and the youngest known fossil coelacanth is Upper Cretaceous in age. Thus, before 1938 this group was thought to have been extinct for about 65 million years. The earliest coelacanth fossils are Middle Devonian in age, and by the later Paleozoic the group was widespread, with over 50 extinct genera. Although the living coelacanth is a fairly deepwater marine form (caught mostly between 150 and 300 meters, or 490 and 990 feet), the fossil varieties include freshwater and nearshore forms.

▼ *The earliest known group of true jawed fishes, the acanthodians are sometimes called spiny sharks but are currently classified with the bony fishes. Climatius possessed the characteristic spines on its fins, with accessory pairs of spines between its pectoral and pelvic fins, and was probably an active swimmer that fed on invertebrates by filter feeding.*

Rhipidistians The rhipidistians are a number of extinct early lobe-finned fishes, some of which are thought to be more closely related to tetrapods than to other fishes. The earliest known rhipidistians are from Lower Devonian sedimentary rocks of China, and the youngest is early Permian (245 million years ago). There are nearly 100 described genera of rhipidistians, and they were geographically widespread by the middle of the Devonian. During the late Paleozoic these fishes were the major bony-fish predators living in fresh water. Many paleontologists believe that some component of this group gave rise to the earliest tetrapods (amphibians). Two major monophyletic groups that have been identified within this "catchall" group are the Porolepiformes

Ray-finned fishes
The largest group of fishes is the Actinopterygii, or ray-finned fishes. This group contains about half of all living vertebrates. Like so many of the other fish groups discussed, the earliest known actinopterygians are late Silurian or early Devonian. The late Silurian record is based on isolated scales, and is therefore not as definitive as the more complete skeletons found in Devonian sedimentary rocks.

The actinopterygians have been classified into two groups: lower actinopterygians (a non-monophyletic group) and neopterygians. The evolutionary relationships of the earliest actino-pterygians with their few living survivors is very poorly understood and more research is needed.

Lower actinopterygians Living members of this non-monophyletic group include bichirs (family Polypteridae), sturgeons, and paddlefishes (order Acipenseriformes). The largest lower actinopterygian group is the extinct so-called Palaeonisciformes, a non-monophyletic group of more than 30 families and over 120 genera. This group includes the earliest actinopterygians, and is known from the late Silurian to the early Cretaceous. Although there are many well-preserved fossils in this group, such as the middle and late Devonian *Cheirolepis*, much work is still needed to clearly delineate even the most basic evolutionary relationships of these fishes. The bichirs are somewhat enigmatic: there is a huge gap in the fossil record and we therefore cannot tell when they diverged from the other actinopterygians. It is possible, however, that this seemingly large gap is due to our lack of knowledge about the evolutionary relationships of lower actinopterygians.

Neopterygians This brings us to the neopterygians, which includes most of the ray-finned fishes. Neopterygians are characterized by a number of unique features indicating that they form a natural (monophyletic) group. Several early neopterygian groups are now extinct; the oldest is the Semionotidae (first known from late Permian deposits). The largest group of neopterygians outside the Teleostei are the halecomorphs, with only a single surviving species, the bowfin *Amia calva*. Halecomorphs are first known from Lower Triassic deposits, and include about 50 genera. Other major groups of neopterygians that are now extinct include the order Pycnodontiformes (late Triassic to Eocene with over 40 genera) and the order Macrosemiiformes (late Triassic to early Cretaceous with about a dozen genera). Both groups have a number of very well-preserved species. The only other surviving non-teleost neopterygian group are the gars (order Lepisosteiformes), with 7 living species. Fossil gars are first known from Lower Cretaceous deposits. Their distinctive shiny rhombic-shaped scales are some of the most common identifiable fish fossils in Tertiary freshwater sedimentary rocks of North America (65 to 2 million years ago).

TELEOSTS

The largest neopterygian group, the Teleostei, contains over 96 percent of all living fishes. Living teleosts number about 23,000 living species in more than 4,000 genera, although estimates vary. These estimates will never be entirely accurate because in some areas (for example, the Amazon and Congo basins) new species are probably becoming extinct faster than they can be described by scientists. The earliest known teleosts (about 200 million years ago) are small Late Triassic and Early Jurassic fishes in the order Pholidophoriformes. Teleosts appear to have undergone a major diversification in the Cretaceous, which seems to have continued up to today where they now dominate the world's fish fauna. Scientists are still trying to understand how living species of some of the larger teleost groups, such as the perches and allies, are related to each other in order to group them with closely related fossil species.

Most of this chapter has emphasized extinct groups of fishes, but each kind of fish has its own history. Vast numbers of well-preserved fossil teleosts and other fishes are awaiting scientific analysis. The most productive way to study fishes through the ages is to include these fossils with their living relatives and analyze them all together. Only then can we evaluate the significance of ancient fishes and discover detailed information about the origin and evolution of modern fishes.

LANCE GRANDE

▼ *The Palaeonisciformes were a large group of now extinct lower actinopterygians whose relationship to the living sturgeons, paddlefishes and bichirs is poorly understood. Many of the best-preserved fossils belong to the genus Cheirolepis, a group of predatory fishes which possessed a single dorsal fin, a shark-like tail, small, tightly fitting ganoid scales, and a bony plated head with a large, tooth-studded mouth.*

HABITATS & ADAPTATIONS

Virtually every aquatic habitat is home to some fishes, from the torrential streams and lakes of the highest mountains (Lake Titicaca is at an elevation of 3,800 meters or 12,350 feet in the Andes) to the deepest freshwater lakes and widest rivers, from desert springs over 40°C (104°F) to supercooled Antarctic waters of -1.8°C (28°F), from coral reefs and open-ocean waters to the deepest ocean waters at the bottom of the Marianas Trench, where Piccard and Walsh saw through the porthole of their bathyscaphe *Trieste* a sole on the bottom at 10,933 meters (35,800 feet).

▼ *Sockeye salmon Oncorhynchus nerka leap rapids on their journey back to their spawning grounds. Many fishes migrate between different habitats. Some, for example, migrate from salt to fresh water and back. The chemical behavior of salts in water is such that water tends to flow inward through a fish's skin in freshwater, outward in the sea—an inexorable trend controlled by numerous physiological adaptations. The net result is that radical adjustments of body chemistry are necessary to enable a migratory fish to move between rivers and the sea.*

IN AND OUT OF WATER

A few fishes have even moved out of the water, at least temporarily. The oceanic flying fishes (family Exocoetidae) and freshwater butterflyfish (family Pantodontidae) fly or glide above the water for brief periods, while the mudskippers (family Gobiidae) make more extended forays out of water. Some lungfishes (genera *Protopterus* and *Lepidosiren*) and the salamander fish (genus *Lepidogalaxias*) are capable of aestivation—literally a summer hibernation, buried in the bottom sediments of dried-up ponds. The walking catfish (family Clariidae) is a well-known example of a fish that can leave the water for short periods, as are some freshwater eels (family Anguillidae). And of course those primitive fishes that evolved into amphibians some 360 million years ago in the Devonian left the water permanently.

Then there are those fishes that bury themselves partially or completely, at least temporarily, in bottom sediments. These range from the spectacular garden eels with waving heads, to rays and flatfishes with only eyes protruding, to some completely buried wrasses sleeping in safety.

FRESH WATERS

It is surprising that almost 40 percent of the total fish species (more than 8,000) live in fresh waters that cover about 1 percent of the Earth's surface and include only 0.01 percent of the Earth's water. Not so surprising is the fact that the two driest continents—Antarctica, with no living freshwater fishes (despite numerous fossils), and Australia, with only some 200 species—have the fewest freshwater fishes. Most freshwater species, like their marine counterparts, are found in the tropics. The poorly studied Southeast Asian region probably has the most freshwater fishes. The largest rivers of the world are home to a wide variety of fishes, and some 1,500 different species are found in the Amazon.

Streams and rivers

The headwaters of a mountain stream, with rocky bottom and cold, clear, and highly oxygenated waters that lack plankton, are vastly different from the typically warm, slow-moving, often turbid, plankton-rich and oxygen-depleted waters of a floodplain river near its mouth. Not surprisingly, the fishes of these two habitats are different, with differing adaptations. The trout (family Salmonidae) of mountain streams is a fast-swimming predator with keen eyesight and a need for low temperatures and high oxygen. The carp (family Cyprinidae) of rivers and ponds is a slow-swimming detritus feeder that utilizes taste buds on its barbels and has a tolerance for higher temperatures and lower oxygen levels.

The torrential mountain streams of Asia are home to the hillstream loaches (family Balitoridae), with both pectoral and pelvic fins formed into sucking disks to hold onto rocks. Other hillstream inhabitants include members of the carp family (family Cyprinidae) and a number of catfish families. Large sucking mouths and belly skin with folds and grooves are two of the adaptations for this extreme environment.

Ponds and lakes

Unlike rivers, most still waters are ephemeral as sooner or later, depending on size, they fill with sediment. As a result the fish species inhabiting most of these habitats are the same or very similar to those in the running water habitats. However, some lakes are older, providing enough time for distinct species, genera, and even families to evolve. The two striking examples are the Great Lakes of Africa and Lake Baikal in Siberia.

Lakes Malawi, Victoria, and Tanganyika in tropical Africa have perhaps 1500 different species and numerous genera of cichlids (family Cichlidae) that have evolved between 500,000 and 2 million years ago. The much older Lake Baikal, with an estimated age of 20 to 25 million years, is the world's oldest and deepest lake, with a depth of 1,637 meters (5,320 feet). Baikal has the seventh largest surface area and contains the largest volume of fresh water. This coldwater lake has more than 50 endemic species of fishes, including two endemic families (Comephoridae and Abyssocottidae) related to the sculpin family Cottidae.

Hot springs

At the other extreme of fresh waters are desert springs, where temperatures can reach 40°C (104°F) and more. The North American desert pupfishes (family Cyprinodontidae), an African cichlid (genus *Tilapia*), and the Australian spangled grunter (family Terapontidae) live in these extreme environments, usually swimming at the edges of the springs where the water has cooled down.

Caves

The underground waters found in caves are another specialized freshwater habitat. Species of a number of fish families have adapted to this lightless environment, including four catfish families in Africa and North and South America, barbs (family Cyprinidae) in Africa, cavefishes (family Amblyopsidae) and the cave characin (family Characidae) of North America, and the blind gudgeon (family Gobiidae) and blind cave eel (family Synbranchidae) of Australia. Three species of the marine cuskeel family Ophidiidae have entered cave waters in Cuba and Mexico.

Although they belong to unrelated families, cavefishes are typically blind, lack body pigment (and are a striking pink color), and often have sensory papillae and ridges on the skin. These sensory organs pick up pressure waves in the water and allow the cavefishes to detect both predators and moving food. The cavefishes of different families are a striking example of convergent evolution. It has been shown that the loss of eyes and pigment is not irreversible in some forms, but depends on whether these species develop in a light or dark environment.

Richard Kirby/Oxford Scientific Films

▲ *The golumyanka Comephorus baicalensis is the main diet of the nerpa seal, and both fish and seal are endemic to Russia's Lake Baikal. Oldest and deepest of the world's lakes, Baikal's total volume is so enormous that it stores about one-fifth of the world's fresh water. Two fish families are entirely restricted to the lake.*

▼ *Eyeless and lacking skin pigment, the blind gudgeon Milyeringa veritas is restricted to sinkholes and subterranean waters on Northwest Cape on the coast of Western Australia.*

G.E. Schmida

▲ *The floor of shallow or coastal seas is a rich environment, occupied by hundreds of fish species like this scorpionfish Scorpaena cardinalis nestled in kelp off New Zealand's Alderman Island.*

OCEANS AND SEAS

Marine waters cover 72 percent of the Earth's surface and, with a mean depth of some 3,700 meters (12,000 feet), include 97 percent of the planet's water. More than 13,000 fish species live in the many subdivisions of this vast environment.

Estuaries

Estuaries are the transition zones between fresh water and the sea. Accounting for only a small percentage of the aquatic habitat, estuaries typically have a broad range of both salinity and temperature. Their rich waters carry nutrients from land run-off down the rivers. Many estuaries are the nursery grounds for shore fishes, and the young of these species must have broad environmental tolerances.

Coral reefs

The tropical marine shore habitat has more fish species than any other. And the epitome of species richness is the coral reef habitat, where an estimated 4,000 fish species are found. Many families, like the gobies (family Gobiidae), wrasses (family Labridae), damselfishes (family Pomacentridae), groupers (family Serranidae), blennies (family Blenniidae), and butterflyfishes (family Chaetodontidae), have numerous species in close proximity, each specialized in its preferences for food and microhabitat. Scientists believe that the evolution of this diversity was enhanced by the relatively long period since the development of the modern coral reef environment dominated by scleractinian corals (some 50 million years ago in the early Tertiary),

◄ *Richest of all marine environments, the coral reefs of the tropical seas of the world support extraordinary numbers of colorful species, like these scalefin anthias* Pseudanthias squamipinnis *in the South Pacific.*

Kevin Deacon / AUSCAPE International

here, from the gobies and blennies of the tidepools, to the varied species found in the fish markets of the world.

The open ocean

The surface waters of the sea are a specialized environment where some of our best-known fishes live. Mako (family Lamnidae) and blue (family Carcharhinidae) sharks, tuna (family Scombridae), and billfishes (family Istiophoridae) are found here. These pelagic (oceanic) species are fast-swimming and well-muscled, important adaptations in an environment where there is no place to hide. A further specialization in some of the oceanic tunas and lamnid sharks is an arrangement of the blood vessels in the muscles in what is termed a counter-current heat exchange system to raise the body temperature and therefore muscular efficiency. All other fishes, like amphibians and reptiles, are poikilotherms—that is, they take on the temperature of their environment, unlike birds and mammals which can maintain a constant body temperature.

Although these large oceanic fishes have no place to hide, most have protective coloration—blue or dark gray on top and white or silvery underneath—to make them inconspicuous to potential predators. Other typical high-seas fishes include the oarfishes and allies (Order

▼ *Those fishes that inhabit the sun-lit upper levels of the open ocean are generally large, sleek, stream-lined and fast-moving, like this jack or trevally.*

D. Parer & E. Parer-Cook / AUSCAPE International

coupled with the many rises and falls in sea level associated with global warming and cooling and resultant ice cap fluctuations.

The vast majority of coral reef fish species are in the perch order Perciformes. This group is characterized by a highly protrusible mouth and anteriorly placed pelvic fins, allowing a specialized type of feeding behavior that involves hovering in place and picking small individual food items. Such coral reef fishes as damselfishes, butterfly-fishes, and some groupers feed in this way.

Marine shallows

Other shallow marine habitats include surf beaches, rocky headlands, and the shallow waters of the continental shelf down to 200 meters (660 feet). Most of the well-known marine fishes live

Lampridiformes) and the huge ocean sunfishes (family Molidae). Some smaller fishes, like the juveniles of some jacks and trevallies (family Carangidae) and the bluebottlefishes (family Nomeidae), have found a place to hide in the open ocean. They swim among the tentacles of jellyfish where they are protected by its stinging cells. Like the anemonefishes, they are not harmed by the lethal weapons of their hosts.

Antarctic seas

The freezing seas surrounding the continent of Antarctica are home to over 260 species of fishes. Dominant among these, and including more than half of the shore and bottom species, are five families of Antarctic fishes, the notothenioids, placed in their own suborder within the perches. Thought to be related to the blennies, they include the thornfishes (family Bovichtidae), plunderfishes (family Harpagiferidae), Antarctic dragonfishes (family Bathydraconidae), icefishes (family Channichthyidae), and the speciose notothens (family Nototheniidae).

Their adaptations are striking, and include the presence, in some of the midwater species, of large sacs of lipid (fat) to obtain neutral buoyancy (notothenioids lack a swim bladder). As sea water has a lower freezing point (-1.8°C / 28.8°F) than fish blood (-1.2°C / 29.8°F), a number of species have evolved glycoprotein antifreezes to allow them to survive in the coldest water. The icefishes have lost their hemoglobin, the oxygen-carrying pigment cells in the blood. The amount of oxygen dissolved in water increases with decreasing temperature; specialized cells like hemoglobin, which would increase blood viscosity, are not required in this oxygen-rich environment.

The deep sea

The deep sea can be defined either by the edge of the continental shelf or the depth in the clearest oceanic waters where so much light is absorbed that photosynthesis does not support phytoplankton. Both definitions place the upper limit of the deep sea at 150 to 200 meters (490 to 660 feet). With the deepest bottom at 10,900 meters (35,425 feet) and the average depth at 3,700 meters (12,000 feet), it is by far the largest environment on Earth. Yet only some 2,500 species—just about 10 percent of all fish species—live there.

Deepsea fishes are of two basic types: the pelagic fishes living in the midwaters (some 1,150 species) and the benthic fishes living on the bottom (about 1,500 species). Midwater fishes between 200 and 1,000 meters (660 to 3,300 feet) live in the twilight or mesopelagic zone where the remaining blue-green light is slowly absorbed by sea water. Those midwater fishes below 1,000 meters (3,300 feet) live in the total darkness of the true deep sea, the bathypelagic and abyssopelagic zones. Only some 300 species are known from this extremely harsh environment.

Lanternfishes (family Myctophidae), lightfishes (family Gonostomatidae), and dragonfishes (family Stomiidae) predominate in the twilight zone, while the deepsea anglerfishes (11 families) and whalefishes (family Cetomimidae) are the largest of the bathypelagic fish groups. Rattails (family Macrouridae) are the dominant family of bottom deepsea fishes, while a cuskeel (family Ophidiidae) is the deepest fish yet trawled at 8,370 meters (27,200 feet)—the flatfish seen from the *Trieste* has yet to be captured.

Most deepsea fishes belong to primitive groups, like sharks, eels, and the less advanced bony-fish families. The environment is harsh, with low levels of food, light, and temperature. Deepsea fishes have presumably been forced into this habitat by competition from the more advanced groups of bony fishes. The lack of light and low level of food probably have been the most important evolutionary forces molding their striking adaptations.

▼ *The open sea supports some of the most formidable of all predators, such as sharks, the tigers of the sea. Many, like the blue shark* Prionace glauca, *are extremely wide-ranging: individuals tagged off England have been recovered off Brazil, and others tagged off New York have been found off Spain.*

Marty Snyderman

Some of these adaptations include the development of light organs, the development of large lateral-line systems that pick up pressure waves created by swimming animals to compensate for the loss of sight in the dark, and bizarre modifications in reproduction like the parasitic males of some anglerfishes. Many twilight-zone fishes undertake a migration from the daytime depths of 500 to 1,000 meters (1,625 to 3,300 feet) to the upper 200 meters (660 feet) at night, to feed in the rich surface waters. The weakly muscled bathypelagic species do not migrate, but have to rely on large gapes and specialized teeth to ensure success with infrequent meals.

A specialized deepsea community found only in the last 20 years is that around deepsea vents at depths below 2,000 meters (6,600 feet). Hot water is warmed by lava-filled chambers associated with midocean ridges. Hydrogen sulfide is dissolved in the hot water and, where it escapes from the bottom, rich communities of invertebrates are found around the vents. Recently three species of vent fishes in the families of viviparous brotulas (family Bythitidae) and eelpouts (family Zoarcidae) have been described. Nothing is known of the biochemistry of these vent fishes, so it is not yet clear if or why they are restricted to these communities.

MIGRATIONS

Not all fishes can be categorized easily by habitat. A number are diadromous—that is, they migrate between fresh water and the sea. Some, like the well-known salmon (family Salmonidae), are born in the headwaters of freshwater rivers and streams, migrate downstream as developing young to the estuaries, and then head to sea as juveniles, to live and grow to adulthood. Maturity sees a reverse migration to the home stream for spawning (and death in some species). This type of freshwater–ocean–freshwater migration, where most of the life is spent at sea, is termed anadromy and is typical of both salmon and searun lampreys (family Petromyzontidae). Much research has been done on the adaptations required for finding the home stream, which may include both celestial navigation and recognition of the water of individual streams or rivers.

The opposite type of migration—described as catadromous—is typical of freshwater eels (family Anguillidae) that spawn in the sea. The developing larvae are carried near the mouths of freshwater rivers by oceanic currents, where the juveniles migrate upstream to live and grow to adulthood. After up to 10 years, the reverse migration downstream and to the oceanic spawning grounds is followed by death. Eels can travel immense distances. The European freshwater eel, for example, swims more than 5,000 kilometers (3,000 miles) to the Sargasso Sea to spawn. The larvae can drift for up to three years in the Gulf

Norbert Wu

Stream before reaching a river in southern Europe. The freshwater eels of southeastern Australia are thought to breed near New Caledonia.

Major changes take place in migrating fishes in the estuaries. Because of osmosis, marine fishes have a problem of water loss, which they overcome by drinking sea water, passing concentrated urine, and secreting salt from the gills. Freshwater fishes have the opposite problem of water retention, so they never drink, pass much dilute urine, and take up salt in the gills. Diadromous species must undertake these fundamental changes in physiology in the estuary as they pass from one habitat to another. Perhaps

▲ *Even amid the vastness of the open ocean, protective environments can be found: these juvenile jacks (family Carangidae) find shelter among the stinging tentacles of a floating jellyfish in the Atlantic's Gulf Stream.*

◄ *Cold and dark, the ocean depths are scarce in food and other resources. Here fishes have evolved a variety of ways to make do with less. In the anglerfishes, for example, males are often dramatically smaller than females, and in some species live permanently attached as parasites on their mates. This female deepsea anglerfish* Haplophryne mollis *has two such consorts.*

Peter David/Planet Earth Pictures

Jeff Foott Productions/Bruce Coleman Limited

▲ *Sockeye salmon, fresh from the sea, about to spawn on the gravel shallows of a high country stream in western Canada. Cold water is richer in oxygen than warm water, so pure snow-fed mountain streams—further aerated by vigorous turbulence—serve as ideal nurseries for the young of several fishes, notably the various species of salmon, genus Oncorhynchus.*

this is why all these diadromous species (Pacific salmons and eels for example) spawn only once in their lives, and also why so few of the total fish species are diadromous. The vast majority of fish species are either strictly freshwater or strictly marine, presumably unable to make the physiological changes necessary for diadromy.

Migrations also take place within one major habitat, such as from lake to stream or from one place to another in the ocean. Often this is associated with mature adults undertaking a breeding migration. Those of the commercial species of the North Sea, such as the herring (family Clupeidae) and cod (family Gadidae), are the best known. The evolution of spawning migrations must have involved one area where the developing larvae have the richest feeding grounds and another area where the adults have the richest feeding grounds. With breeding migrations typically an annual event at the same time and place, the potential for overfishing is considerable. One of the most important limits to species distribution is water temperature. While warm-water fishes will not freeze in cooler waters,

their metabolism may be so slowed that they are easy prey for local predators. Temperature may be most limiting during reproduction, when eggs and larvae are developing. Clearly, many other factors can also limit species distribution.

The maps in each chapter show natural distributions only. Many species have been introduced to new areas, often to the detriment of the native species, but these introductions are not mapped.

LARVAE

The developing larvae of most fish species are so different from the adult that many were originally described as different species or even genera. Only the live-bearing fishes with internal fertilization, like some sharks, rays, viviparous brotulas (family Bythitidae), guppies (family Poeciliidae), surfperches (family Embiotocidae), eelpouts (family Zoarcidae), and kelpfishes (family Clinidae), give birth to essentially miniature adults. Spawned eggs have a much shorter development time and the hatchlings have a completely different ecology than the adults, eating different food and often living in a different environment. Most larvae are pelagic, living in surface waters where food is most abundant. Many have larval specializations that are lost upon metamorphosis to the juvenile stage, which is often associated with a change in habitat, particularly for bottom-dwelling species. Such features as eyes on the end of long stalks, long trailing fin rays and even trailing intestines, extended head and body spines, and larval pigment patterns are all lost at metamorphosis.

COLORATION

Body pigment has a number of functions. It protects the internal organs of shallow-water fishes from sunlight, and is particularly noticeable over the brain of larval fishes. The various patterns and colors enable members of the same species to recognize one another; their colors often become brighter during the breeding season. It is also necessary not to be conspicuous to predators, so colors often blend with the background; this is why fishes are dark on top and light underneath. The ability of flatfishes to change their pattern to match that of the bottom is well documented. One of the most noticeable parts of a fish is its black eye, so many have a dark line through the eye to camouflage this feature. Others have false eye spots on the tail or back of the fins, which can trick a predator to strike away from the head.

Protection from predators can involve mimicking dead leaves, as done by the freshwater leaffishes (family Nandidae) and the marine juvenile tripletail (family Lobotidae). Other forms of mimicry, involving two or more different species of fishes, are described in the chapter on blennies. Camouflage sometimes includes various flaps and strands resembling pieces of seaweed, as found in such unrelated species as carpet sharks

Kevin Deacon/Dive

(family Orectolobidae) and sea dragons (family Syngnathidae).

LUMINESCENCE

Living light, or bioluminescence, takes the place of color in the dark environment of the deep sea. Surprisingly, it is also present in a number of shallow-water marine forms like the flashlight fishes (family Anomalopidae), pineapple fishes (family Monocentrididae), ponyfishes (family Leiognathidae), and some cardinalfishes (family Apogonidae) and bullseyes (family Pempherididae). These have a relatively large light organ, external on the head of flashlight and

◄ Camouflage, or crypsis, is common among fishes as a means of blending with the environment and evading notice from both predators and prey. The complex surfaces and color patterns of this leaf scorpionfish Taenianotus triacanthus render it almost undetectable against the weed-clad rocks it inhabits.

▼ Luminescent organs of various kinds have evolved independently in many families of fishes. The brooch lanternfish Benthosema fibulatum, like other species of the family Myctophidae, has distinct sets of light organs on the side of the body. This fish was photographed at night in 20 meters (65 feet) of water at Bali, Indonesia. Here the shore drops so steeply that these migrating deep-sea fishes can be found in coral habitats at night.

Rudie H. Kuiter

Norbert Wu

◄ The larvae of most deep-sea fishes, like this anglerfish, develop in the food-rich surface waters. A balloon-like envelope of inflated skin is typical of these anglerfish larvae and may be related to maintaining neutral buoyancy.

pineapple fishes and internally associated with the digestive tract in the others, usually with a colony of luminous bacteria that produce the light. A very few shallow water species like an anchovy (family Engraulididae), a croaker (family Sciaenidae), and some midshipmen (family Batrachoididae) have numerous external light organs or photophores in the skin. This is more typical of deepsea species, particularly those in the twilight zone, like the aptly named lanternfishes and lightfishes.

This living light has several functions. In twilight zone fishes, the silhouette will be obliterated to predators hunting from below if the light produced matches the intensity of the downcoming sunlight. Most lanternfish species have distinct patterns of light organs that may be used in recognition for schooling or mating. Large light organs near the eyes presumably help the fish to see food. The elaborate luminous filaments on the heads of anglerfishes and the chins of dragonfishes must lure prey near their enormous mouths. Some fishes have the ability to flash brightly or squirt a luminous cloud to startle or divert a predator.

VENOM AND POISON

One of the most effective ways to discourage a predator is to inject it with a painful venom. Many unrelated fishes have evolved this protective device, which in its most elaborate form involves a venom gland at the base of a grooved fin spine. The most lethal (at least to humans) is the stonefish, described in detail in the chapter on scorpionfishes. Venomous spines are known in at least 40 different species, including some sharks, stingrays chimaeras, catfishes, toadfishes, scorpionfishes, weeverfishes, old wifes, rabbitfishes, and stargazers.

The most famous poisonous fishes are the puffers or toados (family Tetraodontidae), which have caused numerous human fatalities. They are described in the chapter on triggerfishes. Another type of poison is the concentration through the food chain of a toxin, produced by a minute marine plant, so that the largest predators become poisonous to eat. Termed ciguatera poisoning, it affects a number of tropical species of fishes. This condition is the result of the fishes' ecology and what they have eaten, rather than a long period of evolution, as with the other features described.

JOHN R. PAXTON

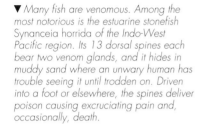

▼ Many fish are venomous. Among the most notorious is the estuarine stonefish Synanceia horrida of the Indo-West Pacific region. Its 13 dorsal spines each bear two venom glands, and it hides in muddy sand where an unwary human has trouble seeing it until trodden on. Driven into a foot or elsewhere, the spines deliver poison causing excruciating pain and, occasionally, death.

Carl Roessler

CLOSE ASSOCIATIONS

The best-known fish symbiosis ("living together") is that of the clown or anemonefishes and their anemones, described in the chapter on damselfishes. A similar form of symbiosis exists between floating jellyfishes and various types of jacks (family Carangidae) and driftfishes (family Nomeidae). In most cases only juveniles take advantage of the shelter provided by the lethal tentacles of the jellyfish. During this stage of their life cycle, they are highly vulnerable to predators whereas the larger, fast-swimming, and elusive adults no longer require this protection. In similar fashion, a tiny white pipefish lives among the tentacle-like polyps of mushroom corals on reefs of the western Pacific. The venomous spines of sea urchins (usually belonging to the genus *Diadema*) frequently form a refuge for other pipefishes (family Syngnathidae), clingfishes (family Gobiesocidae), and cardinalfishes (family Apogonidae).

There are many other, equally interesting, examples of symbiosis between fishes and invertebrate animals. Numerous types of gobies (family Gobiidae) in the tropical Indo-Pacific region share a sandy burrow with shrimps (belonging to the family Alpheidae). The shrimp provides labor, continually excavating the burrow. The fish serves a "watchdog" role, signaling to its partner with a flick of the tail when the coast is clear to emerge from the lair with a load of debris.

Perhaps the most unusual case of symbiosis involves certain types of pearlfishes (family Carapidae) that live inside the body cavity of sea cucumbers (Holothuria). They freely enter and exit via the anal opening and receive part of their nourishment from feeding on the gonadal tissue of their invertebrate host. Other types of pearlfishes live in the

Marty Snyderman

▲ *Intimate relationships between different species of fishes are not unusual in the sea. Remoras Remora remora, for example, hitch rides with larger fish by means of a powerful sucker on their heads, dashing forward to snap up scraps from the larger fish's meal.*

mantle cavity of oyster shells or within the body cavity of starfishes and sea squirts.

Another form of symbiosis involves cleaner fishes that feed on tiny parasitic organisms (often crustaceans such as copepods) that attach to the body, fins, mouth, and gill chamber of other fishes. The best known example of this behavior is provided by the cleaner wrasses of the genus *Labroides* . These fishes usually occur solitarily or in pairs, establishing permanent cleaner stations scattered over the reef. Competition for the attention of the cleaners may be intense among the parasitized clients and the stations are marked by queues of fishes awaiting their turn. Scientific studies have shown that the presence of cleaner fishes is vital to the general good health of the overall fish community. Many other fishes act as cleaners, including certain gobies and the juveniles of various butterflyfishes (family Chaetodontidae) and other types of wrasses.

A more general type of species association is that between the remoras (family Echeneidae) and large pelagic fishes like sharks and billfishes, and even turtles. The suckerfish, as the name indicates, has a large sucking disk on the head, modified from the first dorsal fin, that allows it to attach tightly to its larger host. The attachment is so strong that local people are able to tie a line to the suckerfish and use it to retrieve turtles to which the fish become attached. The suckerfishes gain shelter, protection, and sometimes scraps of food from their hosts.

GERALD R. ALLEN AND JOHN R. PAXTON

David Fleetham/Oxford Scientific Films

▲ *A scribbled pipefish Corythoichthys intestinalis on a coral reef on Palau, Micronesia. Coral reefs represent environments of such complexity that a bewildering array of interspecific relationships have evolved, often between fishes and entirely different groups of animals such as corals, anemones and shrimps.*

FISH BEHAVIOR

A ll fishes are built on the same basic body plan, but they display an impressive range of sizes, shapes, and color patterns. While fishes also display a broad diversity of behavior, there are also common threads running through the behavior of all fishes. Fish behavior, like all animal behavior, is governed by the need to find food and mates, balanced by the necessity of avoiding predators. Fishes accomplish these goals by using specialized sensory systems, some of which are unavailable to other vertebrates.

▼ A moray eel (genus Muraena) displays its unusually prominent, tube-like nostrils, or nares. Most fishes have an extremely well-developed sense of smell.

Marty Snyderman

SENSORY SYSTEMS

As with most vertebrates, vision plays a key role in enabling fishes to process information about their surroundings. However, aquatic life puts strict limitations on the use of eyes. Although sunlight penetrates to 1,000 meters (3,300 feet), 75 percent of the ocean is sunless, and well before this depth (below about 15 meters or 50 feet), the bright reds, yellows, and oranges have dropped out of the spectrum. Most water contains particulates, like mud, which makes the water cloudy and further reduces the depth to which light can penetrate. Most fishes can see at light levels humans would experience as darkness. However, many are particularly vulnerable at dawn and dusk when light levels are changing quickly. At dusk and dawn, when the day-active (or diurnal) denizens are ceding their territories to the night-active (or nocturnal) species, many predators attack.

Perhaps because vision is so often difficult, fishes have extremely well-developed senses of smell and hearing. Although they do not have a nose as we know it, fishes do have nostrils, or nares with which they can detect even small concentrations of chemicals. Salmon, for example, locate the streams they were born in by smelling their way back from the open ocean and following odors they imprinted on as young fry. The sense of smell is intimately connected with the sense of taste. As fishes do not have tongues, their "taste buds", or chemoreceptors, are spread out around the face and sometimes along the entire body. Catfishes and goatfishes have localized sets of chemoreceptors on whiskerlike projections known as barbels. They use these appendages to probe the ocean bottom for tasty meals hidden from view.

Although light does not travel very far under water, sound does. In fact, sound waves travel much farther in the water than they do in the air, allowing fishes to hear noises from a considerable distance. Fishes use sound to locate prey, keep track of predators, and communicate within the species. Boaters in Sausalito, California, are seasonally kept awake by the froglike grunts and groans of midshipmen broadcasting for mates at night.

In addition to these basic sensory systems, fishes have an extra sense not present in other vertebrates. This is the acoustico-lateralis system, or the sense of "distant touch" located on the fish's lateral line. Somewhat like humans feel wind against their cheeks, fishes can feel the wake left by a swimming neighbor. Schooling fishes use these lateral-line senses to regulate the distance between neighbors in muddy water or darkness when vision is not possible.

Sharks, skates, rays, and chimaeras have a very specialized sensory system not possessed by other fishes. Using a network of cells arrayed around the head, these fishes can sense the minute electric fields that project from every living animal. Sharks use this electroreception to orient their downward-projecting jaws toward their prey in the last few seconds of capture. By contrast, electric fishes, including the electric eel, project an electric field within which they sense disturbance. These pulses are used to detect and stun prey, as well as for communication.

These senses of sight, chemoreception, hearing, "distant touch", and electroreception allow fishes to perceive fully their watery three-dimensional world, and respond to the needs of foraging, predator detection, and reproduction.

FORAGING

Fishes eat a bewildering array of food, made up of almost any plant or animal found, even momentarily, in the water. Many species have specialized jaws, which allow them to eat a particular type of food. Halfbeaks, named because of their long and slender, protruding lower jaw, neatly slurp up pieces of floating seagrass like spaghetti. The more careful spinynose sculpin selects a specific size of snail, orients the shell in its mouth, and then punches a hole in the shell with the special teeth that are located on the roof of the mouth.

Fishes forage for invertebrates not only on the bottom, but also up in the water column. Plankton, tiny algae, and invertebrates that drift along in the water are food for many schooling fishes such as anchovies and sardines. Of course, fishes also eat each other.

Because fishes have fins instead of limbs equipped with claws or hooves, they can usually eat things only about the size of their head. This means that most prey items will be up to one-third the length of the predator. For herbivores—fishes that eat plants—this is not a problem as bite-size pieces are easily taken. However, for piscivorous (or fish-eating) fishes, mouth-size strictly limits the upper size of prey taken.

Some fishes, such as the cookie-cutter shark, have managed to circumvent this rule by biting into a much larger animal and twisting out a mouth-sized plug.

Paradoxically, the largest fishes do not eat the largest prey, but the smallest. Whale sharks and manta rays feed on tiny plankton, eschewing the vagaries of hunting down fishes.

G.I. Bernard/Oxford Scientific Films

PREY CATCHING

Finding a meal is often a difficult task for any animal, and fishes are no exception. With each specialized kind of food comes an equally specialized method of food gathering. Archerfish spit well-aimed jets of water at insects hanging from overhead foliage, knocking them into the pond where they are quickly consumed. Foraging is often an energetically expensive task and minimizing search time helps keep costs down. When the density of food is high, planktivores such as anchovies swim through the water with their mouths wide open, straining the plankton out. When the density of plankton is low, these fishes switch feeding modes and pick individual plankters (members of the plankton) from the water. Moreover food is usually not evenly spread across a habitat, but occurs in patches. This means that a hungry fish must

▲ Archerfishes catch much of their insect prey by spitting at them, bringing them down from their perches on overhanging leaves and branches with a bullet-like stream of water droplets.

electroreceptors

▲ Fishes find prey using an unusually wide range of senses, ranging from chemoreception to the sophisticated electrosensory systems of sharks.

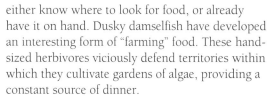

▲ *Many fish evade predators by hiding. The parrotfishes (family Scaridae) are especially prone to seek shelter in coral crevices when threatened or alarmed.*

Norbert Wu

Kathie Atkinson/AUSCAPE International

▲▶ *A striped anglerfish Antennarius striatus shows off the flexibility of its lure, which is waved temptingly about to draw unwary fishes closer. Once within range, the prey is swallowed in a rapid lunge-and-engulfing action. Sometimes anglerfishes are found with missing lures, suggesting that some unusually cool customer has taken the bait and got away with it.*

either know where to look for food, or already have it on hand. Dusky damselfish have developed an interesting form of "farming" food. These hand-sized herbivores viciously defend territories within which they cultivate gardens of algae, providing a constant source of dinner.

While a large number of fish species eat plants or invertebrates, there are always larger piscivores in the area. Fishes that specialize in eating other fishes have developed a wide range of hunting methods. Sit-and-wait predators do just that. Many of them, such as flounder, are cryptic and can blend into the background, relying on camouflage to trick unsuspecting prey into swimming too close. Anglerfish have a fleshy projection resembling a worm, the esca, which they dangle invitingly in front of their waiting jaws, tempting small fishes to inspect the deadly lure.

Not all predators are so sedentary; many actively stalk their prey. Pike slowly move toward feeding minnows, edging forward as they aim for a selected victim. Only when the minnow is within range will the pike suddenly bend its body into an S-shape and lunge. Many predators are extremely fast swimmers and do not need stealth to get close to their prey. Jacks and tunas have streamlined, bullet-shaped bodies which they accelerate into fleeing schools of herring, easily capturing victims in their path. The ensuing mêlée is made worse by the fact that many of these fast-swimming predators hunt in groups. While tuna or barracuda do not hunt socially, as do terrestrial predators such as lions or jackals, they will frequently attack schools of prey en masse, creating such a breakdown in school structure that individual prey

are easily chased and captured. The largest piscivores, however, hunt alone. Marlin slash through schools of tuna with their bill, or rostra, returning to pick off injured victims.

AVOIDING PREDATORS

The aquatic world is full of predators waiting for prey to make an ill-considered move. How do fishes survive? Fortunately, in the predator–prey evolutionary arms race, prey have made just as many advances, allowing them to stay, for the most part, one step ahead.

Many fishes seek the shelter provided by the cracks and crevasses of the sea floor. At night, for example, triggerfish erect their spiny dorsal fin, locking it into place as they wedge themselves into narrow openings in the coral reef. When fishes are young, they are especially vulnerable, as almost everything is bigger than they are. Often this means a conflict between finding dinner and being safe. When largemouth bass are present, young bluegill sunfish prefer to remain in the weeds by the side of the pond even though they can find more food in the open water. Many juvenile fishes have found interesting places to hide. Young cardinalfish, for example, hunker down among the sharp quills of the long-spined sea urchin when predators threaten.

Adult fishes have fewer places to hide but, fortunately, they have a battery of defenses to help them deter predatory attack. Some fishes use chemical defenses to ward off predators. Parrotfish secrete and sleep in a translucent mucous cocoon which may prevent predators from detecting them. At the approach of danger the beautiful lionfish

Kathie Atkinson/AUSCAPE International

spreads out its brightly colored fins containing deadly poison. Bright coloration is sometimes an advertisement of danger. Spines, scales, and tough skin can also deter predators. Porcupinefishes are covered with an interlocking layer of spines. Normally laid flat, these spines are erected when the fish senses danger and inflates into a thorny ball.

Although many prey fishes do not have extraordinary chemical or mechanical means of defense, they have evolved gregarious behavior, or schooling. Over 50 percent of the world's fishes school as juveniles, and roughly 25 percent school as adults. The sheer size of a large anchovy school swamps the effect of potential predators, thereby diluting the individual risk of being attacked and eaten. Faced with a moving morass of identical fishes, predators find it hard to "lock on" visually to a single target. This confusion may cause the predator to miss, delay, or even abort an attack. Schooling prey also have an impressive repertoire of avoidance maneuvers in which they engage when a threatening predator comes too close. Schooling herring will clear the path of an approaching barracuda in a chorus-line wave of movement. Although this might seem simple, it is actually a deadly ballet where one wrong move by a herring can lead to immediate attack.

MATING BEHAVIOR

Fishes display an impressive range of mating and parental care systems, unparalleled in the animal world. While some species collect in large mating aggregations to spawn on full-moon nights, allowing their gametes to drift unattended in the current, others rigorously select and defend nest sites, attract mates with elaborate courtship dances, and tenaciously guard their young. Mating

in fishes can be divided into phases of courtship, mating, spawning, and parental care. In most species mating and spawning are virtually synonymous, but in species with internal fertilization, such as sharks and guppies, spawning is replaced in the sequence by birth.

In the schooling spawners such as anchovies, males and females appear identical and do not engage in courtship displays. However, in those species in which eggs are deposited in nests, or are actually picked up and cared for, elaborate courtship and mating rituals may evolve. Because the males of many species must attract females in order to enter into courtship, they have developed bright coloration patterns, accessory structures on their fins and tails, and occasionally increased body size. These traits, known as secondary sexual

▲▼ *Many fishes use chemical means to deal with predators. The warty prowfish* Aetapcus maculatus *(above), which inhabits coastal waters of southern Australia, responds to threats by expelling a smoke-like cloud of presumed toxic fluid from a vent near the gills. These anemonefishes (*Amphiprion *species, below) use chemistry in a rather different way. They produce a skin secretion that mimics the chemical action of the substance that inhibits anemone tentacles from stinging themselves. They can shelter amid the tentacles safe from harm, taking advantage of the anemone's own formidable defenses. This is the two-banded anemonefish* Amphiprion akindynos.

Alby Ziebell/AUSCAPE International

▲ *Blue-and-gold fusiliers Caesio teres, Great Barrier Reef, Australia. Many fish congregate in schools like these, avoiding predators not by flight nor by hiding but by diffusing the target presented to any potential predator, making it difficult to select a particular victim.*

▼ *Chromis damselfish Chromis cyaneus, Caribbean. Studies have revealed that these algae-feeders defend territories open to most fishes, except that the damselfish can somehow distinguish algae-feeding species like itself from non-algae-feeding species. The latter it ignores, the former it vigorously chases from its territory.*

Norbert Wu

characteristics, not only allow males to compete with each other for access to females, but also allow males to advertise themselves to females, who are trying to choose the best male. This process of sexual selection can lead to the evolution of extremely flashy males, which are noticed not only by females, but also by predators. Thus the process of natural selection keeps secondary sexual characteristics from becoming too extreme.

Although pair-bonding is rare among fishes, some species, such as the butterflyfishes, form monogamous pair-bonds which may last for several years. In habitats rife with potential egg predators, monogamy and shared parental care may be the only way to raise young successfully. Before spawning, cichlids entice mates with elaborate courtship which may last up to several days, before a suitable nest site is selected. The nest, in which the eggs are deposited and the fry are raised, is guarded by both parents. Often there is a division of labor—females tend to the eggs while males patrol and defend the boundaries of the territory.

Members of the same sex sometimes develop conflicting mating strategies. Female salmon, returning from the sea, select and defend nesting sites from other aggressive females. Males compete with each other for females, with the largest, most aggressive male winning. However, some males, known as jacks, return from the ocean several years earlier than their larger counterparts. Instead of fighting for females, jacks sneak in to spawn with ripe females while the larger males are still engaged in contests of strength, or even when the large male and female are mating! In sunfish, sneaky males adopt the coloration and behavior of females, thereby tricking unsuspecting males into yielding access to the real female.

Single male–multiple female mating systems (polygyny) are common in habitats where a male can successfully defend the resources, such as food or protective cover, that females want. Bright orange male garibaldi secure small crevices or overhangs in their rock-reef habitat. Female garibaldi choose to spawn with males that hold the best nest sites, depositing their eggs on the rock surface and leaving their care to the male.

Successful males have large nests that contain the developing eggs of several females. Males must continuously defend the eggs from marauding egg predators (including other garibaldi) and also fan them with their fins, to keep them free of sediment and supplied with oxygen.

In species such as the cleaner wrasse, which do not use benthic (sea-bed) nests but release gametes directly into the water column, polygynous mating systems are similar to harems. A single male controls reproductive access to a group of females, and mates with each of them in turn. This mating system is further complicated by the fact that these fishes have the ability to change sex. When the male is removed from his territory, the dominant female transforms into a male, and takes over the role of her/his predecessor. Many species of fishes change sex, usually from female to male. The process may take one to two weeks. Some species, such as the blue-headed wrasse, have two types of males: large, brightly colored "terminal" males that control access to harems of females, and small, drab "satellite" males that are excluded from the females by the terminal male. Terminal males may mate more than 40 times per day, while satellite males must rely on sneaking fertilizations. However, when the number of satellite males is high, they may band together to overcome the defenses of the terminal male. If the terminal male dies, either a satellite male will transform into a terminal male, or a dominant female will change sex and become the terminal male.

A few species of fishes—some of the guppies and the silversides—are all-female. In this interesting twist on evolution, eggs already contain the necessary genetic material and do not need to combine with sperm. However, sperm of related or "sister" species is sometimes used to initiate development mechanically. In this situation, the males contributing sperm are deceived into believing they have mated with a female of their own species.

PARENTAL CARE

Fishes display a complete range of parental care, from none to extensive, although never as developed as that displayed by birds and mammals. In general there is a trade-off between the number and size of eggs produced, and the amount of parental care provided. Many schooling and open-water species broadcast spawn. Each female may release thousands of small eggs which, carried by the currents, float away from the mating aggregation. Because so many eggs are fertilized at once, egg predators are swamped, leaving some proportion of the eggs to hatch. Some species, such as the coho salmon, make crude nests or redds; a single female may lay more than a thou-sand eggs. However, no parental care is provided other than a light dusting of gravel over the eggs.

Species with benthic nests and parental care usually produce larger, yolky eggs in much smaller numbers than the broadcast spawners, as many fewer are lost to predators. Where egg predation is more intense, nest building and egg guarding become necessary. Stickleback males construct nests out of algae in which they entice a female to deposit her eggs. Males then guard the nest from hungry egg predators, including gangs of spawned-out females and non-territorial males. A guarding male may try to disguise the proximity of his nest by picking at substrate nearby, feigning the presence of eggs in an effort to distract the hungry group.

Some cichlids have circumvented the need for nests entirely by brooding the eggs in their mouths. After the eggs have hatched into tiny fry, the parent releases the brood into the water column. However, if an egg predator should threaten, the parents quickly snap up their offspring and protect them in the safety of their mouths until the danger passes. Pipefish and seahorse males have external brood pouches, in which they place and nurture the fertilized eggs.

Kevin Deacon/Dive 2000

▲ Parental care in fishes is not necessarily restricted to the female. In the weedy seadragon Phyllopteryx taeniolatus of Australian southern coastal waters, the male carries the eggs—sometimes 120 at a time—safe on his tail until they hatch.

Marty Snyderman

Although many fish species fertilize their eggs externally, internal fertilization and live birth do occur in a surprising array of fishes. Live-bearers have dispensed with the need for nest building and early parental care entirely, as the young are born as fully developed, active fishes. Egg and fry losses are minimized, but the number of young produced is also limited. Some sharks may give birth to a single young. However, some live-bearing females, such as guppies, display superfetation, that is, the simultaneous gestation of more than one brood of differing ages. Females store sperm and can fertilize batches of eggs in stages.

JULIA K. PARRISH

▲ Fish species occupy points along the entire spectrum of possibilities from intensive parental care to almost none. Many build nests and guard their eggs and young. Some, like the sergeant major Abudefduf saxatilis, even nest in colonies. This fastidious homemaker is moving a sea star away from its nest.

This book lists endangered species throughout Part Two, Kinds of Fishes (from page 54). The first page of each chapter features a colored Key Facts panel, which includes the Conservation Watch heading.

The conservation information in the Key Facts panels is based on the *1996 IUCN Red List of Threatened Animals*, a co-publication of the IUCN World Conservation Union and Conservation International.

The level of threat is indicated by the symbols below.

!!! Critically endangered
!! Endangered
! Vulnerable
■ Other information

ENDANGERED SPECIES

Fishes are the largest and most diverse group of vertebrates, not only in their form and size but also in their biology and ecology. With a few exceptions, however, fish conservation has been sadly neglected. All kinds of fishes, big and small, are at risk, and over 700 species and 49 subspecies and populations are currently known to be in danger. Unfortunately, this low figure does not represent the true picture worldwide and as our awareness of the plight of fishes increases, the number found to be at risk rises. In the past, research has focused on freshwater species, because they were easier to study. In early 1996, however, the status of marine fishes was examined for the first time. More than 100 fishes from a wide range of marine habitats were listed, clearly indicating the fragility of these ecosystems.

A GLOBAL PROBLEM

If fishes were as easy to see and approach as birds and mammals are, their ecological and biological values would be better understood. Our knowledge of the conservation status of fishes is rather limited, but we do know that they are under pressure right around the planet, with threatened species listed from every continent. Threatened representatives were found in 32 of the 58 orders of fishes, belonging to four of the six classes. The Cyprinidae (carps and their allies) is the most threatened family (157 species listed) with representatives throughout the world. The Cichlidae follows with 86 threatened species, of which 74 (86 percent) are found in mainland Africa; the remaining species occur in Madagascar (about 8 percent) and Mexico (nearly 6 percent). In contrast to the cyprinids, the vast majority of these beautiful and highly specialized cichlids are endemics found in restricted environments such as lakes. Cichlids have attracted attention among

▼ *The cichlid* Copadichromis verduyni *of Lake Malawi. One of the most interesting of all groups of fish are the species swarms of cichlids that inhabit Lakes Tanganyika, Malawi, Victoria and a few other large lakes of East Africa. Hundreds of species are restricted to these lakes, but many have been put at risk by the introduction of the predatory Nile perch for commercial fisheries.*

evolutionary biologists because of the existence of species flocks, such as the African Rift Lake species. Because their habitats are confined, they are particularly vulnerable. Other families with important numbers of threatened species include Atherinidae (41 species), Syngnathidae (37), and Cyprinodontidae (31). The order Acipenseriformes (sturgeons and paddlefishes) deserves special mention as all its members are threatened (6 critically endangered, 11 endangered, 8 vulnerable, and 2 near threatened).

Fishes from all kinds of environments—rivers, lakes, caves, estuaries, coastal areas, coral reefs, and the deep sea—are threatened. Even in the forests, traditionally thought of as terrestrial environments, fishes are at risk because the loss of forest cover affects the amount of water available in the area. Temperate forests (New Zealand and

Chile) and tropical forests (South America and Asia) have their own specific fish faunas.

WHY FISHES ARE THREATENED
Restricted distributions

Many fishes have naturally restricted distributions such as lakes, sinkholes, and desert pools. Restricted habitats include the Great Lakes of North America and the Rift Lakes of East Africa, the Valley of Mexico, the Parras Valley, and the cenotes (sinkholes) in southern Mexico. Enclosed basins are often rich in endemic species, especially in isolated environments, such as desert pools. Members of the families Cyprinodontidae (pupfishes) and Cichlidae are examples of fishes that have adapted to these limited environments. These fascinating families have been compared to Darwin's finches in the Galápagos Islands because they, too, have

▼ A salt creek pupfish in a drying pool, Death Valley, USA. Some freshwater fishes have extraordinarily restricted distributions—notably the pupfish (family Cyprinodontidae) of Death Valley, some species of which are restricted to a single desert pool.

William M. Smithey, Jr/Planet Earth Pictures

◀ Largest of all the world's fishes, the whale shark Rhincodon typus grows to at least 12 meters (nearly 40 feet), and probably much larger. It is a pantropical species, but it is uncommon and its total numbers are thought to be very low.

Rudie H. Kuiter

▲ *The spotted seahorse Hippocampus kuda is broadly distributed in tropical waters of the Indian Ocean and the west-central Pacific. Like all seahorses of this area, it is now targetted by local fishers who dry their catch and sell it for the lucrative Chinese traditional medicine trade. Because seahorses have specialized reproduction with limited numbers of young produced, populations are declining as numbers caught exceed births and survival. Seahorse farming may be one solution to the problem.*

Adrian Davies/Bruce Coleman Limited

▲ *Fishes like the blind cave fish Astyanax mexicanus of Mexico that inhabit underground cave systems can be put at risk by surface water pumping for irrigation or other purposes, lowering the water table and altering the flow of subterranean rivers and streams.*

evolved remarkable adaptations rather quickly in geologic time. Cichlids have developed specialized feeding strategies and the cyprinodontids have adapted to isolated and harsh environments. The Devil's Hole pupfish in Death Valley in Nevada, USA, is probably the most extreme example of a species with a restricted habitat—the entire population is found in a limestone shelf that overhangs the Devil's Hole pool and measures 3 by 5.5 meters (10 by 18 feet). In Lake Victoria, haplochromine fishes may occupy distinctive small niches despite the great size of the lake. The asprete *Romanichthys valsanicola*, found only in the upper tributaries of the River Arges (Danube Basin), is probably the most threatened fish in Europe. Surprisingly, restricted distributions have also been noted for some marine species such as coral reef toadfishes and butterflyfishes.

Forest fishes
In the tropical rainforest of the Amazon and Malaysia, there are fishes that eat fruits, seeds, and leaves that will suffer from the loss of this environment. Between 1970 and 1975 the fish catch in the Amazonian rivers declined by 25 percent as a result of the deforestation around the fishes' breeding grounds. Already several commercial species in the Amazon Basin are at risk from overexploitation and habitat destruction.

Natural changes
Natural environmental changes do occur, and fish distribution and abundance may vary according to factors such as decreased water levels resulting from lack of rain, or increased salinities or temperatures. Extinction and evolution are natural processes and many fishes have followed these two paths for centuries. In addition, species life cycles play an important role in survival: while some, like the Indonesian ricefishes, complete their cycles in a year, slow-maturing species such as the great white shark may take up to 10 years to mature, and the sturgeons up to 14 years.

Human intervention
In recent years human intervention has threatened an increasing number of fishes as aquatic habitats have been altered to provide for agricultural and other development activities with little, if any, regard for the fishes. Habitat alterations by humans have been recognized as one of the main causes of fish losses and it is believed that almost all fish habitats have been affected by urban development, pollution, water extraction or diversion, dam construction (which blocks fish migration paths), canalization of rivers and streams, international trade (including cyanide fishing and dynamiting of coral reefs for the aquarium trade), and overexploitation. The draining of wetlands and streams has resulted in the loss of fish spawning, feeding, and nursery grounds. The IUCN has identified habitat loss or degradation as a major reason for the threatened status of 55 percent of endangered fishes.

Pollution and habitat alteration
Pollution of many kinds is affecting an increasing number of habitats. The Great Lakes of Canada and the United States, one of the world's largest freshwater systems, illustrate the terrible effects of industrial, agricultural, and urban pollution: together with urban development and habitat alteration, this pollution has had a drastic effect on the water quality and productivity of the lakes. Resident fishes such as the lake sturgeon *Acipenser fulvescens*, the shortnose cisco *Coregonus reighardi*, and the shortjaw cisco *C. zenithicus* have been adversely affected.

Rivers that flow through several countries are often subjected to serious impacts along their courses. The beautiful Danube in Europe, passing through 12 countries with a combined population of more than 70 million, has been polluted by more than 1,700 industries and subjected to overfishing, water extraction, and damming. Pollution has been worsened by the low water levels and it is the area of its delta, in Rumania, where the effects are most dramatic. Similarly, pollution and water extraction in the Rio Colorado, which flows through the USA and Mexico, are critically endangering the totoaba *Totoaba macdonaldi*. This marine fish, the largest member of the family Sciaenidae (size up to 2 meters or 6½ feet), is endemic to the extreme north of the Gulf of California. The Mekong Basin is another important and little known system that is under increasing pressure from large-scale dam development along its course.

Introduced species
Introduced species are a major threat to many fishes worldwide and their impact can be devastating, especially in enclosed or very specialized habitats. These species compete with the native fishes for food and living space, as well as preying on them.

They may even breed with the native fishes, and place them at further risk by changing their genetic makeup. The mosquitofish and the Nile perch are well-known examples of the negative effects of introduced species. The Nile perch, a large predatory fish which may grow up to 2 meters (6½ feet) and weigh 200 kilograms (440 pounds), together with four tilapiine species, was introduced into Lakes Victoria and Kyoga. Here the Nile perch fed on the small cichlids; the resulting disaster is detailed in the cichlid chapter (see page 204).

International trade

Many fishes are commercially exploited for food, for either human or animal consumption. Sturgeons and paddlefishes are heavily exploited for caviar. Although fisheries dominate the exploitation of wild fishes, the growing trade in tropical aquarium fishes has become an important source of income for local populations in many parts of the developing world. As many as 800 species are listed as "commonly available" in the aquarium trade. Although a large part of the market is already supplied by farm-bred freshwater species (90 percent supposedly being captive-bred), a considerable portion is still supplied by wild-caught specimens from South America and Africa, especially those with uncommonly attractive varieties. Some individual fishes may be sold for huge amounts: the Asian bonytongue, for example, can fetch up to US$18,000. In contrast, roughly 90 percent of the marine species in the trade are believed to be taken directly from the wild in countries such as the Philippines, Singapore, and Indonesia. In the Philippines alone, 386 species of coral reef fishes belonging to 79 families are utilized in the aquarium trade, supplying up to 80 percent of the market. Although few of the known threatened species are of interest to the ornamental trade,

many more are at risk because of it. Trade may adversely affect wild populations via the introduction of non-native organisms (escapees from fish farms that compete with or prey on the native species and introduce parasites and disease) and the direct depletion of wild stocks. Another detrimental aspect of the international and intercontinental movement of fishes is the threat to the genetics of the species and to the ecology of their habitats. Nevertheless, the aquarium trade has kept some species alive. The Mexican goodeid has not been found in the wild for some time now, but has been kept in aquaria. Aquaculture practices have also been responsible for species translocations: the African tilapias are striking examples.

Sharks and other elasmobranchs are subjected to more gruesome forms of trade. These majestic

▲ *Ocean litter presents a rapidly escalating threat to many fishes in a variety of ways, for example in this moray's lethal predicament, hopelessly entangled in a torn fishing net.*

▼ *Asia's high total is a reflection of the increased knowledge of its fish faunas. Main causes for fish decline are habitat loss or degradation and widespread introductions (deliberate and accidental) of exotic species, which often compete with, or prey on, native species.*

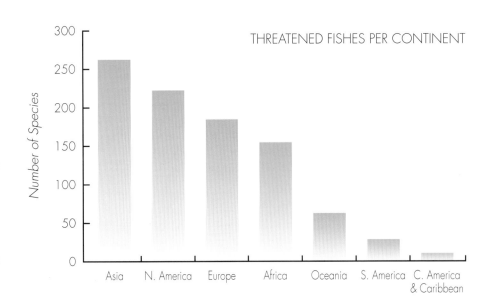

THREATENED FISHES PER CONTINENT

Number of Species

Neville Coleman

Richard Cook/Planet Earth Pictures

▲▶ *A shark victim of human persecution (above) and shark fins drying in the sun (right). Shark numbers worldwide are dropping alarmingly as a result of direct human persecution, over and above the semi-legitimate but appallingly wasteful annual harvesting of thousands of tonnes of shark-fins used in the preparation of the famous Oriental soup.*

KNOWN FISH EXTINCTIONS

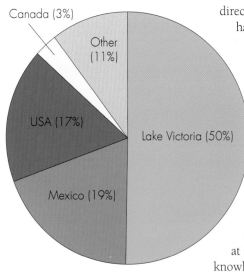

Canada (3%)

Other (11%)

USA (17%)

Lake Victoria (50%)

Mexico (19%)

▲ *Known extinctions of fish species (93) can be apportioned as in this pie-chart, but it seems reasonable to suppose that this pattern may be merely an artifact arising from the fact that, potentially, many more extinctions may have gone unrecorded in developing nations. The proportions shown in this pie-chart will most likely change in the future as our knowledge of the status of fishes throughout the world improves.*

animals are caught for their fins (to supply the market for dishes such as shark-fin soup); their liver oil (used in cosmetics, health food preparation, and as a low-temperature lubricant for high-altitude aircraft); their skins (used in the leather industry); and their meat and jaws (sold as trophies). These goods are not by-products of established fisheries, and it is feared that new industries are being set up to cater for specialist products. The most damaging form of trade is in fins; in the majority of cases, only the fins are utilized and the rest of the carcass is discarded. Species such as the great white shark and the basking shark are under pressure from direct fisheries. Sport fishers also pay handsomely for the opportunity to hunt the great white shark. Traditional Chinese medicine is another form of trade that has recently reached alarming levels. Investigations on seahorses and pipefishes showed that fishing to supply this market has placed many species under serious threat, resulting in 37 threatened syngnathid species.

THE TIP OF THE ICEBERG?

Until recently, fewer marine than freshwater fishes were known to be at risk but this simply reflected our knowledge of their status. In the past, it was believed that marine species were safe because of the great size of the oceans and the pelagic nature of their eggs and larvae. As our knowledge of their situation grows, we are finding alarming signs of deterioration of the various marine habitats and of their species and populations. This is the case with the totoaba, several species of sharks and seahorses, and a multitude of coral reef fishes. Migratory freshwater and marine species such as

the Laotian shad, sturgeons, bigeye and southern bluefin tunas, and the related billfishes are also at risk as they are preyed upon, overfished, and exposed to perils along their paths. Coastal species like the sawfishes are also affected.

One hundred and eighteen marine species, representing 40 families and 18 orders, are now known to be globally threatened. These include both wide-ranging species—such as some of the sharks (blacktip shark, dusky shark, basking shark), the Queen triggerfish, the humpback wrasse, and the jewfish—and more restricted species—such as the Tayrona blenny (Colombia), the Easter Island butterflyfish, the splendid toadfish (Mexico, Belize), the Santa Helena dragonet, and some of the seahorses. Even large species are at risk—the mighty basking and whale sharks, the two largest fishes in the world, are at risk, although the status of the whale shark is still uncertain. Whale and basking sharks feed on plankton and inhabit warm and cool seas respectively. These impressive creatures are popular with divers and yachtsmen who send information on them. Other sharks are also under pressure. Entanglement in fishing nets and selective hunting (which the great white has been subjected to since the *Jaws* films) are two of the causes for the demise of these fishes.

Encouragingly, several sharks—such as the great white, the great nurse, and the basking shark—and some rays—such as the Devil's ray in the Mediterranean and the spotted eagle ray—are now receiving legal protection in some countries. Fishing quotas are also being imposed. However, under the present evaluation system, it is not easy to assess the status of marine fishes because some widespread species, such as sharks and tunas, have separate populations under varying degrees of threat. The system is currently being tested for other marine species.

The impressive coelacanth (see page 73) has both a very small population and limited distribution. It is currently listed as endangered by the IUCN. Local fishermen have caught most coelacanths on handlines while fishing for the oilfish *Ruvettus pretiosus*. However, many past scientific expeditions have encouraged and rewarded coelacanth captures. The Coelacanth Conservation Council was established in 1987.

KNOWN EXTINCTIONS

Despite the difficulties of assessing the status of fish populations in the wild, we are aware of the loss of 93 species from 11 countries, of which the African countries of Uganda, Kenya, and Tanzania jointly account for 52 percent (50 species); Mexico and the USA follow with nearly 20 percent (19 species) and nearly 18 percent (17 species) respectively. In North America, 68 percent of the 39 extinctions that have taken place in the past 100 years were affected by the introduction of alien species. Fish extinctions have been reported from most continents, although they are still unrecorded in many localities worldwide.

Three species of cisco have vanished from the Great Lakes as a result of overfishing, pollution, habitat degradation, and hybridization with introduced species. Increased pumping of groundwater for urban development, and pollution by herbicides, are believed to have caused the extinction of the San Marcos gambusia in the springs on the San Marcos River in Texas, USA. The mosquitofish is believed to be responsible for the decline of many native North American fishes such as the Pahranagat spinedace, known only from two localities in the Pahranagat Valley in Nevada. In Mexico, three species of dace from the Valley of Mexico, where Mexico City now stands, were lost probably as a result of persistent water extraction from the valley's floor for agricultural and urban development.

BACK FROM THE BRINK

Some fishes have persisted in extremely low numbers in remote habitats and have been found only after the populations have recovered sufficiently to be noticed again. This was the case with the Owens pupfish, which was rediscovered in a remote corner of Fish Slough, California, USA, in 1964. The Shoshone pupfish was similarly rediscovered in 1986, but the population had dwindled to such a small size that the genetic variation had been lost. Being rediscovered does not guarantee the survival of the species, for in most cases, the factors that caused population decline are still operating.

FUTURE CONSERVATION

In many parts of the world, such as Asia and Africa, humans depend on fishes for their livelihood and survival, as fishes are often their only source of protein. They are not only important ecological components of aquatic habitats and food webs, but are an indispensable part of human culture and economy and, therefore, must be saved. Unfortunately, the number of known threatened species increases

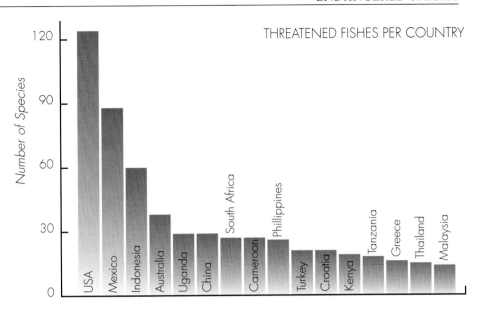

THREATENED FISHES PER COUNTRY

Number of Species

(Bar chart with y-axis marked 0, 30, 60, 90, 120; countries listed: USA, Mexico, Indonesia, Australia, Uganda, China, South Africa, Cameroon, Phillippines, Turkey, Croatia, Kenya, Tanzania, Greece, Thailand, Malaysia)

▲ *Countries such as the USA and Mexico have a greater number of threatened fishes than those in many African or Asian countries. This is likely to be influenced by the amount of information presently available, expecially on species that share aquatic systems in the border between these countries.*

every year and international listings do not necessarily guarantee their survival. Long-term programs to raise public awareness must be combined with a genuine commitment to conserve them. Many countries already provide protective legislation for their fishes but only a few enforce them. Unique environments such as the Danube delta and the Amazon Basin are currently the focus of international efforts. Sadly, unless urgent action is taken, such conservation programs may be too late for many of the species inhabiting systems like the Mekong River Basin. Captive breeding for conservation has also become an important activity in the last few years, with species like the African cichlids and the desert fishes of the United States and Mexico already receiving attention.

Conservation needs may differ according to the species but, in general, a combination of habitat preservation and management, control of pollution, elimination of introduced species, bans on poaching and overfishing, and stricter trade controls will go a long way to ensure the survival of the species now at risk.

Species conservation alone may not be enough to ensure fish survival, and a more recent approach has been to use threatened species as indicators of key environments in need of protection, such as the African Rift Lakes and coral reef systems, where more than one species is under pressure. Indeed, this strategy has unveiled a large number of coral reef fishes at risk and it is feared that the same may occur with other environments. Similarly, an ecosystem approach is also being taken in which scientists attempt to evaluate the services that the ecosystem in question is providing, both ecologically and economically. These "hotspots for conservation" may include a combination of animal and plant species. It is in such important conservation sites that our contribution to maintaining global biological diversity will be greatest.

PATRICIA C. ALMADA-VILLELA

▼ *With a total population possibly as low as three figures, the impressive coelacanth* Latimeria chalumnae *is known primarily from the Comoros Islands near Madagascar, where it is occasionally taken inadvertently by local fishermen as well as sometimes for museums and other research purposes.*

Peter Scoones/Planet Earth Pictures

PART TWO
KINDS OF

FISHES

JAWLESS FISHES

The first fossil record of a vertebrate dates back to the late Cambrian, over 500 million years ago. The first fishes did not possess jaws and hence are placed in a group termed the Agnatha. These agnathans were largely represented by ostracoderms—small fishes with a bony external skeleton—which flourished during the late Ordovician (about 450 million years ago) and Silurian (435 to 408 million years ago) periods, before dying out by the end of the Devonian (360 million years ago). In contrast, the acanthodians, placoderms, and chondrichthyans, the first representatives of the jawed vertebrates (Gnathostomata), have not been found prior to the early Silurian. For this and other reasons, the early jawless vertebrates and the gnathostomes are frequently regarded as forming an evolutionary sequence, during which hinged jaws developed from a more primitive jawless condition.

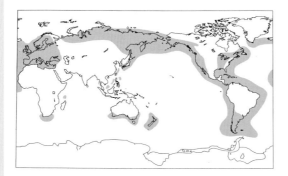

LAMPREYS AND HAGFISHES
The sole survivors of the Agnatha are the hagfishes (order Myxiniformes) and lampreys (order Petromyzontiformes) which, unlike the ostracoderms of the Upper Ordovician, do not have a bony external skeleton. They have an eel-like body with pore-like gill openings, and lack scales, paired fins, and internal ossification. Hagfishes and adult lampreys also possess a multicuspid tongue-like structure (piston). Early taxonomists considered these similarities so fundamental that they grouped the lampreys and hagfishes together in a single class, the Cyclostomata, which literally means "the round mouths".

Although there is still some value in retaining the term "cyclostome" for the two surviving agnathan groups, most scientists now regard these groups as only distantly related. Hagfishes are

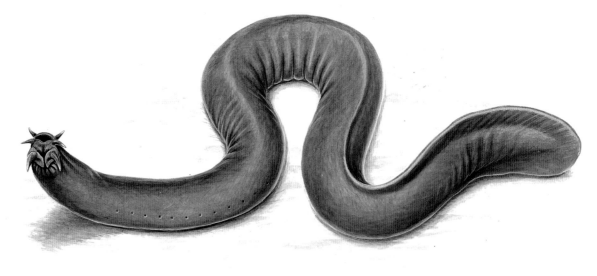

▶ Unlike lampreys, hagfishes such as this Pacific hagfish *Eptatretus stouti* (top), feed on carrion. When feeding on fish, the hagfish enters the host either through the gill region or through the ventral body wall. Many lampreys are parasitic as adults but some, such as the European brook lamprey *Lampetra planeri* (right), become adults long enough to mate, without feeding at all after the larval stage.

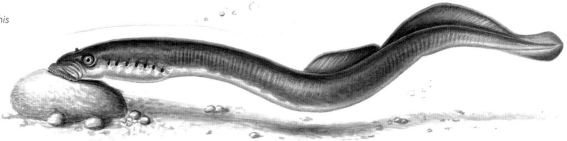

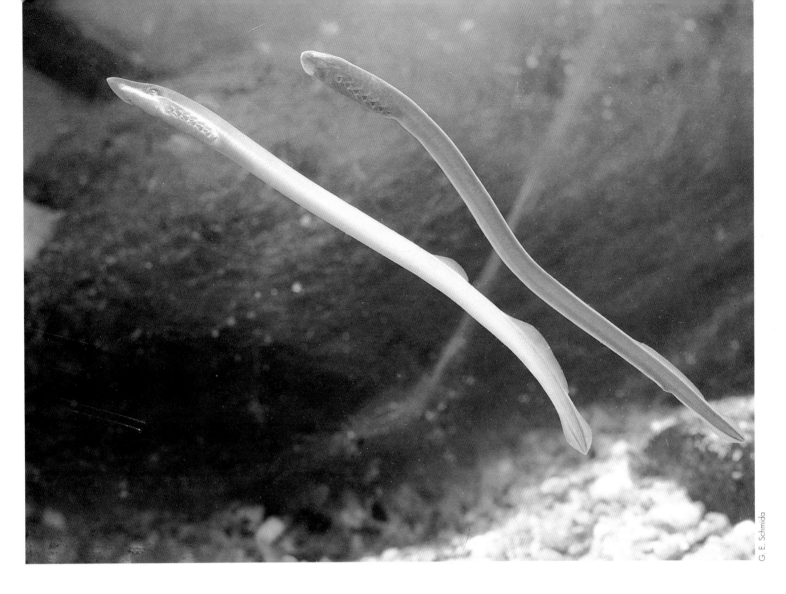

now thought to represent an early and primitive offshoot of the vertebrates, with an origin dating back to the Cambrian (more than 500 million years ago), whereas the lampreys are regarded as having evolved later and as being more closely related to the jawed fishes than to the hagfishes.

It was not until 1968 that the first fossil of either a lamprey or hagfish was described. The beautifully preserved *Mayomyzon pieckoensis* from about 280 million years ago bears a striking resemblance to contemporary lampreys. Although this demonstrates that lampreys pursued a conservative course of change during the ensuing years, the inside of the buccal funnel of *Mayomyzon pieckoensis* did not possess the well-developed teeth that are found in that position in contemporary adult lampreys. Since those teeth are used to aid attachment to their hosts during feeding, *Mayomyzon pieckoensis* may have obtained its food by a different method to modern adult lampreys, possibly by scavenging.

TAXONOMY AND DISTRIBUTION

The hagfishes and lampreys are found in the temperate waters of the northern and southern hemispheres, and in cool deep waters in certain regions of the tropics. A total of about 50 species of hagfishes and 38 species of lampreys have been described. There is disagreement, however, as to whether each group is most appropriately represented by a single family and how many genera should be recognised.

Although some taxonomists place all species of hagfish in the family Myxinidae, others separate them according to whether the ducts from each of the gills join internally, opening to the exterior by a single aperture (Myxinidae), or whether they remain discrete, opening separately to the exterior (Eptatretidae). Representatives of the "myxinids" and "eptatretids" are found in both the northern and southern hemispheres.

While few taxonomists would argue that the living lampreys constitute three discrete groups, some place all lampreys in the Petromyzontidae, whereas others restrict this family name to those groups found in the northern hemisphere and separate those of the southern hemisphere into two families, Géotriidae and Mordaciidae.

LIFE CYCLES

Hagfishes are confined to marine environments and do not have a larval phase. In contrast, lampreys are found in both fresh and salt water and do have a larval phase.

The larval lamprey (called an ammocoete) spends most of its time burrowed in the soft substrata of streams and rivers, feeding on algae, detritus, and micro-organisms drawn from the surrounding water. Although the larval phase of all species typically lasts for at least three years, the

▲ *A photograph showing the fully metamorphosed (left) and ammocoete (right) stages of the Australian short-head lamprey Mordacia mordax. The eyeless ammocoete lies burrowed in the soft bottoms of streams and rivers, filter-feeding on detritus and algae, for around three years before undergoing metamorphosis into a young adult, which has fully developed eyes and a sucking disk. The adult lamprey feeds on fish at sea before returning to rivers to spawn and die.*

▼ *Unlike lampreys, hagfishes do not have a larval stage. The eggs of this female Pacific hagfish Eptatretus stouti will therefore hatch into miniature versions of the adult.*

Ken Lucas/Planet Earth Pictures

G. E. Schmida

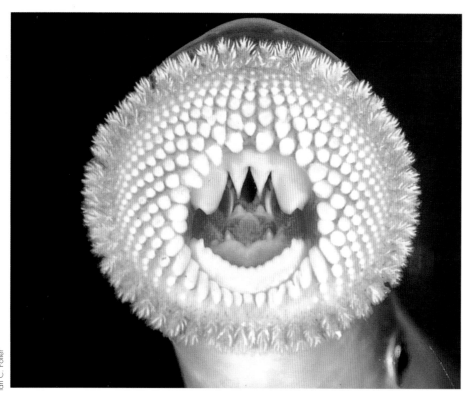

Ian C. Potter

▲ The mouth and oral disk of a lamprey. The arrangement of horny teeth lining the disk and mouth, used by the lamprey to attach itself to and feed from the host fish, is an aid to identification of the numerous species.

13 centimeters (5 inches) and 3.5 grams (⅛ ounce) to about 80 centimeters (32 inches) and 900 grams (2 pounds) in two years. When the marine phase is complete, the lamprey ceases to feed and goes back to the river, where it embarks on a migration to upstream spawning areas. This migration can be as short as a few weeks, as in the sea lamprey, or as long as 15 to 16 months, as in the pouched lamprey of the southern hemisphere.

Some anadromous species of lampreys (those species which migrate out to sea for the main period of feeding and return to the river for spawing) have developed landlocked forms in which the adult feeding phase is confined to lakes or rivers. A notorious example is provided by the sea lamprey which, during the middle part of this century, spread throughout the upper Great Lakes of North America and, by killing its hosts, made a major contribution to the collapse of certain commercial and recreational fish stocks in those waters.

There are also a few parasitic species whose life cycle is confined to fresh water, but which do not appear to have been derived from forms similar to surviving anadromous species. They include three species of *Ichthyomyzon* in large water bodies in North America, *Tetrapleurodon spadiceus* in the Río Lerma system in Mexico, and *Eudontomyzon danfordi* in the Danube River.

Most of the anadromous and freshwater parasitic species have given rise to species which never feed after they have completed the larval phase. Unlike their ancestral parasitic species, metamorphosis is accompanied by rapid development of the gonads, with the result that sexual maturation is reached 6 to 9 months later.

precise duration varies with species and the growth rate of the ammocoete.

Metamorphosis into the young adult usually takes place when the larva is 75 to 160 millimeters (3 to 6 ½ inches) long and takes 3 to 6 months to complete. When metamorphosis is completed, the young adult migrates downstream to the sea. Growth during the marine parasitic phase is rapid; the sea lamprey, for example, increases from about

▶ A sea lamprey Petromyzon marinus feeding. Sea lampreys are parasitic, attaching to the host by means of a sucking disk, rasping a hole in its side with a toothed tongue, and consuming blood and body fluids. As seems to be the case here, this frequently results in the death of the host.

H. Berthoule/Scott/Jacana/ AUSCAPE International

FEEDING

Larval lampreys feed on small algae and detritus that are brought into the branchial chamber with the water current that conveys oxygen to the gills. This food is trapped in mucus and passed down into the intestine to be digested.

Adult lampreys feed by using their sucker disk to attach themselves to their hosts and then, by rocking their toothed piston backwards and forwards, they rasp away the host flesh. Some species feed on blood, others on muscle tissue.

In contrast to adult lampreys, hagfishes do not attach to their hosts. They are the scavengers of the deep sea and feed on dead and dying fish and bottom-living invertebrates. The two plates on their tongue-like piston are everted and protracted and then retracted, thereby cutting and tearing away the flesh. Large numbers of hagfishes can be caught in certain regions by punching a hole in a 5-gallon tin, baiting it, and setting it in deep water.

HAGFISH KNOTS

Feeding in hagfishes is helped by their ability to tie themselves in a knot, and they use this as a form of leverage to tear off flesh. This knotting behavior helps to compensate for the absence of jaws and is facilitated by the fact that the backbone consists only of a rod-like notochord, with no cartilaginous or bony vertebrae, as in jawed fishes.

The ability of the hagfish to tie itself in a knot and then run the knot through the length of the body is also used for other functions. For example, it can be used to free the fish from predators that have grabbed part of its body. This freeing mechanism is helped by the production of copious amounts of mucus, making it difficult for any potential predator to maintain a firm grip. Later, the knot may again be run down the body to remove any excess mucus.

SPAWNING

Since so little was known about the reproductive behavior of hagfishes, the Copenhagen Academy of Sciences offered in 1854 an award for information on this aspect of the biology of one of the local species. We are little the wiser today and this handsome prize has still not been claimed. It has been suggested that hagfishes may burrow in deep water during breeding. The eggs of hagfishes are very large and enclosed in a tough case, from which extend filaments with anchor-like terminals.

The spawning behavior of some species of lampreys is well known. Typically, the maturing lampreys congregate in shallow waters, often in tributary streams, where the substrate is made up of pebbles, gravel, and sand. The lampreys create a depression in the stream bed by their activity and by using their sucker disk to move any larger pebbles to the rim of this "nest" area. The fully mature female uses her disk to attach herself to a large pebble or rock. The male attaches himself by

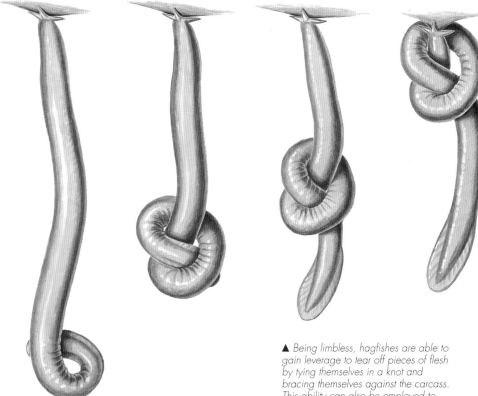

his disk to the female and then coils himself around her. His pressure causes the female to release eggs, which are immediately fertilized by the sperm that are released by the male at the same time. This flurry of activity stirs up the sand in the bottom of the nest, so that the eggs become buried. The eggs take just under two weeks to hatch into larvae.

IAN C. POTTER

▲ *Being limbless, hagfishes are able to gain leverage to tear off pieces of flesh by tying themselves in a knot and bracing themselves against the carcass. This ability can also be employed to escape from the grip of a predator or to rid themselves of excess mucus.*

▼ *Sea lampreys Petromyzon marinus spawning in a Nova Scotia river. After the nesting site is cleared by the male, the female sea lamprey anchors herself to a rock, the male attaches himself to the female and wraps his body around her, then they both vibrate rapidly and release eggs and milt simultaneously.*

Gilbert Van Ryckevorsel/Planet Earth Pictures

SHARKS, RAYS & CHIMAERAS

CLASS CHONDRICHTHYES
• 14 orders • 50 families
• 165 genera • *c*. 960 species

SMALLEST & LARGEST

Dwarf lantern shark *Etmopterus perryi*
Total length: 20 cm (8 in)
Weight: 15 g (½ oz)

Whale shark *Rhincodon typus*
Total length: at least 12 m (39 ft)
Weight: 12,000 kg (26,455 lb) +

CONSERVATION WATCH
!!! The Ganges shark *Glyphis gangeticus* and the largetooth sawfish *Pristis perotteti* are critically endangered.
!! 3 species of the sawfish genus *Pristis* are endangered.
! 4 species including the great white shark *Carcharodon carcharias* are vulnerable.

Sharks, rays, and chimaeras make up the class Chondrichthyes. They have a skeleton that is almost entirely cartilage and are generally referred to as the cartilaginous fishes. With the Osteichthyes, or bony fishes, they form the two major groups of living fishes. While cartilaginous fishes are relatively insignificant in terms of the total number of living fish species, some of their members, mainly the sharks, have gained undeserved notoriety for their attacks on humans, leading them to be seen as primitive, stupid, vicious eating machines. However, the results of recent research means that sharks and their relatives are now being viewed in a much better light, and appreciated as very successful and important high-order predators in marine ecosystems.

▼ *The blue shark* Prionace glauca *is a wide-ranging hunter of the open ocean.*

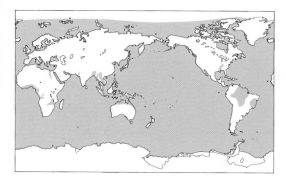

EVOLUTION

Sharks evolved from a group of bony fish ancestors during the Silurian period, about 450 million years ago. Chimaeras probably derived from prehistoric sharks in the late Carboniferous period (about 350 million years ago), with the first skates and rays appearing in the Triassic era, over 200 million years ago. These fishes evolved a cartilaginous skeleton from one of bone, which reduced weight and assisted in buoyancy control.

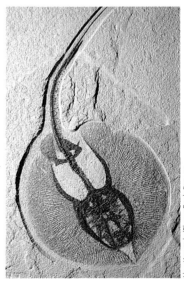

What features separate chondrichthyans from other living fishes? Unlike the jawless agnathans (lampreys and hagfishes), which also have skeletons of cartilage, these fishes have well-developed jaws, paired nostrils, and paired pectoral and pelvic fins. In addition to their cartilaginous skeleton, they differ from the Osteichthyes in having dermal denticles, and teeth that are either replaced serially throughout their life or which are fused into bony plates and grow continuously.

The 14 orders of living chondrichthyans can be classified into two subclasses: a large group that includes sharks, skates, and rays (Elasmobranchii); and the primitive chimaeras (Holocephali), which includes the spookfishes, ghost sharks, and elephant fishes. Holocephalans are a rather bizarre group of mainly deepwater bottom-dwelling fishes. They differ from members of the Elasmobranchii in having only a single gill cleft and four gills on each side of the head, mainly naked skin, teeth fused into plates, and the upper jaw fused to the skull. The elasmobranch orders contain a diverse array of forms which are adapted to a broad range of habitats.

SHARKS WITH SIX AND SEVEN GILLS
Whereas most chondrichthyans have five gill openings, members of one shark group have either six or seven. The order Hexanchiformes, which includes the frill, sixgill and sevengill sharks,

contains two families and five species of mainly deepwater fishes. As well as having extra gill slits, they have a single spineless dorsal fin and an anal fin. These sharks are generally considered to be among the least specialized of living cartilaginous fishes, resembling primitive extinct forms in the structure of their skeleton, digestive, and excretory systems. However, they sometimes show advanced specializations, such as the bizarre body form of the frill shark *Chlamydoselachus anguineus*, with its eel-like body and reptile-like head. The frill shark's shape is a likely adaptation for hunting prey in caves or crevices on the continental slope, while its highly expandable jaws allow it to take unusually large prey. The group includes two large species which grow up to 4.8 meters (15 ¾ feet), including a shallow-water omnivore which eats other sharks, carrion, and even seals.

HORN SHARKS
The horn or bullhead sharks belong to the order Heterodontiformes which has one family and eight species. They have an ancient lineage, being closely related to the extinct hybodontid sharks which date from the Devonian period (408 to 360 million years ago). They can be easily identified, as they are the only living sharks with dorsal fin spines and an anal fin: other living "spiny" sharks lack an anal fin. They also have large, blunt heads with bony crests above the eyes, and a small

▲ *The Port Jackson shark* Heterodontus portusjacksoni *(top) of southern Australia shows the fin spines and bulbous head characteristic of bullhead sharks. A member of the carpet shark group, the epaulette shark* Hemiscyllium ocellatum *(below) is a small tropical species which is active mainly at night and feeds on invertebrates.*

▼ *Sharks arose during the Silurian period some 450 million years ago, long before the first vertebrates invaded the land. This fossil stingray is from early Eocene deposits in Wyoming, USA.*

Alex Kerstitch/Planet Earth Pictures

▲ *Largest of all fish, the whale shark* Rhincodon typus *roams surface waters of the Indian, Pacific, and Atlantic oceans. Despite its awesome appearance, it is a slow-moving plankton-feeder presenting little danger to humans.*

▶ *A filter-feeder that is second only to the whale shark in size, the basking shark* Cetorhinus maximus *favors cooler waters of the Atlantic and Pacific oceans. Its extraordinarily large gill slits almost encircle the head.*

almost terminal mouth. Their highly specialized dentition is differentiated into small, sharp teeth in front and large, blunt and flattened molars at the back. Horn sharks are medium-sized, up to 1.6 meters (5 ¼ feet), mainly shallow-water, bottom-living fishes which have somewhat restricted distributions in warm temperate and tropical areas. They feed primarily on invertebrates such as sea urchins, mollusks, marine worms, and crustaceans at night, resting during the day in caves or crevices. Despite being sluggish swimmers, the adults of one Australian species, the Port Jackson shark *Heterodontus portusjacksoni* make long seasonal migrations, returning each year to their breeding sites. Horn sharks survive well in aquariums, where they have contributed much to our knowledge of shark courtship and mating behavior.

CARPET SHARKS AND THEIR ALLIES

A diverse group of 7 families and about 33 species make up the order Orectolobiformes. It includes the gigantic whale shark *Rhincodon typus*, the colorful carpet and zebra sharks (families Parascyllidae and Stegostomatidae), the flattened wobbegongs (family Orectolobidae), and the nurse sharks (family Ginglymostomatidae). Although differing widely in appearance, they are all closely related, and share a subterminal mouth placed in front of the eyes, nasal barbels, a groove connecting the mouth to the nostrils, and spiracles. They usually live on or near the bottom in shallow tropical parts of the Indo-Pacific, and are most diverse off Australia. The group has adapted to a wide range of life history strategies. Some, such as the carpet and zebra sharks, are relatively inactive with elongate bodies and weakly

developed tails. They feed mainly at night on bottom-dwelling invertebrates and small fish. Specialist feeders, such as the tawny shark, use the gullet as a suction pump to extract prey from their hiding places in reef crevices. The well-camouflaged wobbegongs ambush their prey, seizing them with their dagger-like teeth and extensible jaws. The whale shark of the open sea feeds near the surface, sucking plankton and small fish into its enormous mouth before sieving them through filter screens on its gills.

VARIABLY SHAPED MACKEREL SHARKS

There is a wide range of body shapes within the 7 families and 16 species which comprise the order Lamniformes. Ranging from 1 to 10 meters (3 to 33 feet) in length, they form a rather loose assemblage of species not immediately recognizable by a few distinctive characters, as are some other shark orders, but rather by a lack of them. The group, which is generally widely distributed in both oceanic and coastal parts of tropical and temperate seas, includes many specialized feeders. The bizarre deepwater goblin shark *Mitsukurina owstoni* uses its overhanging snout for electrodetection of prey. The huge basking sharks (genus *Cetorhinus*) and megamouth sharks (genus *Megachasma*) feed on plankton, while the thresher sharks (genus *Alopias*) use their enormously long tail to herd and stun fish. Several species have raised body temperatures and hydrodynamic specializations for rapid swimming; these include the mako, which is the fastest of all sharks. Well-known sharks, such as the gray nurse and crocodile sharks, are in this group. The powerful, super-predatory white shark is at the top of the marine food chains, feeding on large prey such as seals, dolphins, whales, and other sharks.

GROUND SHARKS

The order Carcharhiniformes, collectively known as ground sharks, has the most species of living sharks, containing 8 families and more than 200 species. They include what are widely considered to be the "typical" sharks, such as the whalers or requiem sharks (family Carcharhinidae), which have streamlined bodies, two spineless dorsal fins, an anal fin and five gill slits. There are also other more complex features which characterize the group. Ground sharks are mainly found in tropical and temperate coastal waters, but some catsharks are bottom-dwellers in deep water, some whalers are oceanic, and the highly successful bull shark *Carcharhinus leucas* occurs both in the sea and in freshwater lakes and rivers. Catsharks are mainly small (less than 1 meter or 3 ¼ feet), rather slow-moving, and live near the ocean bottom at depths ranging from very shallow to more than 2,000 meters (6,600 feet) depths. The houndsharks include several important commercial species such as the school and gummy sharks. Most whaler sharks are found over the continental shelves where they feed mainly on small fish and squid, but one large species, the omnivorous tiger shark, includes sea snakes, turtles, and other sharks in its diet. The strange-looking head of the

▲ The leopard shark *Triakis semifasciata* is one of a group of ground sharks known as hound sharks. Because of its relatively small adult size and attractive markings it is a popular display fish for public aquaria.

▼ The swellshark *Cephaloscyllium ventriosum* is a member of the catsharks—with nearly 100 species, much the largest family of sharks. Almost a scavenger rather than predator, it is a sluggish, nocturnal bottom-feeder that feeds on crustaceans and dead or sleeping fish.

Howard Hall/Oxford Scientific Films

63

hammerhead sharks (family Sphyrnidae) gives them increased maneuverability for catching agile prey such as squid, and may also improve their sensory capabilities.

DOGFISHES

Dogfishes and their allies (order Squaliformes) form a large group of mainly deepwater sharks which occur in tropical, temperate, and even Arctic and Antarctic seas. There are 3 families and over 90 species which share the common characters of 5 gill slits, no anal fin, and two dorsal fins that are usually preceded by spines. Among them are the luminous deepsea lantern sharks (genus *Etmopterus*), one species of which grows to only 20 centimeters (8 inches) and is the smallest living cartilaginous fish, and the huge Greenland shark *Somniosus microcephalus*, which grows up to 6.4 meters (21 feet). It inhabits Arctic seas and includes carrion and seals in its diet. The most bizarre feeding method of any shark is employed by the aptly named cookie-cutters (genus *Isistius*), reaching only 50 centimeters (20 inches) in length. They ambush their prey, latch onto, and carve out plugs of flesh from large fish, dolphins, seals, and even whales. Other fascinating dogfishes include the strangely shaped prickly dogfish (genus *Oxynotus*), the bramble sharks (genus *Echinorhinus*) with their thorn-like denticles, and the small, oceanic pygmy sharks, which migrate from near the surface at night to below 400 meters (1,310 feet) during the day. One species, the white-spotted spurdog *Squalus acanthias*, which is fished commercially in some areas, has the longest gestation period (18 to 24 months) of any known shark. It also lives the longest, and has been recorded as reaching up to 70 years.

ANGEL SHARKS

The order Squatiniformes consists of a single family and about 15 species of flattened, ray-like fishes. They have no anal fin, a terminal mouth, nasal barbels, eyes on top of the head, and a tail with

▼ *The sawfishes, family Pristidae, number about five species distinguished by their flat, elongated, and blade-like snouts. Rimmed with slender triangular or needle-like teeth, the saw is used to burrow in bottom sand or mud, and as a weapon to slash among schools of small fish to maim and immobilize them.*

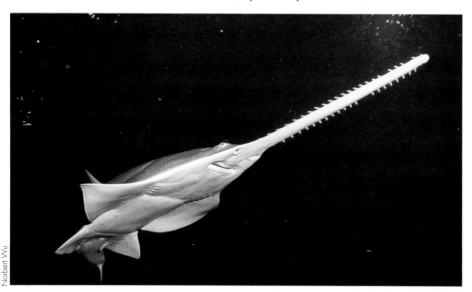

Norbert Wu

REPRODUCTIVE SYSTEMS

All living chondrichthyans have internal fertilization, and produce a low number of large yolky eggs relative to bony fishes. They have a range of reproductive strategies varying from simple oviparity (egg-laying) to advanced viviparity (live-bearing) where the embryos are nourished via a placenta analogous to that in mammals. Copulation is achieved by means of the penis-like claspers of males, following complex courtship behavior which may involve fairly severe biting of the female. Male chimaeras also have accessory claspers on the head and before each pelvic fin.

About 43 percent of chondrichthyans are egg-layers, including bullhead sharks and catsharks, skates, and chimaeras. The eggs are encased in a tough horny case known as a "mermaid's purse". Although the eggs usually have tendrils to anchor them to weed on the bottom, some have instead screw-like flanges which serve to wedge them into sand or rocky crevices. In most oviparous species, the embryo develops inside the eggcase and outside the mother, with hatching taking up to 15 months after the egg is laid. In a few species the eggs are retained for a period inside the mother so that the embryo is more advanced and hatches more quickly once the egg is laid.

In some live-bearing species, such as the sixgill and sevengill sharks, dogfishes, most houndsharks, and some rays, the embryo is nourished until birth from its yolk sac, with few additional nutrients provided by the mother. Many species of rays have a more advanced form of viviparity, where the embryo's yolk reserves are supplemented later in pregnancy by the transfer of "uterine milk" via specialized fingerlike projections in the uterus, called villi. The mackerel sharks, which produce more numerous, smaller eggs than other chondrichthyans, have evolved a bizarre form of viviparity called oophagy. In one type of oophagy, embryos of makos and threshers are nourished by feeding on unfertilized eggs which the mother continues to ovulate. This form of reproduction is taken to its extreme in the gray nurse sharks where the strongest embryos actively hunt and consume their siblings inside the uterus, so that only one pup from each uterus survives. In the whalers and

the lower lobe longer than the upper lobe. They are medium-sized fishes, less than 2 meters (6 ½ feet) in length and are found in tropical and temperate zones mostly in shallow water, although one species occurs down to 1,300 meters (4,250 feet). Angel sharks often lie half-buried in the sand, where they are perfectly camouflaged to ambush approaching fish and bottom-dwelling invertebrate prey. They have extensible jaws and

Eric M. LeFeuvre

hammerheads, the embryonic yolk-sac becomes modified in mid-term into a placenta which implants on the uterine wall. For the remaining gestation period the embryos receive maternal nourishment through an umbilical cord. Gestation periods in viviparous chondrichthyans vary from about 6 to 22 months and litter sizes range from 2 to 300. Females of different species may breed every year, every other year (or even less frequently) or occasionally twice a year, and birth may occur in distinct seasons or throughout the whole year.

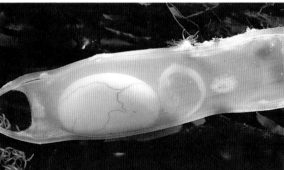

G.I. Bernard/Oxford Scientific Films

▲ *Mating in sharks usually follows elaborate courtship rituals. During copulation the male inserts one of his claspers into the female's cloaca while maintaining a bite-hold on one of her pectoral fins.*

◀ *An embryo with its yolk in the eggcase of a smallspotted catshark Scyliorhinus canicula.*

spike-like teeth, and can snap upwards or forwards when supported by their pectoral fins. Some species are fished commercially.

FISHES WITH SAWS
In saw sharks (order Pristiophoriformes) and sawfishes (order Pristiformes), the head is modified to form a long, flattened blade with sharp, tooth-like projections along its sides. This

"saw" is used to stun and kill prey, which may range from small, mud-dwelling invertebrates to quite large fishes. Members of both orders are essentially shark-like in shape but show distinct adaptations for life on the sea floor: their undersurface is somewhat flattened and their eyes are located on top of the head. The smaller saw sharks, comprising 1 family, 2 genera and 7 or more species, rarely exceed 1.5 meters (5 feet).

Georgette Douwma/Planet Earth Pictures

▲ *The fiddler ray* Trygonorhina fasciata *of southern Australia is a member of the family Rhinobatidae, which includes the guitarfishes and shovelnose rays and is allied to the skates, stingrays and electric rays.*

▶ *Measuring about 90 centimeters (3 feet) across, the southern stingray* Dasyatis americana *favors sandy flats under warm shallow water, where it often lies half-buried. As in other stingrays, venom delivered through spines at the base of the long whiplike tail can inflict painful and occasionally fatal wounds to unwary humans.*

a disk, and the extent to which the disk is developed varies between groups. In the guitarfishes or shovelnose rays (order Rhinobatiformes), the disk is absent or poorly developed compared with the tail. Nevertheless, in some species the head is highly modified to be broadly flattened and oval, triangular, or shovel-shaped. The group contains 3 families and 50 or so species which live on the continental shelves in warm temperate and tropical seas. Most species are small but some, such as the sharkfin guitarfishes, may grow quite large, attaining a length of 3 meters (10 feet).

SKATES

The highly successful skates (order Rajiformes) are the most diverse group of rays, with approximately 200 species worldwide. Our knowledge of their taxonomy is far from adequate and the exact number of species and genera is unknown; there are even several differing views as to which families are valid. They live near the bottom in all oceans (apart from much of the Pacific), from shallow estuarine to deepsea habitats. Most have narrow distributions and many are only found in relatively small geographic regions. Skates may grow up to 2 meters (6 ½ feet) in length but most are much smaller, usually less than than 1 meter (3 ¼ feet). They have a distinctive body shape consisting of a large, flattened, circular to rhomboidal disk and a small tail with 0 to 2 (mostly 2) small dorsal fins near its tip. The skin surface is usually partly covered with thorns and fine granulations, the arrangements of which vary from one species to another. Males of this order have extra patches of specialized thorns on the outer disk which are used to grip the female during copulation.

ELECTRIC RAYS

The electric rays and their allies (order Torpediniformes) have electric organs located either side of the disk behind the eyes, capable of producing quite high voltage. The order consists of 4 families and more than 43 species worldwide in temperate and tropical seas. All are sluggish bottom-dwellers living from inshore to depths exceeding 1,000 meters (3,280 feet) on the continental slope. Their general form is similar, although their disk varies greatly in size relative to the tail from one genera to another. In extreme cases, such as in the coffinray, the tail is smaller than the pelvic fins. Unlike many other rays that use a undulating motion of the disk to move around, propulsion comes from the tail and caudal fin. Consequently, those with very short tails are particularly slow.

To catch prey, these rays rely on lying concealed beneath the substrate, then stunning unsuspecting passers-by, such as small fishes and bottom-dwelling invertebrates.

Most occur offshore in the Indo-Pacific, although one species is found in the Caribbean.

Although members of the two groups are similar in form, sawfishes are rays rather than sharks. Apart from major skeletal differences, the gills of sawfishes lie on the undersurface (instead of on the sides of the head, as in sharks). Also, unlike saw sharks, they lack sensory barbels on the undersurface of the snout. Sawfishes are widespread inshore in tropical seas, but at least one species lives primarily in estuaries and fresh water. Some of the seven species of the single family which make up the group are among the largest chondrichthyans, reported to reach in excess of 7 meters (23 feet).

Members of both orders are live bearers. In order to overcome the inherent difficulties of giving birth to an offspring with a sharp-toothed blade on its head, pups are born with a flexible saw and soft (often raked-back) teeth which become firm soon after birth.

SHOVELNOSE RAYS

Adapting to life on the sea floor has resulted in the evolution of major structural changes to the head and bodies of some chondrichthyans. In many rays the head and pectoral fins have become joined to the body to form a structure resembling

Marty Snyderman

▲ *Devilrays (genus Mobula) like these photographed off Costa Rica, differ from the closely related mantas (Manta) in having the mouth on the underside, not the front, of the head. They "fly" through the water with slow, graceful flaps of their enormous "wings".*

STINGRAYS AND THEIR ALLIES

Not all rays live primarily on the bottom. Three families of the order Myliobatiformes (devilrays, cownose, and eagle rays) are adapted to life in open water. The order contains more than 150 species, most of which live on the continental shelf in tropical regions. They are powerful swimmers, with wing-like disks. These include the largest of rays, the plankton-feeding manta, which may attain a width in excess of 6.5 meters (21 ⅓ feet). The head of these rays protrudes slightly forward of the disk and they usually have long, whip-like tails. The other five families of this order consist mostly of bottom-dwellers. The stingrays, round stingrays (stingarees), sixgill rays, and river rays have a large circular to rhomboidal disk and a

▶ *Among the most unusual of the chimaeras, the plownose or elephant fish Callorhynchus milii inhabits cool, deep waters off southern Australia, but sometimes enters coastal waters and river estuaries. The fleshy trunk-like snout is used to probe bottom ooze for food.*

slender tail usually armed with a serrated stinging spine. This spine, which is a modified body scale, has a venom gland and grooves for passing toxin to its tip. The remaining family, the butterfly rays, have a large, extremely flattened disk and a minute thread-like tail. Like other members of the group, they use the pectoral fins rather than the tail to propel themselves.

GHOULS OF THE SEA
Chimaeras and their allies belong to the order Chimaeriformes which has 3 families and about 34 species. Their ghostly appearance and unique body form has given rise to a wide array of common names, such as spookfishes, ghost sharks, and elephant fishes. All have rather elongate, soft, scaleless bodies with a bulky head covered in prominent sensory canals, a single gill-opening, a prominent spine before the first dorsal fin, and only three pairs of large, often beak-like, teeth in the mouth. The snout is highly modified in some of the genera. The elephant fish, for example, uses its hoe-shaped snout to probe the substrate for invertebrates and small fishes on which it feeds. The snouts of spookfishes are extended to form a conical or paddle-shaped appendage, used for sensory perception. In addition to the claspers typical of sharks and rays (reproductive organs on the ventral fin used for injecting sperm), male chimaeras have retractable sexual appendages before the pelvic fins and on the forehead. These spiny structures are used to clasp or stimulate the female during copulation. Chimaeras are small to moderate in size, attaining up to 1.5 meters (5 feet) excluding the fleshy filament that is sometimes present at the tail tip. Most live near the bottom on the continental shelf, to a depth of at least 2,600 meters (8,500 feet), but a few occur inshore near the coast in temperate regions.

INTERACTIONS WITH HUMANS
Each year about 30 people are killed by sharks—and some 700,000 tonnes (771,400 tons) of sharks and rays are killed by humans. Chondrichthyans are important for food and pharmaceutical and medical products but, being susceptible to over-fishing, they are difficult to manage economically on a sustained basis. With more responsible media reporting and better public education, we are learning to appreciate sharks and rays and the role they play in marine ecosystems. Their conservation has now become an important issue.

JOHN STEVENS & PETER R. LAST

▼ (Below) Although its scientific name means "water rabbit", the other common name of the blunt-nosed chimaera Hydrolagus colliei is spotted ratfish, because of its long, tapering tail. The spine in the dorsal fin is grooved and venomous. Although most sharks and rays are marine, the ocellated freshwater stingray Potamotrygon motoro (bottom) belongs to a family of rays confined to freshwater in South America and Africa. Like their marine cousins, river rays are armed with a grooved, venomous spine and in South America are feared more than piranhas.

SUBCLASS SARCOPTERYGII
• 3 orders • 4 families • 4 genera
• 7 species

SMALLEST & LARGEST

East coastal lungfish *Protopterus amphibius*
Total length: 44 cm (17 in)

East African lungfish *Protopterus aethiopicus*
Total length: 2 m (6½ ft)

CONSERVATION WATCH
!! The coelacanth *Latimeria chalumnae* is listed as endangered in the IUCN Red List.
■ One African lungfish, *Protopterus dolloi,* has narrow habitat preferences.

LUNGFISHES & COELACANTH

Lungfishes and the coelacanth are classified together in the subclass Sarcopterygii, or fleshy-finned fishes. They are relics of ancient fish groups that were near the evolutionary stem of amphibians, reptiles, birds, and mammals. They therefore occupy an important position in vertebrate evolution.

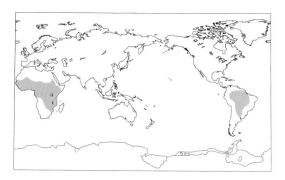

ANCIENT SURVIVORS

Lungfishes first appeared in the fossil record about 380 million years ago, during the Devonian period. They are now represented by the family Protopteridae with one genus and four species in Africa, the South American family Lepidosirenidae with one genus and one species, and the Ceratodontidae with one genus and one species in Australia. The swim bladder of lungfishes has developed into a highly vascularized lung that enables them to breathe air. The three families show a gradation in tolerance of seasonal environmental desiccation, some species of *Protopterus* aestivating (passing the dry season in a dormant condition) in a dry cocoon, *Lepidosiren* exhibiting partial aestivation in a moist chamber,

and *Neoceratodus* not aestivating at all. The African and South American lungfishes are more closely related to one another than to the Australian lungfish.

Coelacanths were first recorded as fossils over 380 million years ago and were previously widespread in marine and freshwater habitats. Their fossil record ends abruptly about 65 million years ago (the beginning of the Paleocene epoch) and it was therefore a great surprise when a living coelacanth was caught off South Africa in 1938. The coelacanth *Latimeria chalumnae* is now known to occur mainly off the Comoros Islands in the western Indian Ocean, although a single specimen was recently caught off Mozambique.

AFRICAN LUNGFISHES

African lungfishes are elongate fishes with soft, partially embedded scales; a continuous dorsal, tail, and anal fin; thread-like pectoral and pelvic fins; and two lungs. *Protopterus annectens*, which reaches 1 meter (3 ¼ feet) in length, occurs in West and Central Africa and ranges southwards to the Zambezi and Limpopo rivers. *P. aethiopicus* occurs in Central and East Africa, *P. dolloi* in the Zaïre basin, and *P. amphibius* in coastal East Africa. They inhabit floodplain lakes, swamps, small streams, large rivers, and some of the rift valley lakes. Their preferred habitats are the shallow

▶ One of the many groups of fishes whose distribution can best be explained by continental drift, lungfishes are strictly freshwater. The South American lungfish Lepidosiren paradoxa (above) and the African species, such as the spotted African lungfish Protopterus dolloi (below), are more closely related to one another than to the Australian lungfish.

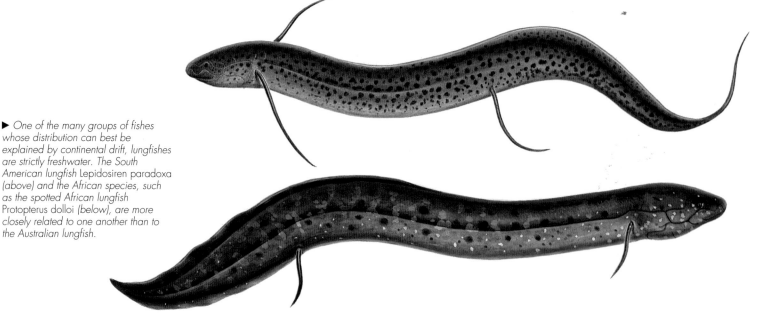

◄ Most primitive of the living lungfishes, the Australian lungfish Neoceratodus forsteri differs from the African and South American species in that it has large scales, flipper-like fins, well-developed gills and only one lung. The larvae of this large species bear a strong resemblance to amphibian tadpoles.

edges of deeper bodies of water and temporary pools with low oxygen levels.

African lungfishes are carnivores that feed on mollusks, crustaceans, and larval insects. Larger specimens also eat fishes and may be cannibalistic. They approach their prey by slow stalking or ambush, and capture it with a powerful suction action. After capture the food is chewed, spat out, and sucked back into the mouth so that it is thoroughly broken up when swallowed. Their main predators are crocodiles, otters, large fishes, storks, and humans.

African lungfishes use two methods of movement: anguilliform (eel-like) swimming, and crawling with the paired fins. They swim when scavenging in midwater or moving to the surface to breathe air. They crawl when in very shallow water or when scavenging along the bottom. When crawling, the buoyant body is raised and the thin paired fins are used alternately to propel the body forward.

With the onset of the dry season, P. annectens prepares to aestivate. It digs a vertical burrow in the submerged mud of the swamp. The hole is excavated by body movements and by biting the soil and expelling the mud through openings in its gills. The depth of the burrow is about equal to the length of the fish. When the base of the burrow is reached, the fish turns through 180°, coming to rest with the snout pointed upwards, so that a bulb-like chamber is formed. The fish makes regular visits to the mouth of the burrow in order to breathe, then sinks back into the chamber. Once the water level in the mud falls below the level of the fish's snout, these visits cease and the fish remains coiled in the chamber and secretes large quantities of mucus. As the water dries up, this mucus dries and forms the closely fitting cocoon that envelops the dormant lungfish and prevents it from drying out. The cocoon wall is very thin and its top has a small hole leading to a funnel that enters the fish's mouth and through which air is drawn into the lungs. P. annectens aestivates for several months depending on the length of the dry season. In captivity they may remain dormant for several years. During

aestivation the lungfish undergoes marked metabolic changes that include a reduction in heart rate, a decrease in blood pressure, and a decline in oxygen consumption. Fat reserves and muscle protein are used as sources of energy, and waste products are stored. They are sometimes fished with spears while aestivating.

In the East African species P. aethiopicus cocoon formation is relatively rare, as they live in more permanent bodies of water. Their burrows are horizontal or diagonal and usually remain flooded or moist. P. dolloi adult males burrow into the bottom or re-occupy existing burrows or nests at the onset of the dry season. The females and immature males leave the swamps and spend the dry season in open water. The males in the burrows are not dormant but frequently swim up the burrow to breathe at the surface of the water. They remain in the nests as the swamps refill with the next rains and spawning takes place there. Although aestivating P. dolloi have not been observed to secrete cocoons in the wild, they can, like P. aethiopicus and P. amphibius, be induced to form cocoons under laboratory conditions. In all four species environmental conditions determine the actions taken to cope with their habitat drying out, but the "classical" pattern of cocoon formation may only be common.in P. annectens.

African lungfishes are seasonal spawners that breed during periods of protracted local rainfall. The breeding nest of P. annectens is usually a broadly U-shaped tunnel extending about 40 centimeters (16 inches) into the substrate with the two openings about 50 centimeters (20 inches) apart. Between the vertical arms of the nest there is an enlarged chamber in which the eggs are deposited and the larvae are guarded. The openings of the nest may be submerged or exposed to the air. The nest of P. aethiopicus consists of a vertical or diagonal burrow under roots and plant debris or a clearing in shallow water. The male vigorously defends the larvae in the nests.

Female P. aethiopicus spawn over 5,000 eggs each about 5 millimeters (⅕ inch) in diameter. The embryos hatch 11 to 15 days after

Ken Lucas/Planet Earth Pictures

▲ The East African lungfish Protopterus aethiopicus is the largest of the four African species of lungfishes. These four species are similar to the South American lungfish in having a pair of lungs; the Australian lungfish has only a single lung.

▶ *The South American lungfish* Lepidosiren paradoxa *spawns in burrows, and the male guards the developing young. Small supplementary gills develop on his pelvic fins, which are thought to supply additional oxygen to the young by enriching the muddy waters in the burrow.*

▼ *The Australian lungfish now has a restricted distribution in east-central Queensland but fossil specimens show that it was once far more widespread across the continent. Very similar species occurred in Australia during the Cretaceous period some 100 million years ago.*

Ken Lucas/Planet Earth Pictures

G.E. Schmida

fertilization. The larvae are relatively inactive at first and are attached to the nest wall by their thoracic suckers. They begin airbreathing at a length of about 2.5 centimeters (1 inch) at this stage while their external gills are still well developed. They remain in the nest for about 55 days by which time the yolk supply has been exhausted. The free-living juveniles then forage actively among roots and debris for insects and crustaceans.

THE SOUTH AMERICAN LUNGFISH

Very little is known about the South American lungfish *Lepidosiren paradoxa* despite its wide distribution in swamps and floodplain lakes of the extensive Amazon and Parana river systems. In common with African lungfish, they have reduced gills, two lungs, and thread-like pectoral and pelvic fins, but they are more elongate and the paired fins are shorter. They can only breathe air and are capable of partial aestivation, surviving for months in a resting chamber of moist mud and mucus. South American lungfish spawn at the beginning of the rainy season in a burrow excavated by the male who guards the young. The pelvic fins of the males bear supplementary respiratory capillaries during the spawning season that may release oxygen from the blood to the water surrounding the young. Adults reach a length of 1.2 meters (4 feet). They capture their prey of bottom-living crustaceans, mollusks, and small fishes using a sudden powerful sucking action; they masticate their prey with their toothplates before swallowing.

THE AUSTRALIAN LUNGFISH

The Australian lungfish *Neoceratodus forsteri* differs from the African and South American lungfishes in appearance and habits. It has a wide flat head, a heavily scaled body, paddle-shaped paired fins, and a pointed tail. The mouth is on the underside of the body and the eyes are small. The snout is covered in pores and the sensory canals on the head and along the lateral line are prominent. The body is dull brown or olive green dorsally and pink below. They reach a size of 1.8 meters (6 feet) and 45 kilograms (100 pounds).

Australian lungfish mainly inhabit large water bodies such as deep pools and reservoirs and the mainstreams of large, slow-flowing rivers. Their natural range includes the drainages of the Burnett and Mary rivers in south-eastern Queensland but they have been introduced into the Brisbane, Fitzroy, and adjacent river systems in southern Queensland.

It mainly feeds at night on prawns, crabs, insect larvae, mollusks, and small fishes, although plants are also eaten. The prey is captured by active pursuit and suction and the strong toothplates are capable of crushing and grinding the shells of snails and crabs. They swim using slow, sinuous movements of the tail and fins, but when alarmed they move rapidly using the tail only. The paired fins are used to brace the body during feeding.

Unlike African and South American lungfishes, the Australian lungfish is not restricted to only breathing air. The lung is a supplementary respiratory organ that is used during intense activity, when oxygen tensions in the water are low, or when the gills become clogged with mud. The juveniles likewise can breathe air but they also use skin respiration. Australian lungfish do not have the ability to survive complete habitat desiccation but they can survive for several months in a dried-up pool, provided that moist leaves and mud are present.

Australian lungfish spawn after a protracted courtship, usually before the summer rains. The eggs are laid in well-oxygenated water at temperatures of 20 to 26°C (68 to 79 °F). Batches of 50 to 100 eggs are deposited on water plants in flowing water. No nest is made and the adults do not exhibit parental care. The newly laid eggs, which are 3 millimeters (1/10 inch) in diameter and heavily yolked, adhere to aquatic plants. Hatching occurs after 21 to 30 days and a size of 25 centimeters (10 inches) may be reached after 6

months and 50 centimeters (20 inches) after 20 months, although growth may be much slower. The larvae start external feeding at 41 to 56 days, once the yolk has been absorbed. There is no obvious metamorphosis and the larvae gradually assume the adult form. The hatchlings do not have external gills like other lungfish and rely on skin and gill respiration. Predators of the young fish include insect larvae, fishes, and ducks.

THE LIVING COELACANTH

The coelacanth *Latimeria chalumnae* is the sole living representative of a long line of ancient fishes that reached the peak of their diversity during the Triassic period (about 240 million years ago). Coelacanths are distinguished from other living fishes by their lobed fins, the absence of a fully formed vertebral column, an extra lobe in the tail fin, and a joint on the skull which allows the front part of the head to be lifted during feeding.

Coelacanths are colored a deep metallic blue with irregular white spots, an effective camouflage against the dark, oyster-shell-encrusted reefs that they inhabit. Single coelacanths have been caught off South Africa, Mozambique, and Madagascar, but most have been found among the Comoros Islands, where they inhabit rocky reefs at depths from 150 to over 700 meters (490 feet to 2,300 feet). They prefer steep-sloping rocky shores that are scoured by currents and have numerous caves, such as the west and south coasts of Grand Comoro and the coasts of Anjouan. No coelacanths have been caught off the geologically older islands of Moheli and Mayotte, where the deeper rocky reefs are covered by sand.

Reg Morrison/AUSCAPE International

▲ *An Australian lungfish embryo nears hatching. In this species there is no parental care; the eggs are fixed to water plants where they hatch unattended in about 21 to 30 days.*

◀ *Coelacanths were believed to have been extinct for 65 million years until a specimen of* Latimeria chalumnae *was trawled up near South Africa's Chalumna River in 1938 and spotted in the catch by Marjorie Courtenay-Latimer. Found mostly around the Comoros Islands, this species is associated with deep lava caves, where its white speckling serves as camouflage.*

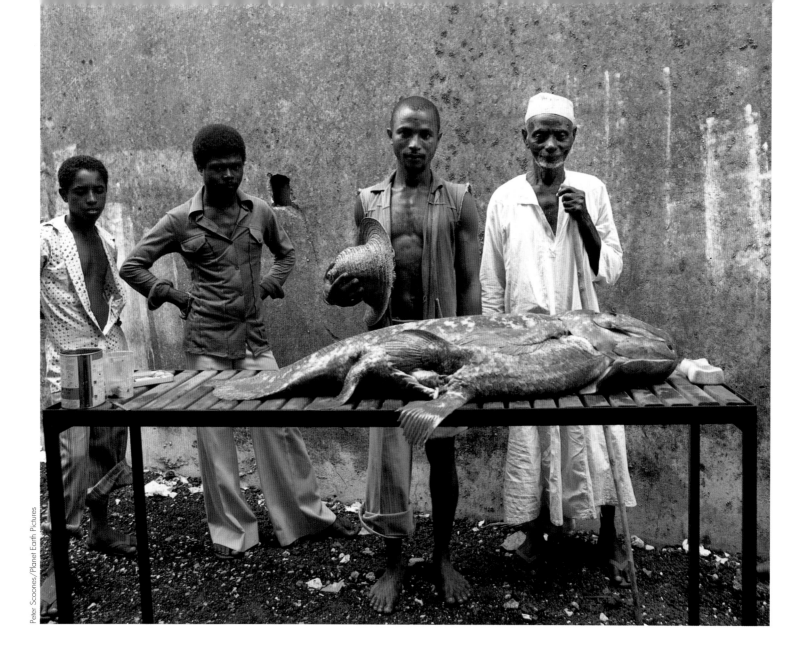

Peter Scoones/Planet Earth Pictures

▲ *Comoros Islands fishermen pose with a coelacanth. They are sometimes captured by accident when fishing for oilfish* Ruvettus, *which occurs in the same habitat.*

During the day coelacanths seek refuge in caves, often in groups, to escape predation from large sharks and to shelter from strong water currents. They live in cool water at less than 18°C (64°F) in permanent twilight. At night they forage for reef or midwater fishes and squids.

The unique swimming method of the coelacanth is both graceful and economical. The lobed fins generate thrust, while the large tail fin is used for fast starts to capture prey and escape predators. The paired fins are used together in a sculling action to propel the fish forward in a drifting glide with the current. While foraging, coelacanths sometimes perform a curious headstand during which the body is raised perpendicularly to the substrate. This behavior may be related to prey detection.

Coelacanths are live-bearers that carry about 20 to 65 developing eggs. The fully formed eggs are large (9 centimeters or 3 ½ inches in diameter) and hatch internally. The developing embryos receive nutrients from the large yolk sac and possibly by direct transfer from the mother through a placenta-like organ. From 5 to 26 young are born after a gestation period of about 13 months. The young are about 38 centimeters

(15 inches) long when born and resemble the adults. They initially grow at a rate of about 6.5 centimeters (2 ½ inches) per year until the twentieth year, after which the growth rate decreases. Females, which can reach 1.8 meters (6 feet) and 98 kilograms (216 pounds), grow larger than males, which can reach 1.55 meters (5 feet), and 65 kilograms (143 pounds).

Observations from an underwater craft of the number of coelacanths off Grand Comoro give estimates of between 210 and 420 adults. The catch rate of Comoran islanders, who are mainly targeting the oilfish *Ruvettus pretiosus* and catch the coelacanth by accident, is usually less than five coelacanths per year. The islanders do not present a major threat if they continue to use their traditional fishing gear. Increased fishing efficiency and effort, and a high price for captured fishes despite an international ban on their trade may, however, threaten the coelacanth population. Their restricted geographical range, narrow habitat preferences, rarity, and susceptibility to capture by humans combine to make the coelacanth vulnerable to extinction, and have prompted a worldwide campaign to conserve this extraordinary species.

MICHAEL N. BRUTON

BICHIRS & THEIR ALLIES

KEY FACTS

ORDER POLYPTERIFORMES
• 2 genera • 10 species
ORDER ACIPENSERIFORMES
• 4 genera • 25 species
ORDER LEPISOSTEIFORMES
• 2 genera • 7 species
ORDER AMIIFORMES
• 1 genus • 1 species

SMALLEST & LARGEST

Polypterus sp.
Total length: up to 30 cm (12 in)

Beluga *Huso huso*
Total length: up to 4.2 m (14 ft)

CONSERVATION WATCH
!!! Yangtze sturgeon *Acipenser dabryanus*; common sturgeon *Acipenser sturio*; Alabama sturgeon *Scaphirhynchus suttkusi*; Syr-Dar shovelnose sturgeon *Pseudo-scaphirhynchus fedtschenkoi*; small Amu-Dar shovelnose sturgeon *Pseudoscaphirhynchus hermanni*; and Chinese paddlefish *Psephurus gladius* are critically endangered.
!! 11 species are endangered.
! 8 species are listed as vulnerable.

B ichirs, sturgeons, paddlefishes, gars, and the bowfin are the living remnants of a diverse assemblage of extinct fishes. They have a worldwide freshwater and marine fossil record dating from the Permian period (285 to 245 million years ago) and, together with the teleost fishes, they make up the subclass Actinopterygii. Primitive actinopterygians usually have rhombic-shaped scales covered with an enamel-like substance called ganoin, a single dorsal fin (bichirs are the exception), a spiral valve in the gut, and a swim bladder connected to the gut which is frequently used to breathe air.

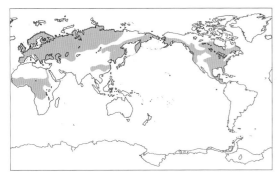

BICHIRS AND THE REEDFISH
Together, bichirs and the reedfish are a small group of 11 species placed in their own family, Polypteridae, in the order Polypteriformes. Fossils are known from the Cretaceous period (145 to 65 million years ago) of Africa. Although fossils have been found in North Africa, the family is now confined to the freshwaters of tropical Africa and the Nile River system.

Bichirs are elongate fishes with a distinctive series of dorsal finlets along the back and pectoral fins with fleshy lobes. Superficially these lobes resemble the fleshy lobes of coelacanths and lungfishes, but the internal structure is different. In fact bichirs are similar to lungfishes in a number of other ways, and some ichthyologists prefer to classify them with the sarcopterygians.

There are ten species of bichirs, all in the genus *Polypterus*. They are found in the rivers and flood plains of Central and West tropical Africa and throughout the Nile River system. They are similar in size and general appearance. Their color tends toward green and yellow-brown, with occasional darker spots and stripes; some species are mottled. They grow to about 0.3 to 1 meter (1 to 3 ¼ feet) in standard length (that is, body length without the tail).

The reedfish *Erpetoichthys calabaricus* is restricted to the coastal lowlands of tropical West Africa between the mouths of the Niger and Congo rivers. The reedfish is more commonly known as *Calamoichthys* in the aquarium trade. It is much more elongate than the bichirs and is frequently known as the ropefish. At least one authority suggests that the reedfish is paedomorphic, that is, the adult reedfish retain characters seen in the larvae and juveniles of the related genus *Polypterus*.

There are no detailed studies of the ecology of this group. Bichirs usually inhabit lake and river margins, flood plains, and swamps. Since they can breathe air, they have access to low oxygen environments not available to other fishes. Nocturnal predators, bichirs eat a variety of prey including fishes, amphibians, crustaceans, and insects. Within its range, the reedfish prefers marginal shore environments with vegetation that

▼ The ornate bichir Polypterus ornatipinnis *inhabits the middle and upper reaches of the Congo River, West Africa.*

Hans Reinhard/Bruce Coleman Limited

STURGEONS AND PADDLEFISHES

Sturgeons and paddlefishes are members of the order Acipenseriformes. The sturgeons, family Acipenseridae, and the paddlefishes, family Polyodontidae, are both confined to the northern hemisphere. They are united by a number of characters of a technical nature. Relatively few genera of fossil sturgeons and paddlefishes are known. They occur from the late Jurassic period (about 150 million years ago) to the Pleistocene epoch (2 million to 10,000 years ago) and are confined, like the living species, to the northern hemisphere.

Both families have a strongly heterocercal tail in which the vertebral column turns upward and extends into the upper lobe, resembling that of sharks. The skull and vertebral column are largely composed of cartilage.

Sturgeons

Sturgeons are unique among actinopterygians as they have five rows of bony plates on the body. They are elongate fishes with flattened snouts, a protractile mouth located under the head, and fleshy tentacle–like projections (barbels) used for searching the sediment for prey.

Sturgeons are found in the fresh waters and coastal marine waters of Europe, Asia, and North America. The genus *Acipenser* includes some 19 species. The majority are European or Eurasian, two are confined to China and Japan, and four are North American. Several of these species are traditional sources of European caviar (for example, *A. ruthenus* and *A. sturio*) and one (*A. medirostris*) is found on both sides of the North Pacific. The white sturgeon *A. transmontanus* of Pacific coast river systems is the largest fish taken in North American fresh waters.

The genus *Huso* includes two species of sturgeons. The beluga *Huso huso*, of the rivers of southeastern Europe and the coasts of the Black, Caspian, and Adriatic seas, is the largest fish found in fresh waters, said to reach a length of 8 meters (26 feet) and a weight of 1,300 kilograms (2,865 pounds). It is the most famous of the caviar sturgeons, both for the quantity and quality of the caviar it produces. A 4-meter (13-foot) female has been recorded as yielding about 180 kilograms (400 pounds) of caviar. *Huso dauricus*, of the Amur Basin of Asia and the coastal waters of the Amurskiy Liman Bay, is also an impressive fish, growing to over 4 meters (over 13 feet). Unfortunately, the numbers of both species are sadly depleted.

The genus *Scaphirhynchus* is made up of the smaller shovelnosed sturgeons of North America and Asia. They are characterized by having longer and flatter snouts. The two North American species, *S. platorynchus* and *S. albus*, have distributions centered on the Mississippi River and its major tributaries. The three Asian species, *S.*

Yves Lanceau/AUSCAPE International

▲ *Characteristic features of the sturgeons, such as Acipenser baeri of Siberia, include the rows of bony plates along the body and the sensitive fleshy barbels dangling below the jaw, used for searching the bottom ooze for small aquatic animals.*

has underwater roots with the stem and leaves above the water. It is a more specialized predator, preferring crustaceans and insects. The reedfish reaches about 1 meter (3 ¼ feet) in length.

Little is known of the courtship and breeding habits of the family in general, but breeding habits have been observed for certain *Polypterus* species. Males and females move into swamps created by flooding during the wet season, from August to September. Courtship has been observed in *P. senegalus*. A pair of males take turns leaping. The smaller male then continuously attends the female as she cruises about. A later phase consists of the male tapping the female's head while throwing his swollen anal fin, cup-like, beneath her abdomen. Eggs are laid a few at a time among rooted vegetation. Upon hatching, the young form schools. There is no information available on possible parental behavior.

kaufmanni, *S. hermanni*, and *S. fedtschenkoi*, are all only found in the Amu-Darya and Syr-Darya rivers that drain into the Aral Sea. These species are frequently placed in their own genus (*Pseudoscaphirhynchus* or *Kessleria*).

Sturgeons are bottom feeders, but they are not scavengers. The outside of the mouth is covered with taste buds and the characteristic fleshy barbels probe the sand and mud searching for insect larvae, snails, worms, and small fishes in fresh waters or for mollusks, shrimps, and other small crustaceans in marine environments. Some species, such as the beluga, specialize on small fishes during the winter months.

Many sturgeons are anadromous (that is, they live part of their lives in coastal waters and ascend rivers to breed). For example, the beluga makes two runs up the Don River in Russia: in the spring and in the fall. Normally anadromous species frequently have land-locked populations. *Scaphirhynchus* species are entirely freshwater but still migrate to favored spawning grounds. Spawning takes place in the spring and eggs are released a few at a time. Spawning habits differ among species. Many *Acipenser* species prefer spawning grounds upstream in smaller tributaries. Belugas spawn in deep holes in the Volga River. The shovelnosed sturgeons of North America prefer swift streams flowing over gravel.

Sturgeons are valuable fishes, and commercial fisheries were common in Europe and North America before declines in numbers became serious this century. Sturgeons are valued for the delicate taste of their flesh and, of course, for sturgeon caviar. There are two ways to collect caviar: dissection of dead females and stripping the eggs from live females so that they can be returned to their habitat to produce more eggs. The latter is preferred by conservationists. After collection, the eggs are separated from the egg-mass membrane, and then lightly salted, so that water is drawn from them. The brine is drained and the eggs packed for consumption. Caviar quality and price are directly related to the degree of salting. The best Russian caviar is very lightly salted and almost liquid.

Females that are killed for their caviar are used in several ways. As mentioned, the flesh is prized, especially when smoked. A fine oil also can be obtained. But the most valuable part of the sturgeon per unit weight is the swim bladder which yields isinglass. Processed isinglass is a sheet of almost pure gelatin. It can be used to make special glues and water-proofing materials, but its traditional use is to clarify white wines. English kings and queens once decreed the sturgeon a royal fish, to be served only at the royal table. Unfortunately even royal protection is not enough. Dams, pollution, silting, and overfishing have greatly reduced both the numbers and the geographic ranges of many sturgeons.

Paddlefishes

Paddlefishes are elongate fishes and are largely scaleless, having only a few scales on the upper lobe of the caudal fin. Their elongate snouts, flattened from top to bottom, make up 36 to 52 percent of the total length of the adult. Unlike that of gars, this elongation involves the beak and the snout and is covered with electroreceptors and supported by star–shaped bones. There are two living species, the American paddlefish *Polyodon spathula* of North America and the Chinese paddlefish *Psephurus gladius*. Only *Polyodon* has a fossil record, from the Paleocene of Montana. Related extinct genera are known from the Cretaceous period (145 to 65 million years ago) to the Eocene (57 to 37 million years ago) of North America.

The American paddlefish grows to about 1.5 meters (5 feet) and a weight of 80 kilograms (176

▼ *Swimming with mouth agape, the American paddlefish* Polyodon spathula *(top) filter feeds on zooplankton. Originally incorrectly classified as a shark, the paddlefish's closest living relatives are the sturgeons. One of the traditional sources of European caviar, the sterlet* Acipenser ruthenus *(bottom) is also one of the smallest species of sturgeon and is consequently often seen in aquaria.*

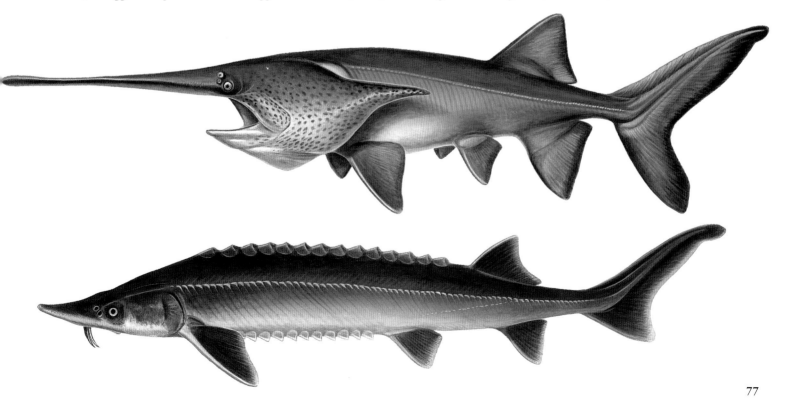

Ken Lucas/Planet Earth Pictures

▲ *A gaping American paddlefish* Polyodon spathula, *showing the characteristically flabby gills and toothless jaws of an adult. The long paddle-shaped snout is a sensory organ.*

pounds). It is so shark-like in appearance that it was originally described as *Squalus spathula* (*Squalus* is a genus of sharks). It has a distribution largely confined today to the Mississippi River system of central North America and adjacent rivers flowing into the Gulf of Mexico. Apparently, natural populations also occur in Lake Erie, between Canada and the United States. Although considered a freshwater fish, there are confirmed reports of it in brackish waters. It lives in large rivers rich in phyto- and zooplankton. The snout is a sensory organ and is not used to dig up the substrate. Paddlefish feed on zooplankton, cruising through swarms with gaping mouths, straining food as they go. They migrate to favored breeding spots, and local fisheries were established to take advantage of this behavior.

Paddlefish breeding habits were not known until the early 1960s when C.A. Prukett observed them breeding in midstream over gravel, bare rock, and boulders, during spring flooding on the Osage River in Missouri. The eggs are adhesive, sticking to the first rock touched. Upon hatching, the young are swept downstream into deep pools. Natural populations of paddlefish have declined because of damming and general loss of quality habitat. Efforts to strip females of caviar and then release them, and to culture young in fish hatcheries for release into the wild have been successful, but population numbers and ranges of naturally reproducing stocks remain on the decline. Unfortunately, this is compounded by an increase in poaching for paddlefish caviar.

By all accounts, the Chinese paddlefish of the Yangtze River system is a true giant, reaching 7 meters (23 feet) in length. The confirmed record is 3 meters (9 feet). Although valued as food, it is a rare fish.

GARS
Gars are a small group of two genera and seven living species placed in a single family, Lepisosteidae, and order Lepisosteiformes. These gars should not be confused with teleost fishes of the families Belonidae and Hemiramphidae whose species are called gars in some countries. Living gars are restricted to eastern North America south to Costa Rica as well as Cuba and the Isla de Pines.

They have an extensive fossil record from the Cretaceous periods (145 to 65 million years ago) in India, Africa, North America, and Europe.

Gars are considered mostly freshwater fishes. However they are common in brackish and coastal marine waters, especially along the coast of the Gulf of Mexico. A Cretaceous fossil is known from the epicontinental Niobrara Sea of Kansas.

Gars are elongate fishes encased in ganoid scales (these consist of an inner bony layer and an outer layer of an enamel-like substance) and they have a semiheterocercal tail. Uniquely, gars have a snout that is elongated in the ethmoid region (between the nose and the eyes). Their nostrils are at the tip of the snout rather than being at the base just forward of the eyes, as they are in paddlefishes and teleost fishes with beaks, such as needlefishes. This is correlated with a series of features which make the identification of fossil fragments easy, hence there is an extensive fossil record.

There are two genera of gars (or one genus with two subgenera). *Atractosteus* gars are heavy bodied, with two rows of enlarged canine teeth as adults, and distinctively ornate gill rakers. *Lepisosteus* gars are more slender, have one row of canine teeth as adults, and have rather simple gill rakers.

The alligator gar *Atractosteus spatula* is a large fish reaching 3 meters (10 feet) and is found primarily in the larger tributaries of the Mississippi River and more western rivers emptying into the Gulf of Mexico. It and the Cuban gar *A. tristoechus* are common in brackish and coastal marine waters. *Atractosteus tropicus* is the most southern member of the family. The longnose gar *Lepisosteus osseus* is the most widespread gar, being found over most of eastern temperate North America. It has a distinctively long snout and is common in brackish as well as fresh waters. The spotted gar *Lepisosteus oculatus* and its close relative the Florida gar *L. platyrhincus* also enter brackish waters. The shortnose gar *Lepisosteus platostomus* is a strictly freshwater fish, inhabiting the Mississippi River and its tributaries.

Most gars mate between April and early June. Several males attend a single female who lays small numbers of green eggs. The young hatch in a few days and attach themselves to vegetation until they are ready to become predators themselves. Gars are largely fish eaters, using their elongate jaws to snap their prey with a slashing motion. Birds are also favored prey, as are crabs in estuaries. Although gars are considered undesirable by many anglers, their flesh is prized by many in the southern United States where a commercial fishery exists for *A. spatula* and larger *L. osseus*. The Cubans culture *A. tristoechus*.

THE BOWFIN

Amia calva is the sole living bowfin belonging to the family Amiidae and the order Amiiformes. The family originated in the Late Jurassic (150 million years ago), and is now restricted to temperate eastern North America. Amiidae, in turn, is related to fossil caturids and parasemionotids whose fossil record stretches to the Triassic period (245 to 200 million years ago). These groups share a peculiar jaw articulation. Although frequently grouped with the gars in the Holostei, *Amia calva* is actually the closest living relative of teleosts, as evidenced by such derived features as a mobile upper jaw.

Bowfins grow to modest size, with females reaching about 75 centimeters (39 inches) and a weight of 3.8 kilograms (about 8½ pounds). They live to about 12 years. They have a distinctive set of features: they have an elongate body and the dorsal fin runs its entire length; the tail is semi-heterocercal, like that of gars, but the scales on the body are cycloid, like that of teleost fishes.

Bowfins are a typical lowland species most common in backwaters, oxbow lakes, and clear, well-vegetated streams. They prefer clear waters. Although usually described as fish eaters, bowfins eat a variety of prey and seem to like freshwater crayfish as much as fish.

Bowfins have complex spawning and parental care habits. Mating usually occurs between April and June in weedy shallow waters. Groups of males arrive first and proceed to construct nests in the form of a circular mat. Where possible, this will be among the roots of trees or under the protection of fallen logs. As a female approaches a nest, the mating ritual begins with nose bites, nudges, and chasing behavior until the female is receptive. The female then lies on the nest while the male positions himself at her side and the eggs are laid. Occasionally more than one couple will use the same nest and a male may attempt to mate with several females. After the eggs hatch, the fry school, and the male guards them and the nest against potential predators, usually basses and sunfishes.

EDWARD O. WILEY

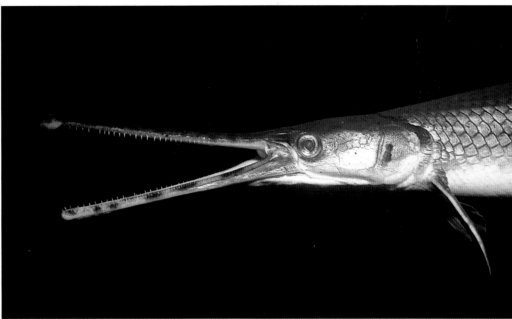

Ken Lucas/Planet Earth Pictures

▲ *The longnose gar* Lepisosteus osseus *of eastern North America. Gars have unusually tough skins and extremely hard enamelled scales formed of ganoin. The scales were sometimes used as arrowheads by Amerindians, and the skins are often used as leather in the production of numerous luxury goods and novelties.*

Ken Lucas/Planet Earth Pictures

◄ *The bowfin* Amia calva *inhabits large rivers in eastern North America. Clutches of up to 1,000 eggs, sometimes contributed by several females, are guarded by the male until they hatch after about 10 days.*

SUPERORDER
OSTEOGLOSSOMORPHA

ORDER HIODONTIFORMES
• 1 family • 1 genus • 2 species

ORDER OSTEOGLOSSIFORMES
• 3 suborders • 5 families
• 28 genera • 215 species

SMALLEST & LARGEST

Pollimyrus castelnaui
Total length: 2 cm (⅘ in)

Pirarucu *Arapaima gigas*
Total length: 2.5 m (8 ft)

CONSERVATION WATCH
‼ The Asian bonytongue
Scleropages formosus is listed as
endangered.
■ The spotted baramundi
Scleropages leichhardti and
featherback *Chitala blanci* are
listed as near threatened.

BONYTONGUES & THEIR ALLIES

Although the name "bonytongues" is usually applied only to species of the suborder Osteoglossoidei, the term is very descriptive of the large and prominent tongue bones characterizing all members of the whole superorder Osteoglossomorpha.

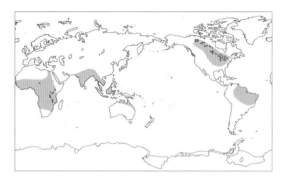

TOOTHED TONGUE BONES

In most osteoglossomorphs the tongue bones are toothed and bite against rows of teeth on the base of the skull. Anatomically, all osteoglossomorph fishes retain many other primitive features, and both their present-day and past distributions suggest that they were widely dispersed before the supercontinent Gondwana broke up some 135 million years ago. Fossils occur in Europe, North and South America, North, East and West Africa, Australia, Sumatra, and China. Living osteoglossomorphs also have a wide range, including North and South America, Africa, Australia, India, Bangladesh, the Malay

Archipelago, Thailand, Borneo, and New Guinea. All bonytongue species are restricted to fresh waters, but some fossil forms may have lived in brackish water.

It is impossible to give a short description of the group on the basis of its external appearance, so varied are the body forms, fin shapes and sizes, scale types, coloration, and even the sizes reached by members of its several genera and many species.

The superorder Osteoglossomorpha is subdivided into two orders: the Hiodontiformes and the Osteoglossiformes.

MOONEYES: ORDER HIODONTIFORMES

Hiodontiformes contains a single family, the Hiodontidae, which are the only living North American members of the Osteoglossomorpha. The Lycopteridae of China and Siberia was once included in the Hiodontiformes, but this exclusively fossil family constitutes the most primitive known osteoglossomorph group and is now classified in a separate order, the Lycopteriformes.

Hiodontids are restricted to the northeastern and north-central regions of North America, where they live in lakes and rivers. The family has a single

► The sole member of the family Pantodontidae, the butterflyfish *Pantodon buchholzi* is a surface dweller, lurking camouflaged among floating vegetation and snapping at insects with its large trapdoor mouth. Although it can leap a fair distance out of the water, reports of it gliding on its large pectoral fins are still unconfirmed.

living genus, *Hiodon,* which consists of two species, the mooneye *H. tergisus* and the goldeye *H. alosoides.*

Both *Hiodon* species feed on small fishes and aquatic invertebrates such as insects. They are egg-layers and are not known to exercise any obvious form of parental care.

Goldeyes can reach a length of 50 centimeters (20 inches). In some areas they are fished commercially, smoked, and sold as a delicacy. The smaller mooneye is of less economic value and reaches a length of 43 centimeters (17 inches).

BONYTONGUES AND THEIR RELATIVES: ORDER OSTEOGLOSSIFORMES

The order Osteoglossiformes is far more diverse than the Hiodontiformes in body shape, life habits, number of species, and geographic distribution. It can be divided into three suborders containing five families with living representatives.

TRUE BONYTONGUES: SUBORDER OSTEOGLOSSOIDEI

The true bonytongues (suborder Osteoglossoidei) have the widest distribution, occurring in South America, Africa, Australia, Malaysia, Borneo, Sumatra, Thailand, and New Guinea. Members of the two families that make up the suborder—the Pantodontidae and Osteoglossidae—are the most primitive living osteoglossiforms.

The freshwater butterflyfish *Pantodon buchholzi,* endemic to western tropical Africa, is the only representative of the family Pantodontidae. These small fish, rarely exceeding 10 to 13 centimeters (4 to 5 inches) in length, have very large, wing-like pectoral fins controlled by a well-developed and complicated set of muscles. It is believed that *P. buchholzi,* which often rests just below the water surface with its pectoral fins raised and expanded, can leap from the water and glide for short distances. Certainly it is able to leap but its gliding

capabilities still require confirmation. The first specimen ever captured had a dragonfly in its stomach, but whether the insect was an aquatic larva or an adult was not noted. Later records indicate that other aquatic insects are an important part of the fish's diet.

The butterflyfish has a large and lung-like swim bladder, branches of which penetrate into the ribs. Judging from its appearance, the swim bladder functions as an air-breathing organ, supplementing the oxygen supply that the fish obtains through its gills.

Aquarium observations indicate that the butterflyfish lays floating eggs that form a raft at the surface. Fertilization may be internal, with the eggs being expelled shortly afterwards.

At the opposite size-extreme to that of the butterflyfish is the giant pirarucu *Arapaima gigas* (family Osteoglossidae) from the Amazon river system. It reaches a length of at least 2.5 meters (8 feet) and is one of the largest freshwater fishes.

The swim bladder of the pirarucu is also lung-like. It seems that this fish is unable to meet its respiratory requirements through gill-respiration alone, and thus needs to take in air at the surface.

Parental care is well developed and includes making a nesting area among submerged plants. About 47,000 eggs are laid in the nest. Here the developing eggs and young are guarded by both parents. Once free-swimming, the young fishes are still guarded by a parent. The fry tend to cluster around the parent's head, attracted by a pheromone secreted from glands in the skin of the parent's head. It has been suggested that this skin also produces mucus that the young feed on.

The pirarucu is a much-favored food fish, and there are indications that in some areas it is now overfished. In northern Brazil the species has been cultured successfully in ponds.

▲ *Among the largest of all freshwater fishes, the pirarucu* Arapaima gigas *of Amazonia has a highly vascularised swim bladder that functions partly as a lung, and the fish can absorb oxygen from either air or water.*

▶ *The arowana* Osteoglossum bicirrhosum *is distinctive because of its unusually shaped head, two forward-pointing barbels, and with the lower jaw opening at a very abrupt angle.*

Max Gibbs/Oxford Scientific Films

▼ *Members of the family Notopteridae have extremely long anal fins, extending almost from chin to tail. With rippling movements of this fin, the fish can swim backwards or forwards with equal facility. Sometimes called the false featherfin,* Xenomystus nigri *is further distinguished by its lack of a dorsal fin.*

The living species most closely related to the pirarucu is *Heterotis niloticus* from West Africa and the Nile. Like the pirarucu, *H. niloticus* is a nest builder, clearing a site among aquatic plants, and later guarding its offspring. Unlike the predatory pirarucu, however, *H. niloticus* feeds on small planktonic organisms and on organic matter in the bottom debris. Food is collected and mixed with mucus as it passes through the mouth and into a complex spiral-shaped organ lying above the gills on each side of the head. From there the mixture is moved into the pharynx, and then swallowed. Earlier ideas that the spiral organs were accessory air-breathing devices are now discounted. The swim bladder, however, is lung-like and serves that function, thus allowing the species to occupy swamps and the relatively deoxygenated water of a river's reeded shores, both regions being its preferred habitats.

H. niloticus rarely exceeds a length of 80 centimeters (32 inches). The species is of some economic importance in West Africa, where it is also cultivated in ponds.

Two other genera, *Osteoglossum* (two species) in South America and *Scleropages* (three species) in Australia and Southeast Asia, are also included in the family Osteoglossidae. Both genera are characterized by their large, near-vertical, trap-like mouths, and by having a pair of prominent sensory barbels on the chin. The five species eat fish and insects, crustaceans, and other aquatic invertebrates.

All five species practice mouth brooding—that is, the fertilized eggs, and later the young, are carried in the mouth cavity of a parent. In *Osteoglossum* species the male performs this duty, but in the genus *Scleropages* the female does. The parents guard the shoal of young fishes for some time after they have grown too large to shelter in the parental mouth. Compared with the eggs of *Heterotis* and *Arapaima,* those of *Scleropages* and *Osteoglossum* are much larger, at around 1 centimeter (⅖ inch) in diameter, and far fewer in number—between 75 and 180.

FEATHERBACKS: SUBORDER NOTOPTEROIDEI
Members of the family Notopteridae (featherbacks), which is made up of four genera

Jane Burton/Bruce Coleman Limited

and eight species, constitute a living tropical family from Asia and Africa. They have a very distinctive shape unlike that of any other of the osteoglossiform fishes. The body is elongate and slender, tapering to a greatly reduced caudal fin that is continuous with the long anal fin, which occupies about two-thirds of the body length. The dorsal fin is short-based and narrow (hence the name featherback) and is entirely absent in one species from Africa.

Xenomystus nigri, Papyrocranus afer, and *P. congoensis* are the only representatives of the Notopteridae in Africa. Both are confined to the rivers of West Africa, including the extensive Zaïre river system. Their Asiatic counterparts are the genera *Notopterus* and *Chitala,* whose six species have a wide distribution encompassing India, Bangladesh, Malaysia, Thailand, Sumatra, Java, and Borneo, where they are found in rivers and swampy areas.

Not a great deal is known about the feeding and breeding habits of the African notopterids, but, like those in Asia, they are probably omnivorous carnivores. All *Notopterus* species are egg-layers; some prepare nests among aquatic plants and one of the parents acts as guardian. No details of breeding habits are available for *Xenomystus* and *Papyrocranus* species.

Xenomystus nigri, the smallest notopterid, reaches a length of only about 15 centimeters (6 inches). *Papyrocranus* grows to a length of about 65 centimeters (26 inches). Both are somewhat smaller than the largest of the *Chitala* species (*C. chitala*), which can grow at least 1 meter (3¼ feet) long.

All notopterids swim in a very characteristic manner. The body is held straight, and the fish propels itself forward, or backward, by a wave-like movement passing through the long anal fin. The stiff body probably results from the large, long, and extensively branched swim bladder. This organ runs from the head almost to the posterior tip of the body. Over most of its length there are numerous, finger-like processes reaching downward between the internal bony supports of the anal fin. At its front end the swim bladder, like that in *Hiodon,* is closely associated with the inner ear. In *Papyrocranus* it actually passes into the skull where it is intimately associated with the inner ear and parts of the brain. The inner wall of the swim bladder is not lung-like as it is in other air-breathing osteoglossiforms. Nevertheless, it seems to be used as a breathing accessory. In some species there are reports of it also being involved in producing underwater sounds, presumably by the expulsion of gases through the narrow passage connecting it with the gullet.

Some notopterid species are of commercial importance in Asia, but in Africa the family is of negligible economic value.

ELEPHANTFISHES: SUBORDER MORMYROIDEI

Africa is the home of the largest group of osteoglossiform fishes, the suborder Mormyroidei,

▼ *(Top) Despite its common name, the elephantnose or ubangi mormyrid Gnathonemus petersi has an enlarged chin, rather than nose, and uses this to probe the muddy bottom for food. Mormyrids navigate in their murky habitat by producing an electrical field and have extraordinarily large brains for their body size. (Bottom) Like the unrelated South American knifefishes, the Asiatic featherback or clown knifefish Chitala chitala moves by undulating its long anal fin. By reversing the undulations this fish can also swim backwards, allowing it to maneuver with ease in heavily vegetated backwaters.*

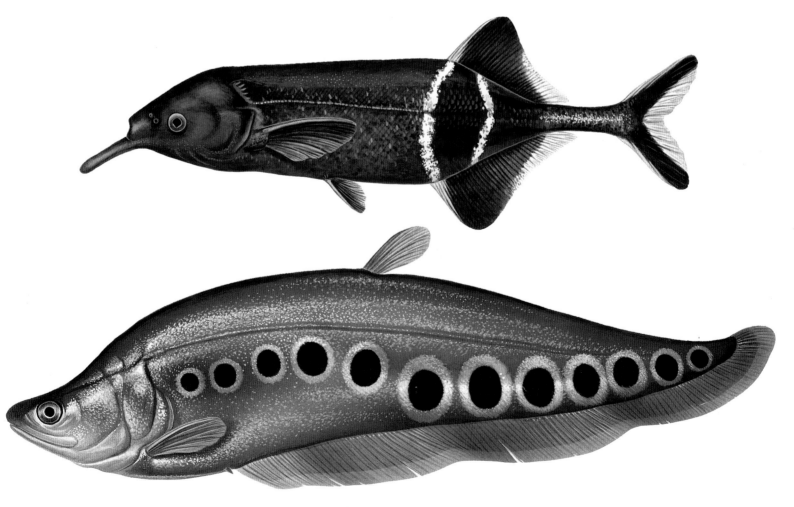

Ken Lucas

organs must be of considerable survival value. The system is also used as a means of communication between individuals, for recognizing its own and other species, and for keeping a shoal together.

In relation to body mass, mormyroid brain size is equaled only by that of humans. This hyper-development, especially of the cerebellum, may be associated with the interpretation of electric signals, and possibly explains the fact that, at least in aquariums, mormyroids are quick learners.

Within the Mormyridae (elephantfishes), one of the two families making up the suborder, there is considerable diversity in mouth form and shape. Some species have a small mouth situated at the tip of a long and slender, down-curved, proboscis-like snout. Others have a bluntly rounded head with a slit-like mouth on its underside. Still others have a larger mouth in a more usual position, or it may be at the end of short, but still obvious and barrel-like, snout. Many species feed on the aquatic larvae of insects or on other bottom-living invertebrates, such as mollusks. Other species, reaching a large adult size, eat small fishes.

Little is known about mormyrid breeding habits. Because the shape of the anal fin differs in the two sexes, it has been suggested that the fins of a spawning pair could be brought together to form a cup into which the eggs and sperm are ejected.

The family has a wide distribution involving most of the lakes, rivers, swamps, and streams of Africa, from the Nile to the warmer parts of South Africa. Mormyrids are absent or rare, however, in highly alkaline waters, possibly because the ionic concentration could interfere with the transmission and reception of electric signals.

Opinions vary on the palatability of mormyrid flesh, and in some areas there are tribal taboos on its consumption, especially by women, since the flesh is thought (without any pharmacological foundation) to induce abortion.

The second mormyroid family, the Gymnarchidae, contains only one species, the aba *Gymnarchus niloticus*, which differs from all other mormyroids in having no caudal, pelvic, or anal fins; it also reaches a larger size, around 1.6 meters (5¼ feet). The dorsal fin is very long, ending just before the narrow, finless posterior part of the body. It is the main source of propulsion.

The aba is a predator. Large individuals feed on fishes, and smaller ones on insects. It breeds in well-vegetated, marginal areas of swamps and rivers (its preferred habitat), where a large, floating nest, about 1 meter (3¼ feet) in diameter is constructed. Here the eggs are laid and later the young are guarded by one of the parents.

The species has a broad longitudinal, but narrow latitudinal, range across Africa between about 5° and 18°N, and is present in the Nile, Niger, Senegal, and Gambia river systems.

P. HUMPHRY GREENWOOD & MARK V. H. WILSON

▲ *Like the electric eel, the aba* Gymnarchus niloticus *possesses electrical organs along the flanks towards the tail, but their relatively feeble output is used for navigation rather than the immobilization of prey.*

which is made up of about 18 genera and 198 species. In terms of the varied shapes and sizes of those species, the Mormyroidei is also the most diverse osteoglossiform suborder.

In the absence of fossil evidence to the contrary, it seems possible that the Mormyroidei originated and evolved in Africa, and that for geological or biological reasons they have never been able to leave that continent. The few and fragmentary fossils are all from the same area.

Two outstanding mormyroid characteristics are the exceptionally large brain (mostly its cerebellum) and the presence of electric organs and well-developed electroreceptor cells. There are two other unusual features: one is the presence in the skull of a small balloon-like portion of the swim bladder completely isolated from the rest of that organ and lodged within the semicircular canals of the inner ear; the other is the complete separation of the semicircular canals (concerned mainly with balance) from the lower auditory part of the inner ear.

The electric organs are derived from body musculature situated in the region of the tail stalk or caudal peduncle. These produce a continuous, but weak, current that creates an electric field around the fish. Unlike other electric fishes (for example, the electric eel, order Siluriformes), mormyroids do not use the current to shock and stun their prey. Instead it is used as an early warning system. When the field is interrupted by some object, animate or inanimate, the break is detected by electroreceptor cells on the head and front part of the body, and the fish is alerted. Since many mormyroids are nocturnal or live in murky water, and all have small eyes, the electric

EELS & THEIR ALLIES

T he superorder Elopomorpha (eels and their allies) is made up of about 730 species of living fishes, ranging from large-scaled, silvery-sided tarpons to slender, black, and scaleless deepsea gulper eels. Although very different in appearance and behavior as adults, all elopomorphs share a similar larval stage called a leptocephalus (meaning "thin head"). These transparent, leaf-like, or ribbon-shaped creatures drift on ocean currents, apparently feeding on dissolved organic molecules until they develop to the next stage. Then they return to the sea bed, shrink, and firm up their flimsy bodies, and take on the coloring of the juvenile or adult stage.

KEY FACTS

ORDER ELOPIFORMES
• 2 families • 2 genera • 8 species
ORDER ALBULIFORMES
• 1 family • 2 genera
• c. 6 species
ORDER NOTACANTHIFORMES
• 2 families • 6 genera • 25 species
ORDER ANGUILLIFORMES
• 19 families • c. 138 genera
• c. 690 species

SMALLEST & LARGEST

Onejawed eel *Monognathus ahlstromi*
Total length: 5 cm (2 in)
Weight: c. 0.1 g (0.003 oz)

LONGEST
Giant moray *Strophidon sathete*
Total length: 3.75 m (12⅓ ft)
Weight: c. 8 kg (17⅗ lb)
HEAVIEST
Atlantic tarpon *Megalops atlanticus*
Total length: about 2 m (6½ ft)
Weight: 160 kg (350 lb)

CONSERVATION WATCH
■ No eels or allies are endangered.

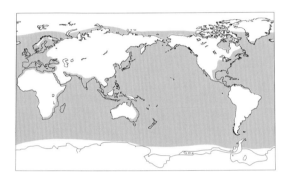

TARPONS AND THEIR ALLIES
Tarpons and their allies (order Elopiformes) are well known in the fossil record, dating back to the Upper Cretaceous period some 135 million years ago. Early tarpon-like fossil marine fishes are identifiable because they have gular plates (throat bones which are not present in most fishes). The two living tarpons are born in the sea but the larvae ultimately end up in estuarine nursery grounds. Tarpons overcome the occasional low oxygen conditions of such lagoons by breathing air at the surface. The Atlantic tarpon *Megalops atlanticus* is found from Massachusetts, USA, to Argentina and along the northwestern coast of Africa. It looks like a monstrous herring, reaching 160 kilograms (350 pounds) and more than 2 meters (6 ½ feet) in length. Its flesh is poor and bony but it is an excellent sport fish which makes spectacular leaps when hooked. The Pacific congener *Megalops cyprinoides*, known from

▼ *Also known as the silver king because of its unusually large, metallic scales, the Atlantic tarpon makes poor eating but is highly regarded as a sport fish, especially in Florida.*

Peter Scoones/Planet Earth Pictures

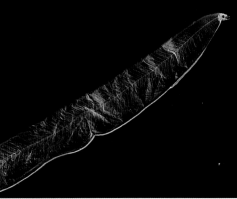

Norbert Wu

▲ *All members of the superorder Elopomorpha have in common a distinctive pattern of development, which includes a larval stage known as a leptocephalus, shown here.*

Southeast Asia, is similar but only reaches 18 kilograms (40 pounds) in weight.

Closely related to the tarpons are the ladyfishes or tenpounders. Belying their name, tenpounders achieve 7 kilograms (15 pounds) in weight and nearly 1 meter (about 3 feet) in length. A related order (Albuliformes) includes the bonefishes (genus *Albula*), which are just that—bony. They were previously thought to have a single form from the tropics, but molecular geneticists have discovered subtle differences which identify at least five separate species of bonefishes. They are much sought after by sport anglers who cautiously stalk these "ghosts of the flats", hoping to entice them by casting shrimp flies and small fish imitations. Those lucky enough to succeed enjoy a heart-pounding run from a mostly inedible creature.

The Notacanthiformes, or deepsea spiny eels, are poorly known. The 25 species, divided among 6 genera and 2 families, are worldwide in distribution and live on the sea bottom, at depths between 120 meters (400 feet) and 4,900 meters (16,000 feet). They are long and scaly, but not eel-like in appearance, and feed on slow-moving or attached invertebrates such as echinoderms, sponges, anemones, and mollusks. *Lipogenys gilli*, the suckermouth spiny eel, lives in the western north Atlantic and western Pacific and probably feeds by sucking up benthic ooze (detritus found on the sea floor) within its toothless mouth.

The leptocephalus larval stage common to all elopomorphs reaches extraordinary dimensions in notacanths. It is so large, in fact, that when a 1.8 meter (6 foot) larva was captured off South Africa it was extrapolated that the adult must grow to 30 meters (98 feet) or more. This was based on the assumption that the leptocephalus of the deepsea spiny eel, like that of the true eel, would generally be about 5 to 10 percent of the total length of the mature adult. These larvae were named *Leptocephalus giganteus* and it was assumed that they would grow into the as-yet-uncaptured sea serpent of mariner's lore. This theory was put to rest when a larval notacanth was captured mid-metamorphosis, and shrank considerably rather than grew excessively in its transformation.

TRUE EELS

True eels (order Anguilliformes) have long suffered from an identity crisis. The general public, and many zoologists as well, assume that most elongate fishes share a recent common ancestry and call them all eels. Included in this are such disparate vertebrates as the slime eel (a myxinoid), the electric eel (a gymnotiform), the blenny eel (an elongate perciform), the spiny eels (applied to both notacanthids and mastecembelid perciforms), and the Congo eel (an amphibian!). Most of these creatures evolved from a fish which looked more like a minnow or a perch, but they have lost appendages, scales, and more, and have taken on an elongate shape as adults to take advantage of the benefits of being eel-shaped.

The vast majority of true eels, 625 of about 690 species, are from three families, the Muraenidae,

▶▼ *Two tropical reef-dwelling eels, the ribbon eel* Rhinomuraena quaesita *(above) and the tessellated or black-spotted moray eel* Gymnothorax favagineus *(below). Male and female ribbon eels are so differently colored they were once thought to represent separate species. Although males (like the one illustrated here) are blue with yellow fins and face, females are yellow with a black anal fin.*

David Fleetham/Oxford Scientific Films

Ophichthidae, and Congridae. The remainder are elongate evolutionary experiments which have filled niches as variable as a freshwater pond or a deepsea trench.

Moray eels

There are more than 200 species of the family Muraenidae distributed among 15 genera which live in all the world's tropical and subtropical oceans and seas, and are common residents of coral and rocky reefs. Large morays are often feared by humans because of their malevolent appearance (which is actually a result of their respiratory behavior) and their occasional habit of attacking human hands which are thrust at them. The bite is more a result of moray myopia than meanness, and the similarity in appearance of our digits to those of an octopus is also presumably contributory.

Although not considered delicious, morays are eaten in many parts of the world. On rare occasions they have caused tropical fish poisoning (ciguatera). The most vivid example of eel ciguatera was that of the Filipino banquet when 57 diners consumed a large yellow margin moray *Gymnothorax flavimarginatus*. All soon became sick, eleven became comatose, and two died. All morays are carnivorous and important participants in the food web. Their abundance and significance within an ecosystem are often overlooked due to their inconspicuous behavior. Scientists in Hawaii were able to determine that as much as 46 percent of the carnivorous biomass of a reef was moray eels—critical components of any energy analysis!

Morays range in size from the 20-centimeter (8-inch) redface eel *Monopenchelys acuta*, of the Atlantic and Indian Oceans, to the giant green moray *Strophidon sathete* of Indo-Pacific reefs, which attains 3.75 meters (12 1/3 feet) in length. The teeth of morays are variable, ranging from the large backward-turned fangs of the dragon morays (genus *Enchelycore*) to the round, marble-like teeth of the species of *Gymnomuraena* and *Echidna*. Morays are also variable in color, ranging from plain to stripes, spots, bands, and bars. Not only is the leaf-nosed moray *Rhinomuraena quaesita* bizarre in anatomy and coloration (males are dark blue and yellow, females are black and yellow), but they also begin life as males and become females. Recent examination of other morays has discovered that such sexual strategies are not uncommon.

Snake eels

The snake eels and worm eels of the family Ophichthidae are the most diverse of the living eels and have the most species. More than 260

▲ Bold color patterns and fearsome teeth characterize many morays, but the dragon moray Enchelycore pardalis of Hawaii also has unusually prominent nose-tubes.

Sophie de Wilde/AUSCAPE International/Jacana

▲ *Common in European ocean waters, the conger eel* Conger conger *hunts mainly at night and is often especially numerous in old sunken wrecks.*

living species are distributed among 57 genera. Snake eels and worm eels occupy all tropical oceans and seas. They have adapted to live in numerous habitats, ranging from the shallow

intertidal areas to depths of 750 meters (about 2,500 feet). Although they are primarily burrowers in the sand and mud, some strikingly marked species swim above the bottom and others occupy

A GARDEN OF EELS

With the advent of modern scuba, divers can now observe curious colonies of conger eels embedded like grass blades above sand and shell bottoms most anywhere in tropical seas. These gardens of eels received their common name from Dr William Beebe, an American scientist and explorer who described his observation of a "swaying garden of eels" in the Gulf of California. A half century later, more than 20 species have been captured and identified. They occur in colonies of juveniles and adults at depths of 1 meter (3¼ feet) to more than 300 meters (1,000 feet), usually where a current flows carrying the plankton upon which they feed.

Their enlarged eyes, binocular vision, and short upturned mouths are adaptations to such a lifestyle; so too are the specialized tail tips, lateral line modifications, and secreting organs which allow them to dig a burrow tail-first, and line it with mucus. The eels extend nearly half their length from the burrows and bob and weave, grasping passing plankton. They mate like cobras, their heads and trunks entwined but their tails firmly anchored within their burrows, thereby allowing a hasty retreat should an unexpected predatory snapper or ray appear. Fertilized eggs hatch and current-borne leptocephali larvae develop and subsequently

Allan Power/Bruce Coleman Limited

transform. The juveniles burrow into the substrate at the periphery of a new colony, and again begin the cycle of one of the world's few vertebrates permanently attached to substrate.

▲ *An animal oddity, permanently attached vertebrates: a colony of garden eels* Taeniconger hassi *waits for currents to bring them food.*

the midwater realm. The family name, Ophichthidae (literally, "snake eel"), refers to their often spotted or striped appearance, a mimicry of the venomous sea snakes meant to intimidate a would-be predator.

Ophichthids are advanced eels which share an increased number of gill pouch support bones, a unique head pore pattern, and in one subfamily, a sharply pointed tail tip which allows it to rapidly burrow tail first into the substrate. Other modifications, such as the loss of all fins and the movement of the posterior nostril to within the mouth, allow these fishes to live an earthworm-like lifestyle. Snake eels range in size from 10 centimeters (4 inches) to nearly 3 meters (10 feet).

Conger eels

The conger eels are second only to the snake eels in diversity. There are approximately 25 genera and more than 115 species, living in all tropical and subtropical oceans and seas. They live in shallow water from tidepools to deep oceanic water. They are excellent food fish and are considered great sport, but are difficult to subdue. The family contains three independent lines of evolution, recognized as subfamilies, and the general lack of distinguishing characteristics among the majority of them makes their identification difficult even by specialists. Included in this family (Congridae) are the garden eels of the subfamily Heterocongrinae (see page 88).

Freshwater eels

Members of the family Anguillidae are unique among eels in that they spend most of their lives in lakes and rivers, returning to the sea to spawn. It had long been known that large anguilla depart to the sea and much smaller juveniles return, but it wasn't until 1777 that developing gonads were recognized in an adult. This discovery challenged the conventional wisdom that variously suggested that eels grew from horse hairs, scrapings of skin, the eggs of beetles, or "the entrails of the earth". Finally, early in this century, the Danish naturalist Johannes Schmidt traced returning leptocephali larvae back to the Sargasso Sea, where adults breed in the warm, saline deep water. The two Atlantic species have also been a source of controversy among biologists who presume that the American and European eels spawn in the Sargasso at the same time, but somehow segregate and return to their shore of origin. The deepwater spawning grounds of most Pacific species still remain to be discovered.

Other true eels

The remaining 15 true eel families are made up of a number of deepsea bottom and midwater forms and a variety of shallow-water bottom burrowers and coral reef dwellers. Each family is worthy of brief mention.

The Moringuidae are a small, worldwide family of noodle-shaped, shallow-water tropical burrowing eels that look much as their name implies—living spaghetti. The Heterenchelyidae are an even smaller family of highly modified burrowing eels, primarily from the West African coast. They are poorly known. The Chlopsidae are small moray-like eels. They live in shallow tropical and subtropical coral reefs, out to the continental shelf. The five species of Colocongridae are bottom dwelling, living between 300 meters (1,000 feet) and 900 meters (3,000 feet) in the tropical Atlantic and Indo-west Pacific. They are absent from the central and eastern Pacific ocean.

The Muraenesocidae are large, heavily toothed eels that are found in all tropical seas except the central Pacific. They are popular with fishers. Rarest of all eels is *Myroconger compressus*. Until recently, it was the sole member of the Myrocongridae, based only on the 50-centimeter (19½-inch) type specimen sent from St Helena Island to the British Museum in 1868, and a head purchased at the Dakar fish market in 1972! Now observations and collections from deep-sea submersibles have identified three more species and many individuals.

Several eel families spend their entire lives in the open ocean, rather than being bottom dwellers. Among them are the widely distributed Derichthyidae, small (to 60 centimeters or 24 inches) eels which differ considerably in their two related forms. One is the narrowneck eel *Derichthys serpentinus*, a blunt-snouted, large-eyed species with large head pores, and the other is the spoonbill eel *Nessorhamphus ingolfianus*, which has an elongated, spatulate nasal region.

Rodger Jackman/Oxford Scientific Films

▲ *After a two to three year migration across the Atlantic, larvae of the European eel Anguilla anguilla are 5 to 10 centimeters (2 to 4 inches) long when they finally reach the river estuaries. After transforming to juvenile elvers, they head upstream in groups to find suitable places to spend most of their adult lives.*

▼ *As adults, European eels Anguilla anguilla live for many years in lakes, ponds, and rivers before ultimately leaving for the depths of the Sargasso Sea to spawn.*

Hans Reinhard/Bruce Coleman Limited

The Serrivomeridae, or sawtooth eels, are common deepwater forms that inhabit all temperate and tropical seas. The Nemichthyidae, or snipe eels, are so-named for the long-billed shore birds with a similar snout. These are elongate, fragile, midwater species. Only the females and young males have the snipe bill. Mature males lose their teeth and bill and the anterior nostrils enlarge into large tubes, probably an adaptation for smelling the pheromones released by receptive females.

The gill openings of some species of Synaphobranchidae, or cutthroat eels, are single, mid-ventral slits. They are worldwide, moderate sized (to 1 meter or 3¼ feet), deepsea (to 3,700 meters or about 12,000 feet), bottom-living forms.

Lastly, the Nettastomatidae, or duckbill eels, are slender-tailed, toothy eels from the outer continental shelves and slopes of the world's tropical and warm-temperate seas. Little is known of their biology except that they live on or near the bottom and eat small fishes and invertebrates.

LOOSEJAWED OR GULPER EELS

The 4 families and 26 species of deepsea gulper eels demonstrate the variety of adaptations that fishes of shallow-water origin must make to life in the deep sea. Feeding in a dark, slow-moving, and sparsely inhabited world is aided by a large mouth, prickly teeth, lures and decoys, and a stomach that can enlarge to make the most of a rare meal. Enlarged surface sensory organs provide some protection from lurking predators.

All four families of deepsea gulper eels are worldwide in distribution, inhabiting the open water of the deep sea. They all possess a leptocephalus larva which lives in much shallower water than the adults.

The Monognathidae, or single-jawed eels (lacking the upper), are small, fragile, nearly transparent, and poorly understood. The Cyematidae, or bobtail snipe eels, at first glance look like a foreshortened snipe eel, but careful examination of the larvae suggests that their affinities are within this suborder. And finally, the deepsea gulpers and swallowers, Eurypharyngidae and Saccopharyngidae, respectively, are perhaps the most curious of any living vertebrate. One can only imagine what their life in the depths of the ocean might be like.

JOHN E. MCCOSKER

GULPER EELS

The anatomical modification and presumed behavior of these deepsea eels place them among the most bizarre of living vertebrates. And among them, the umbrella mouth gulper eel *Eurypharynx pelecanoides* is unique—a slender, elongate black beast to 75 centimeters (30 inches). It has a large swollen head with eyes and small brain near its tip, followed by an enormous mouth, the jaws of which are nearly one-quarter of the body length. The flimsy maw is covered by an elastic black membrane that when opened appears like a cavernous wind sock. The tiny teeth and the stomach contents remaining after trawl capture suggest that it eats small shrimps. Traveling tailward along the scaleless, velvety body, there is little remarkable other than the lateral line organs which, instead of lying protected within a subcutaneous canal, stand out from the skin. In this position they are probably more sensitive to the vibrations of possible prey or lurking predators.

It is the end of the tapering, whip-like tail that is the most curious feature of gulper eels, well developed in genus *Eurypharynx* but best developed in species of *Saccopharynx*. A complex luminous organ consisting of tentacles, papillae, grooves, and mounds emits a continuous pink glow that is punctuated by startling red flashes. Because these rare fish are usually damaged and nearly dead when recovered from a midwater trawl net, no one has yet seen the device in action. It is reasonable to suggest that it might serve as a decoy or lure to

Bruce H. Robison

attract crustaceans or small fishes, although the shape and proportions of the eel's body would require the skills of a contortionist to allure a shrimp to the maw with a "come-hither" flexure.

Consider how strange it would be to see a giant swallower resting peacefully in the gloomy ocean depths, its tail tip contorted full circle into its open mouth, appearing somewhat like a neon cocktail sign in a dark roadside lounge window, waiting to ensnare an incautious crustacean in the gloom of eternal night!

▲ A deepsea swallower Saccopharynx lavenbergi. These strange, deep ocean distant relatives of the river eels have enormous jaws, long filamentous tails, and light-emitting organs used as lures.

SARDINES & THEIR ALLIES

S ardines are known to the modern world as small fish, packed in orderly rows inside tins sold in food stores. Several different species can be found in tins, but usually only one species at a time, depending on the source: sardines from Norway are young sprats *Sprattus sprattus*; from Portugal, adult pilchards *Sardina pilchardus;* and from the Atlantic coast of North America, young herrings *Clupea harengus*. Other species, even anchovies (genus *Engraulis*), are sometimes sold as sardines.

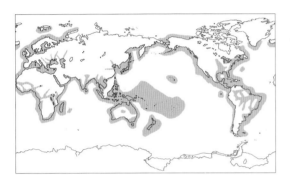

FOUR FAMILIES

Sardines and their allies are known to science as herring-like fishes (clupeoids), members of the order Clupeiformes, which includes four families: two small families (Denticipitidae, one species, and Chirocentridae, two species) and two large families (Clupeidae and Engraulididae). The two large families are commonly known as herrings or sardines (Clupeidae, 66 genera, 222 species) and anchovies (Engraulididae, 16 genera, 138 species).

Otherwise, the term "sardine" has no exact meaning beyond its ancient use as the name of the fish from Sardinia, presumably the European pilchard *Sardina pilchardus*, sometimes reckoned the only true sardine because it is the only species currently in the genus *Sardina*. Before 1929, however, the genus accommodated the five other pilchards, now placed in the genus *Sardinops*, distinguished from *Sardina* by the disposition of rakers on the first gill arch (the posterior lower rakers are overlapped by the upper more prominently in *Sardina*, but are reduced in *Sardinops*).

True sardines have variously been reckoned to include the temperate pilchards, as well as the tropical sardinellas (*Sardinella*, 21 species, locally known as Spanish, Indian, or oil sardines) and other clupeoids.

LIFE HISTORY

Clupeoids are small silvery fishes that swim in large schools; as adults they feed on plankton

◀ *A mixed school of clupeoids and silversides of the family Atherinidae. Members of these two unrelated families often school together in the tropics, feeding on plankton.*

Norbert Wu

▲ *Members of the Clupeiformes occur in huge schools, like these Seychelles glassfish in a coral cave.*

(mainly crustaceans and their larvae) along coasts of temperate and tropical seas, and are themselves eaten by numerous other fishes, birds, and mammals. Low on the food chain, clupeoids tend toward abundance, as if their purpose in life was to be eaten. They usually spawn seasonally or, if throughout the year, in seasonal peaks, often after migration to particular spawning areas. Some seasonally migrate upriver to spawn in fresh water. Large numbers of eggs, 1 to 2 millimeters ($\frac{1}{24}$ to $\frac{1}{12}$ inch) in diameter and up to 200,000 in larger clupeoids, are dispersed near the surface. The eggs hatch into small larvae of 2 to 5 millimeters ($\frac{1}{12}$ to $\frac{1}{5}$ inch) that drift in the plankton for months, until they grow to about 30 millimeters (just over 1 inch). The larvae then change into juveniles that actually resemble adults. This transformation typically involves the forward migration of the dorsal fin and gut.

Larvae that are old enough to have a swim bladder apparently inflate it only at night, and many adult clupeoids are most active, or nearer the surface, during hours of darkness. Some live their entire life in fresh water, but no freshwater species occurs natively in more than one continent. In temperate seas there are few clupeoid species, and in cooler areas perhaps only one or two. In tropical seas there may be some dozens living together. Clupeoids are absent from polar seas and from oceanic depths. Few clupeoids are encountered far from shore in the open ocean.

COMMERCIAL VALUE

No other group of fishes has been as important to humans, who throughout history have exploited clupeoids wherever and whenever marine fishes were sought as human food or as sources of oil, fertilizer, or animal feed. For much of the present century clupeoids comprised 30 to 40 percent by weight of all marine fishes for which records are available. For all marine fishes the total catch during this time has gradually increased from 10.2 million tonnes (11.2 million tons) per year to more than 61 million tonnes (67.2 million tons). Anchovies of the genus *Engraulis* (eight species) and pilchards of the genera *Sardina* and *Sardinops* (six species) occur together in commercial quantity in temperate marine areas of high primary (phytoplankton) productivity throughout the world: in northern Europe and the Mediterranean, South Africa, Australia–New Zealand, Japan, California, and Peru–Chile. These 14 species have comprised most of the clupeoid catch from the world's oceans. Many other clupeoids are fished commercially, and chief among these are western Atlantic menhadens *Brevoortia* (six species), northern herrings *Clupea* (two species), and tropical sardinellas *Sardinella* (21 species).

One anchovy species, the Peruvian anchoveta *Engraulis ringens*, was taken at a rate exceeding 10 million tonnes per year during the 1960s and

1970s before anchovetas declined in abundance, presumably because of overfishing. Even after their precipitous decline in the mid 1970s, anchovetas were taken at a rate exceeding 1 million tonnes per year. Remarkably, one anchoveta weighs in at 20 to 30 grams (about 1 ounce) at its adult size of 15 centimeters (6 inches). No other fish in history has been, or perhaps ever will be, as exploited.

Mature after one year and reaching a fairly small size in a life span of two to three years, anchovetas are typical of temperate and tropical clupeoids. The few species living in cooler seas, such as the Atlantic and Pacific herrings *Clupea harengus* and *C. pallasii* of high northern latitudes, grow more slowly, live up to 25 years, and may reach 40 centimeters (16 inches).

Among the larger clupeoids are the shads of the northern hemisphere. Esteemed for its roe, the American shad *Alosa sapidissima*, a spring spawner in major rivers of eastern North America (introduced in the North Pacific), reaches a length of 60 centimeters (24 inches) and 3 kilograms (6 ½ pounds).

The largest anchovy *Thryssa scratchleyi*, from northern (tropical) Australia and Papua New Guinea, reaches nearly 50 centimeters (20 inches). Not as long as wolf herrings, these clupeoids are nevertheless much heavier.

SILVERY AND STREAMLINED

To many people, clupeoids look like the typical fish, silvery and streamlined, with large scales, no spines in the fins, a forked tail, and small pelvic fins in the middle of the body below a short dorsal fin. Clupeoids are technically recognized by their peculiarities. None except one species (*Denticeps clupeoides*) has a sensory canal (lateral line) on the body. All have a distinctive caudal skeleton (not apparent externally), comprised of bones that support the finrays of the tail. A prominent external feature of many, but not all, clupeoids is the single row of scutes that extends between the paired fins along the belly. Scutes are modified scales, usually sharply pointed to the rear, and clupeoids with well-developed scutes may be recognized by touch, even in the dark. A few species, the Chinese gizzard shad *Clupanodon thrissa*, the Australian river herring *Potamalosa richmondia* and sandy sprats *Hyperlophus* (two species), and the Pacific menhaden of Peru *Ethmidium maculatum*, have scutes along the dorsal surface from the back of the head to the start of the dorsal fin. Curiously, dorsal scutes occurred also in the early clupeoids *Armigatus* and *Diplomystus*, known from fossils of late Mesozoic (Cretaceous 145 to 65 million years ago) and early Tertiary (about 60 million years ago) deposits of Asia, the Americas, Africa, and the Near East.

▼ *Anchovies, like these* Thryssa baelama, *resemble miniature herrings except in their prominent snouts and long, distinctively underslung jaws. Incredibly prolific, some anchovy species have sustained commercial catches exceeding 10 million tonnes (11 million tons) per year.*

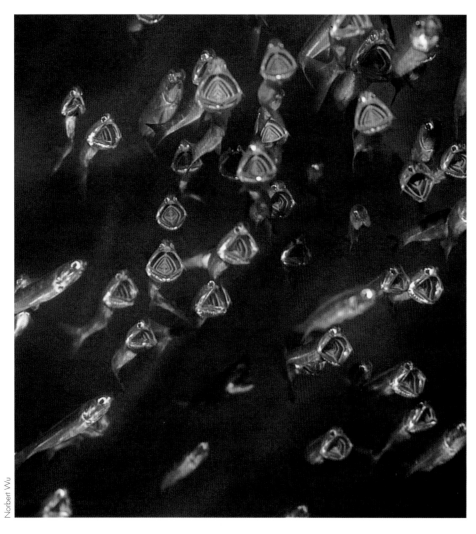

Norbert Wu

▲ *Feeding by night with mouths agape, a school of anchovies filters plankton from waters off the coast of California. Elaborate, close-packed rakers on their strongly flexed gill arches skim off minute animals for swallowing. Anchovies serve in turn as food for tunas.*

including 18 species) lack scutes (other than a peculiar scute wrapping partly round the base of the pelvic fins) and have a body round in cross-section. Pristigasterids (*Pristigaster* and eight other genera, including 34 species) have a long anal fin (more than 30 rays) and a body that tends to be very thin. Shads (*Alosa* and six other genera, including 31 species) and gizzard shads (*Dorosoma* and five other genera, including 24 species) have a notch between the left and right upper jaws. Most gizzard shads have the upper jaw flared outwards, and the last dorsal finray is usually prolonged as a filament that extends along the back almost to the tail. The number of gill-rakers increases into the hundreds with age. Their well-developed pharyngeal pouches (epibranchial organs) are used in feeding on microscopic life, the stomach has developed as a gizzard-like organ, and the intestine is lengthened in a characteristic pattern of complex folds.

The two final groups in this family, freshwater herrings (*Pellonula* and 22 other genera, including 43 species) and herrings proper (*Clupea* and 15 other genera, including 72 species), lack the distinctive features of the other groups. Freshwater herrings have only one supramaxilla, a small bone above the back of the upper jaw, in contrast to the usual two. Most of the six groups approach a worldwide distribution in tropical seas. The small (rarely more than 7 centimeters, or 3 inches maximum) freshwater herrings, including *Pellonula* and relatives, are exceptional in their restriction to Africa and other lands bordering the Indian Ocean.

THE DENTICLE HERRING: A LIVING FOSSIL

Because of its lateral line and tail skeleton, the denticle herring *Denticeps clupeoides* (the only species of its family, Denticipitidae) is considered primitive and distantly related to all other modern clupeoids. Undiscovered until the late 1950s, this small species (4 centimeters or 1½ inches) is confined to four river systems of southwestern Nigeria. A living fossil, it is peculiar in having teeth (odontodes) over the external surface of the head, giving it a furry appearance.

WOLF HERRINGS

Wolf herrings (genus *Chirocentrus*, family Chirocentridae) are peculiar in having large fang-like teeth and an extremely elongate and laterally compressed (thin) body. Lacking scutes, these predatory species are placed in their own family because their exact relationship to other clupeoids is obscure.

HERRINGS, SHADS, SARDINES, AND MENHADENS

All other clupeoids that are not anchovies are placed in one family, Clupeidae, which divides into six major groups or subfamilies. Round herrings (*Dussumieria* and four other genera,

ANCHOVIES

The second large family, the anchovies (Engraulididae), are easy to recognize because of their prominent snout, and long and underslung lower jaw, extending far behind the eye and making a large mouth. The mouth is simply left open as anchovies filter feed by swimming through the water. The anchovy snout contains a rostral organ, unique to the family, formed from the anterior part of the sensory canals that extend forward above and below the eye. Not apparent externally, except for the distinctive snout, the rostral organ presumably has a sensory function not yet discovered. The sensory canals on the head of anchovies are distinctive also in forming an anastomosing (net-like) complex on the cheek below and behind the eye. Behind the head, herrings and anchovies look very similar and are generally indistinguishable.

Anchovies divide into two major groups. Grenadier anchovies *Coilia* (including 13 species) have a body tapering to the rear, ending in a pointed rather than forked tail continuous with the anal fin. Anchovies proper (*Engraulis* and 13 other genera) include 125 species. Grenadier anchovies, of the Indian and Western Pacific oceans, are highly distinctive, often with posterior elongation of the upper jaw bone (maxillary) and several (up to 19) of the pectoral finrays. Two species, *Coilia brachygnathus* and *C. nasus*, are important commercially in the catch from the Yangtse River of China. One species, *C. dussumieri*, is unique among clupeoids in having light organs in several rows on each side of the body. Of anchovies proper there is a primitive assemblage (*Thryssa* and three other Indo-Pacific genera, including 34 species) with scutes along the belly and a single scute (predorsal spine) just in front of the dorsal fin. An intermediate group *Stolephorus* (19 species) tends toward scute reduction. An advanced group (*Engraulis* and nine other genera, including 72 species) has few or no scutes (except for a reduced scute just in front of the pelvic fin) and additional fusion of bones of the tail skeleton.

The advanced group includes the nehu *Encrasicholina purpurea*, which is native to Hawaii and has perhaps the smallest geographic distribution of any marine anchovy, as well as the buccaneer anchovy *E. punctifer* of the Indian, Western, and Central Pacific Oceans; the buccaneer anchovy has the largest geographic distribution and apparently the most pelagic (open ocean) habits. The advanced group also covers all anchovies of North and South America (nine genera, 64 species), including many freshwater species native to the large rivers (Orinoco, Amazon, Parana) of South America. Among these freshwater species occurs the smallest (2.5 centimeters, or 1 inch) anchovy, the Rio Negro pygmy *Amazonsprattus scintilla*, as well as one of the largest, the wingfin *Pterengraulis atherinoides*, a predatory species reaching 30 centimeters (12 inches). Feeding primarily on other anchovies, the few predatory anchovies reach a large size (more than 20 centimeters, or 8 inches), have large teeth, and tend to have solitary rather than schooling habits; they include the sabertooth anchovies *Lycengraulis* (three species) in tropical America, and the sabertooth thryssa *Lycothrissa crocodilus* and hairfin anchovies *Setipinna* (eight species) in Indo-Malaya and China–Japan. In some respects the most advanced anchovies are the two tropical marine anchovetas *Cetengraulis mysticetus* in the eastern Pacific and *C. edentulus* in the Caribbean, separated by the isthmus of Panama. Like the gizzard shads, they have developed extreme adaptations to enable them to feed on microscopic life: large epibranchial organs, and numerous gill-rakers and complex intestinal coiling that increase with age.

Anchovies are delicate fishes that lose all their scales when caught in a net. With the scales goes all their superficial skin, and death soon follows. Despite their success in nature and their importance to humans and other forms of life, anchovies do not survive long in captivity and are the least understood of all the clupeoids.

GARETH NELSON

◀ *A common Australian freshwater representative of the herrings, the bony bream* Nematalosa erebi *is often called the hairback herring from the long filamentous extension to the dorsal fin clearly visible here. The bony bream is one of the most widespread and adaptable of Australian freshwater fishes, common alike in coastal streams, major rivers and desert bores.*

G.E. Schmida

KEY FACTS

ORDER GONORYNCHIFORMES
• 4 families • 7 genera
• *c.* 30 species

ORDER CYPRINIFORMES
• 6 families • *c.* 284 genera
• *c.* 2,050 species

SMALLEST & LARGEST

Danionella translucida
Total length: 1.2 cm (½ in)

Catlacarpio siamensis
Total length: up to 2.5 m (8 ft)

* *Catlacarpio siamensis* was
recorded at about 2.5 m (8 ft) in
length but not in recent decades.
In the 19th century, the mahseer
of India *Barbus tor* or *Tor putitoria*
was recorded with a total length of
almost 3 m (10 ft). Such giants are
no longer found.

CONSERVATION WATCH
!!! 41 species are listed as
critically endangered, including:
Twee River redfin *Barbus
erubescens*; border barb *Barbus
trevelyani*; barred danio *Danio
pathirana*; Charalito saltillo
Gila modesta; Cape Fear shiner
Notropis mekistocholas; bitingu
Ospatulus truncatus; manalak
Puntius manalak; palata
Spratellicypris palata; cui-ui
Chasmistes cujus.
!! 36 species are listed as
endangered.
! 114 species are listed as
vulnerable.

CARPS & THEIR ALLIES

Carps and their allies belong in the great superorder Ostariophysi. Ostariophysans, as they are known, are predominantly freshwater fishes, and are found on all continents capable of supporting fish life. They form about 30 percent of all living fish species. Included in Ostariophysi are the fishes of this chapter, as well as the characins and catfishes of the next two chapters.

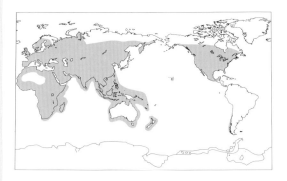

TWO UNITING FEATURES

The superorder Ostariophysi dominates the world's freshwater fishes and almost all its members are united by two features. The first is the Weberian mechanism or its precursor. This is a modification of the first few vertebrae into a series of interconnected levers to transmit sound waves picked up by the swim bladder to the inner ear. This is thought to enable these fishes to hear. The mechanism is in a very primitive state in the order Gonorynchiformes but highly sophisticated in all other ostariophysans. The second characteristic is the "fright" or "alarm" substance. This is a chemical in the skin which is released if the skin is injured. As soon as it is detected by other ostariophysans, they flee. This early warning system allows the others to escape if one fish is taken by a predator.

▶ *One of the hardiest of all fishes, the common carp Cyprinus carpio has been virtually domesticated for thousands of years, resulting in a variety of ornamental types, like these koi carp feeding close to the surface.*

ORDER GONORYNCHIFORMES

This order contains both marine and freshwater fishes showing some of the simpler modifications of the anterior vertebrae. These modified cephalic (head) ribs are thought to fulfill the same function (transmitting sound waves to the inner ear) as the more sophisticated Weberian mechanism in the other ostariophysi. All species are egg-layers.

Milkfish (Family Chanidae)

The only member of the family Chanidae, the milkfish *Chanos chanos* inhabits the rim of the Indian and tropical Pacific Oceans. It is a streamlined, silvery, compressed fish that may grow as large as 1.5 meters (5 feet) long but is seen more frequently at about 1 meter (3 feet). In the wild it lives in the sea and in brackish water, but it can tolerate a wide range of salinity. This has resulted in it becoming an important food fish in many areas of Southeast Asia. It is a filter feeder, its diet consisting of phytoplankton (microscopic floating algae). Around the South China Sea region, the young are caught offshore and raised in ponds in coastal villages. To encourage rapid growth, the algae which it feeds on may often be manured by human excrement.

Family Gonorynchidae

Gonorynchus gonorynchus is presumed to be the only species of the family Gonorynchidae. It is a shallow-water marine fish from the tropical and temperate Indo-Pacific. It lives near the bottom, has a ventral mouth (the snout overhangs the mouth), and lacks a swim bladder. This elongate fish grows to about 60 centimeters (2 feet) long. Fossils of its relatives, about 70 million years old, have been found in Alberta, Canada.

Family Kneriidae

The family Kneriidae are small fishes, less than 15 centimeters (6 inches) long, coming mostly from the streams and small rivers of Africa. Many of the species live in pools in fast-flowing, cool, high-altitude streams. Species of the genera *Kneria* and *Parakneria* are sexually dimorphic (they have distinctly different male and female forms). The male has a strange rosette on the operculum, the function of which is unknown. It is not present on the female. The other two kneriid genera are *Cromeria* and *Grasseichthys*. These come from the Nile River in West and Central Africa. They are unusual in that they are neotenic: they retain the juvenile appearance (that is, they are small, transparent, and scaleless) even after they have achieved sexual maturity.

Family Phractolaemidae

There is just one species in the family Phractolaemidae. This is *Phractolaemus ansorgii*, from the Niger delta and parts of the Zaïre system.

J.M.Labat/AUSCAPE International/Jacana

It is normally found in muddy waters with a low oxygen content. This strangely shaped fish grows to about 15 centimeters (6 inches) long. The body is cylindrical and the protrusile (forward thrusting) mouth is on the upper surface. It has just two teeth, at the junction of the two halves of the lower jaw. The swim bladder is divided into compartments and is used for breathing atmospheric air. *Phractolaemus ansorgii* feeds on small, mud-dwelling animals.

ORDER CYPRINIFORMES

This order has about 280 genera and over 2,000 species. Almost all are freshwater fishes (only a very few enter brackish or coastal waters) and they are widespread in North America, Eurasia, and Africa. They are not native to South America, Australia, or New Zealand although they have been introduced into these areas. A feature of this order is that the third bone of the last gill arch is modified into a stout tooth-bearing bone (the pharyngeal bone). The paired pharyngeal bones grind food against a keratinized pad (a hardened pad made of a substance similar to fingernails) on the roof of the pharynx at the rear base of the skull. Cypriniformes lack jaw teeth and their function is replaced by the pharyngeal teeth. Over 80 percent of all species are in just one of the six families—the Cyprinidae. Very few species show sexual dimorphism and all are egg-layers.

Cyprinids (Family Cyprinidae)

The distribution of the Cyprinidae is similar to that of the order. The other five families are much more restricted.

▲ *Growing to nearly 12 centimeters (4½ inches) in length, the red-finned black "shark" Epalzeorhynchus frenatum is a native of Thailand, where it favors flowing rivers and streams over quiet ponds and lakes. The sexes are virtually identical in appearance.*

Y. Lanceau/AUSCAPE International/Jacana

The absence of jaw teeth has not prevented cyprinids from exploiting almost any food resource. Indeed, it is a source of evolutionary wonder how a simple jaw and mouth plan can be modified to many specialized feeding methods. Even within one feeding strategy, cyprinids have solved the same problem in different ways.

Fish-eating fish can normally be easily recognized by the presence of curved, awl-like teeth on the jaws which are capable of holding slippery prey. Cyprinids cannot adopt this technique but rely instead on others. *Macrochirichthys* is a piscivore (fish eater) found from Thailand, through Java and Sumatra to Cambodia, Laos, and South China. It has a large mouth which is strongly angled downward. The junction of the two sides of the lower jaw is formed into a small dagger-like projection. Specially modified muscles enable it to pull its head back (a rare phenomenon in fish). This helps to increase the gape and enables the fish to seize prey above it.

An equally predatory large cyprinid from China, *Luciobrama*, has a surprisingly small mouth and the part of the head behind the eye (post-orbital) is greatly elongated. In front of the eye the snout is tube-like. *Luciobrama* takes fish by approaching the prey with its mouth closed, evacuating the elongated mouth cavity, opening its mouth and sucking the prey in.

Another predator, *Elopichthys*, from China, is less modified. It just has a large mouth and moves fast. By contrast *Hypophthalmichthys*, also from China, feeds on microscopic algae filtered from the water by the modified gill rakers. Its scientific name reflects the fact that the eye lies below ("hypo") the level of the mouth. It is a large fish, growing to about 1 meter (3 feet) long. At the other end of the size scale is the small *Pectinocypris* from Thailand, which is also a filter feeder. It has over 100 gill rakers on each gill arch to trap algae. The grass carp *Ctenopharyngodon idella*, originally from China, feeds voraciously on large water plants and has been introduced to many parts of the world to clear weeds from canals and rivers. Bottom-grubbing species like the carp *Cyprinus carpio* have barbels around the mouth laden with

▲ *The common carp* Cyprinus carpio *closely resembles the wild goldfish except in the barbels dangling from the gape. A native of Europe, it has been widely introduced into North America and elsewhere.*

▶ *A popular aquarium fish from western India, the zebra danio* Brachydanio rerio *prefers sluggish waters, and is often common in rice paddies. Males tend to be somewhat slimmer than females, and the two sexes maintain strong and long-lasting pair-bonds.*

taste buds in order to detect delectable food items in, or on, the substrate.

Small fishes of the genus *Garra* live in Africa and Asia. They have developed expanded lips which act as a sucker to hold onto rocks in fast-flowing water. *Garra* and *Labeo,* another Afro-Asian genus with expanded lips, show another intriguing phenomenon. In the breeding season, the males of some cyprinid species develop horny tubercles (small hard protrusions like pimples) on the cheeks. Certain species of *Garra* and *Labeo* have solid masses of tubercles on the snout and cheeks, which were once thought to be an exaggeration of the breeding signs. However, recent research has shown that the tubercle clusters have a hydrodynamic effect: tubercle development is greatest in fish from fast-flowing waters and serves to minimize turbulence by smoothing the water flow over the fish.

Almost all food items are consumed by cyprinids. Even slimy algae on rocks are eaten. Some genera (*Varicorhinus* in Africa and *Semiplotus* in Asia) have developed broad, ventral mouths with a sharp-edged covering to the lower jaw, which enables them to scrape off the algae.

The shape of the pharyngeal teeth also reflects the diet: snail-eating fishes have great crushing teeth forming a three-rowed grinding surface; plant eaters have scythe-like shredding teeth; fish eaters have thin, hooked teeth; and omnivores have pharyngeal teeth of all kinds.

Although the specializations just described, achieved through millions of years of evolution, are remarkable, even more outstanding is the variation (or plasticity) in mouth and body form shown within the lifetime of individuals of one species. This is best displayed in some species of *Barbus* from African lakes. *Barbus altianalis* lives in Lakes Victoria, Kivu, and Edward-George. The pharyngeal bones and teeth of fish from Lake Victoria are consistently stouter than those in fish from Lake Edward. Water snails are commoner in Lake Victoria than in Lake Edward. Consequently, the innate plasticity of the fish has enabled it to capitalize on a local food resource.

A more extreme example occurs with *Barbus intermedius* in Lake Tsana, the source of the Blue Nile in Ethiopia. Here, until they are about 10 centimeters (4 inches) long, all fish look alike. From this size, they slowly change into a whole array of specialized feeding forms. They vary from long, thin fish eaters to deep-bodied molluskivores (shellfish eaters). Some develop thick rubbery lips, while others have a wide underslung mouth with a sharp-edged lower jaw. If there weren't intermediates between all the extremes, they could have been (and indeed once were) considered different species.

Humans have exploited the potential for change in form and color shown by some cyprinids by deliberately breeding many bizarre varieties of the goldfish *Carassius auratus* from a fairly undistinguished brownish ancestor.

It is very likely that the variability or plasticity both within an individual species and within the family has been another factor in the success of the cyprinids. There are few freshwater habitats capable of supporting freshwater life where the cyprinids do not live and thrive within their area of distribution.

Eyeless and pigmentless cyprinids have evolved in underwater caves and aquifers in Africa, the Middle East, and Asia. High up in cold streams in the Himalayas live the so-called snow trouts (*Schizothorax* and related genera), characterized by small or reduced body scales but with a row of large, tile-like scales at the base of the anal fin. In the thermal borax lake in Oregon lives the endemic *Gila boraxobius* which can withstand temperatures of over 30°C (86°F). From rapids to lakes, there are cyprinids.

A few can tolerate some degree of salinity. The European roach and bream (*Rutilus* and *Abramis*) have populations in the brackish part of the Baltic Sea. The Japanese *Tribolodon* spends part of its life at sea, as does the Pacific Canadian *Mylocheilus.* These are, however, exceptions.

Most cyprinids are egg-scatterers, where spawning is a communal process. A few species have elaborate breeding techniques. The bitterlings (*Rhodeus* and related genera) of Eurasia lay their eggs inside freshwater mussels. To this end, the female develops a long ovipositor in the breeding season and places the eggs inside the

Peter Gathercole/Oxford Scientific Films

▲ The bitterling Rhodeus sericeus has intriguing breeding habits. The female maneuvers her 5-centimeter (2-inch) ovipositor to lay her eggs inside the mantle cavity of a freshwater mussel. Meanwhile the male stands by to shed his milt over the mussel's syphon, through which it will be drawn to fertilize the eggs as the mussel breathes.

▼ Its head virtually hidden by massive warty growths, the "lionhead" is one of the most bizarre of the exotic breeds of goldfish Carassius auratus.

Y. Lanceau/AUSCAPE International/Jacana

mussel's gill chamber. The male fertilizes the eggs by extruding his sperm close to the mussel's inhalant siphon and the respiratory current carries the sperm to the eggs. As a "payment" for this service, bitterlings frequently carry the parasitic larval stages of the mussels involved.

Many of the small cyprinid species from the tropics are brightly colored and are popular aquarium fishes. The great majority are bred in farms in Asia and Florida. Sadly, the escapees have often replaced the native fauna of those regions.

Psilorhynchids (Family Psilorhynchidae)
This family, from the mountain streams of southern Asia, contains only a few species of poorly known, bottom-dwelling fishes. They never grow more than a few inches long and are cryptically colored to blend with the streambed.

Hillstream fishes (Family Balitoridae)
This family is more widespread in Asia than the former. They also live on the bottom in fast streams. The body is greatly flattened and the pectoral and pelvic fins are greatly expanded to act as suckers which enable the fish to maintain its position in the fast current by hanging onto boulders or stones.

Loaches (Family Cobitidae)
Loaches are mostly bottom dwelling and have a body shape ranging from worm-like to stocky and compressed. The mouth is surrounded by six to eight barbels. The scales are greatly reduced and embedded in the skin. The swim bladder is partially encased in a bony capsule which has led, in at least one genus, to their use as living

barometers. As the degree of expansion of the swim bladder is governed by changes in atmospheric pressure, air is expelled or taken in via the mouth. In parts of Europe, the weatherfish *Misgurnus fossilis* was kept by peasants and its activity used to forecast a change in air pressure.

The algae eaters (Family Gyrinocheilidae)
The members of this family have two unique features: they lack pharyngeal bones; and the gill cavity is divided into two halves. The top half lets the water in and the lower half lets it out. The normal situation is that respiratory water comes in via the mouth and out over the gills. This modification lets the fish adhere to stones and plants with its sucking mouth without having to release its hold to breathe. Gyrinocheilids grow to about 30 centimeters (1 foot) long and feed more on detritus than on algae.

Suckers (Family Catostomidae)
The suckers have thick, fleshy lips. The pharyngeal bone is long and thin with many pharyngeal teeth in a single row. The teeth look much like large gill rakers. Almost all the species live in North America from the Arctic to Mexico, but one species, the widespread *Catostomus catostomus*, also occurs in eastern Siberia, presumably via the Bering land bridge. A second Asiatic species, *Myxocyprinus asiaticus*, from the Yangtse-Kiang has a remarkable body shape. For a fish that lives in fast-flowing water, the body is surprisingly deep. However, the cross-section profile is triangular and contoured in such a way that the water current presses the fish's flat ventral surface onto the bed of the river.

KEITH E. BANISTER

▼ *Growing to about 8 centimeters (3 inches) in length, the slimy loach* Acantophthalmus myersi *(below) inhabits rivers and streams in Thailand. The saddled hillstream loach* Homaloptera orthogoniata *(bottom) lives in rushing mountain streams of Asia, where it clings to rocks with highly modified pelvic and pectoral fins.*

CHARACINS & THEIR ALLIES

KEY FACTS

ORDER CHARACIFORMES
• 13 families • 260 genera
• *c.* 1,400 species

SMALLEST & LARGEST

Bolivian pygmy blue characin
Xenurobrycon polyancistrus
Total length: 17 mm (⅔ in)
Weight: 0.03 g (½ grain).

Tigerfish *Hydrocynus goliath*
Total length: 1.3 m (4½ ft)
Weight: 50 kg (110 lb)

CONSERVATION WATCH
!! The only species listed as
endangered is the naked characin
Gymnocharacinus bergi.
! Sardina ciega *Astyananax
mexicanus jordani* is listed as a
vulnerable subspecies.

F our families of the order Characiformes occur in Africa and ten in South America. Only one of these, the family Characidae, occurs on both continents, although there are anatomical differences distinguishing the American and African species. After the separation of Africa and South America in the late Cretaceous period, about 90 million years ago, characiform fishes on each continent independently evolved different life habits, body shapes, breeding habits, and mechanisms for capturing and eating different prey.

CITHARINIDS

The citharinids, family Citharinidae, are a small group of six species. Deep-bodied and silvery-colored, they feed on small food items filtered from the water and substrate. Citharinids are found in West Africa, the Zaïre River basin, the Nile River, and certain great lakes of East Africa. Their abundance in most of these areas and the large size of some species (84 centimeters or 33 inches and 18 kilograms or 40 pounds) make them important food fishes in many regions.

DISTICHODONTIDS

There are more than 90 remarkably diverse species of distichodontids, family Distichodontidae, and they cover the entire geographic range of the African characiforms. Species of *Nannocharax* are small with the largest growing up to 8 centimeters (3 inches). They are elongate fishes, with enlarged pectoral fins that support them while they search for minute prey on the bottom of sandy rivers and streams. The two dozen species of *Distichodus* include some with striking body pigmentation that are popular for aquarists. Other species in this genus reach a weight of 10 kilograms (22 pounds) and are important in fisheries. The elongate distichodontids related to *Phago* and *Eugnatichthys*

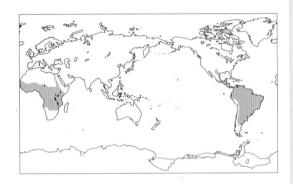

have heavy modified jaws and close-set teeth adapted for their specialized habit of clipping and feeding on the fins of other fishes.

AFRICAN PIKE CHARACIN

The African pike characin, family Hepsetidae, is widely distributed from West Africa through the Zaïre River basin to the Zambezi River of East Africa. Only one species, *Hepsetus odoe*, an elongate fish reaching 70 centimeters (28 inches) and 4 kilograms (9 pounds) is known. This species is a predatory fish, with numerous teeth in each jaw, and is best known as a sport fish.

Unlike other characins, the African pike characin builds a distinctive bubble nest in which the eggs are deposited and guarded.

▼ An American pike characin *Boulengerella maculata. Long, slim, and predatory, pike characins occur in fresh waters in both Africa and South America but, though similar in general appearance and lifestyle, they are not closely related and are usually placed in separate families, Ctenoluciidae and Hepsetidae.*

Hans Reinhard/Bruce Coleman Limited

Max Gibbs/Oxford Scientific Films

▲ *The three-banded pencilfish* Nannostomus trifasciatus *(family Lebiasinidae) occurs in forest streams in upper Amazonia. Complex distribution patterns, active speciation, and variable chromosome formulae make this a challenging genus for taxonomists, and species limits are ill-defined.*

of their silvery golden coloration, which sometimes shows dark markings, and a brightly colored fleshy process at the tip of the upper jaw. The adults are among the larger predatory characiforms in the Amazon basin, reaching over 1 meter (nearly 40 inches) in length. There are two genera and a total of seven species.

VOLADORAS, PYRRHULINAS, AND PENCILFISHES

The family Lebiasinidae consists of two major groups. One of these, the voladoras, includes 2 genera (*Lebiasina* and *Piabucina*) and about 12 species. They are mostly confined to the streams and ponds of north Andean upland areas in Peru, Ecuador, Colombia, Venezuela, and further east in the Guiana Highlands of Venezuela and the Guyanas. They also occur further north in Panama and Costa Rica, often at lower elevations. The cylindrical voladoras have large, somewhat thick scales and strong, well-toothed mouths. Some species reach 18 centimeters (7 inches) in length. They are excellent mosquito-larvae eaters and some or perhaps all species can also breathe air, allowing them to live in rather stagnant waters.

The second group, the colorful pyrrhulinans and pencilfishes, has many species kept in aquariums. The pyrrhulinans, genera *Copeina, Pyrrhulina,* and *Copella,* with about 25 species, are mostly lowland fishes occurring in the Orinoco, Amazon, and Paraguay basins, and the coastal streams of the Guyanas. Some species reach lengths of about 7 centimeters (2¾ inches), and these larger species are also excellent predators of mosquito larvae. There are 15 species of pencilfishes, genus *Nannostomus,* all with cylindrical bodies and tiny mouths. Adults reach lengths of 2 to 4.5 centimeters (1 to 1¾ inches). They are found in the Amazon and Orinoco basins and the Guyanas.

TRAHIRAS AND ALLIES

The trahiras, family Erythrinidae genus *Hoplias*, are a small group of cylindrical-bodied, blunt-headed, partly air-breathing, voracious predators, which feed on other fishes. Three species are widely recognized, but there are many more undescribed. They reach lengths of at least 90 centimeters (35 inches) in some areas. They live in lakes, swamps, and streams and are found from Costa Rica southward to South America and into Argentina.

Two other genera, *Erythrinus* and *Hoplerythrinus*, each with a single species, are also widely distributed in South America. These are smaller fishes, with a length to at least 25 centimeters (10 inches), and are of similar shape and habits to the trahira. They, like the voladoras discussed later, can breathe air and can survive in waters low in oxygen.

AMERICAN PIKE CHARACINS

The American pike characins, family Ctenoluciidae, range over much of the tropics of North and South America. They are most diverse in the Amazon basin and in regions immediately adjoining it. Juveniles of some species (*Boulengerella* and *Ctenolucius*) are interesting aquarium fishes because

AFRICAN CHARACINS

The African characins, family Characidae, occur throughout the range of the Characiformes on that continent, and are most common in the coastal rivers of West Africa and through the massive Zaïre River basin. There are about 18 genera and 130 species, ranging from the diminutive species of *Lepidarchus*, mature at 3 centimeters (1⅕ inches), to the tigerfish *Hydrocynus goliath*, the largest characiform. Most African characins are small to medium-sized silvery, bluish green, or yellowish fishes. Small species are called tetras. The males typically have highly developed anal fins and sometimes dorsal fins, which in species such as the Congo tetra, genus *Phenacogrammus*, have thread-like rays. The small species feed on insects, invertebrates, and some plant materials, while the large species are predatory. One such is the tigerfish, genus *Hydrocynus,* which has massively developed canine teeth. It is internationally known as a sport fish.

▼ *The congo tetra* Phenacogrammus interruptus *is widespread in forest waterways in the Zaïre River drainage system of West Africa. Though it is a popular and widely kept aquarium fish, little is known of its life history in the wild. These are females, slightly smaller and somewhat duller than males.*

J.M. Labat/AUSCAPE International

G.E. Schmida

◀ *The neon tetra* Paracheirodon innesi *is probably the most popular aquarium fish. The species lives in the upper region of the Amazon River on the Brazilian–Peruvian border. The striking coloration is found in both males and females.*

AMERICAN CHARACINS

The American characins, family Characidae, occur throughout the range of the Characiformes in the Americas, being exceedingly common in nearly all the fresh waters of South and Central America. They consist of approximately 170 genera and 900 species. Many are small, commonly 5 to 10 centimeters (2 to 4 inches) long. The group includes the larger piranhas, which grow up to at least 60 centimeters (24 inches) and their relatives (see page 105), and a series of 60 to 80 species of brycons. Brycons are omnivorous fishes that are found over the range of the family with some species also reaching 60 centimeters (24 inches) or more in length. They are used locally as food, especially in the Amazon and Orinoco basins. Members of the genus *Salminus*, especially those in the Rio Paraguay, are internationally known as fine sport fishes. They can reach a length of 1 meter (3½ feet) and a weight of 22 kilograms (49 pounds); weights of 30 kilograms (66 pounds) have been reported.

Characins have diverse feeding habits. Some species are almost completely herbivorous while others feed only on other fishes. Some are fin and scale eaters. This diversity of ways of feeding is reflected in the variety of body shapes, from swift pike-like predators to slow, deep-bodied, compressed omnivorous or insect-eating species that swim in schools.

Many species in this family, such as those in the genus *Astyanax*, are small silvery fishes, much like small minnows in appearance. Usually called

tetras, they are found almost throughout the range of the family. Many species, especially those that live in tea-colored tropical forest streams, have evolved into colorful, even spectacularly brilliant fishes, such as the neon tetras *Paracheirodon innesi*, popular in aquariums.

These forest fishes live in waters that are soft, somewhat acid, and chemically stable. Forest destruction can lead to severe alteration of the chemical nature of the water and many of the forest fishes will become locally extinct if the forest is destroyed.

Most of the American characids scatter their eggs among aquatic plants where the young can grow partly protected from predators. Some species migrate during the spawning season. There are 50 or so species that are relatively small, and some of these are very small. In these species the females are internally fertilized, and the males bear a scent gland on the tail, apparently to attract the females during courtship.

TOOTHLESS CHARACINS

The nearly 100 species of toothless characins, family Curimatidae, exploit an unusual food source: the slimy film of algae, fungi, and microscopic animals that cover underwater surfaces. Some species, used as food fishes, reach lengths of 35 centimeters (14 inches). The smaller toothless characins (genus *Curimatopsis*) show pronounced differences between the male and female: the brightly colored male has the tail fin and its associated bones developed for use in

▼ *The carnivorous tigerfish* Hydrocynus vittatus *(family Characidae) is the commonest and most widespread member of a genus of six species inhabiting African rivers and streams. It grows to a length of 50 centimeters (1½ feet), but its relative H. goliath is the largest of the group, reaching 1.3 meters (4½ feet) in length.*

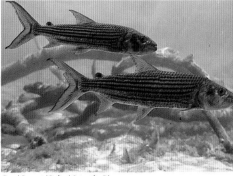

Carol Farneti/Oxford Scientific Films

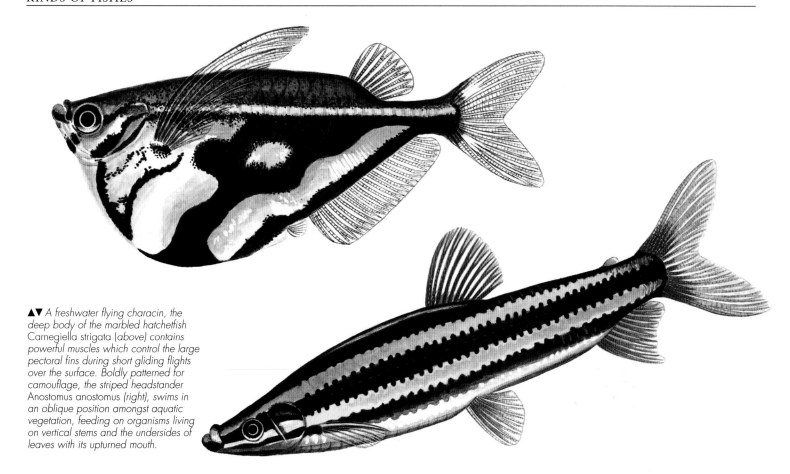

▲▼ *A freshwater flying characin, the deep body of the marbled hatchetfish* Carnegiella strigata *(above) contains powerful muscles which control the large pectoral fins during short gliding flights over the surface. Boldly patterned for camouflage, the striped headstander* Anostomus anostomus *(right), swims in an oblique position amongst aquatic vegetation, feeding on organisms living on vertical stems and the undersides of leaves with its upturned mouth.*

territorial displays. The larger species gather in huge spawning runs that are exploited by fishers throughout the tropics in the Americas.

FLANNEL-MOUTHED CHARACINS

The flannel-mouthed characins, family Prochilodontidae, are among the most important food fishes in South American fresh waters. Some species reach a length of at least 60 centimeters (24 inches). Huge schools of the three genera of flannel-mouthed characins travel up rivers, through rapids, and into smaller streams in spawning runs, making the motorcycle-like grunting sounds characteristic of the family. They were the first fishes to have spawning artificially induced by hormone injections.

ANOSTOMINS AND LEPORININS

The anostomins and leporinins, family Anostomidae, are related to the toothless curimatins, the flannel-mouthed characins, and most closely to the headstanders. The anostomins and leporinins include 10 to 12 genera, the largest being *Leporinus* with over 70 species and *Schizodon* with 10 species. All have small, modified mouths with extended, somewhat curved teeth. The long bodies of most species are rather circular in cross-section, with a tapering head and a small mouth.

Many species reach only 15 centimeters (6 inches) long, but some are up to 60 centimeters (24 inches) long. Some of the smaller, colorful species are kept as aquarium fishes. Many of the larger

species migrate upstream for spawning, and are used as food by humans.

HEMIODONTIDS

The approximately three dozen species of hemiodontids, family Hemiodontidae, are moderate-sized fishes (15 to 50 centimeters or 6 to 20 inches long) widely distributed throughout South America east of the Andes. These streamlined swift swimmers are famous for their jumps to escape predators and fishing nets. Hemiodontids are exploited as food fishes throughout their range and some smaller species are used as aquarium fishes. Some members of the family (genus *Bivibranchia*) have protrusile mouths, an adaptation unknown among other characiforms, and dramatic elaborations of the gill arch nerves and central nervous system.

HEADSTANDERS

Species of the two genera of South American headstanders, family Chilodontidae, are easily recognizable by the distinctive head-down body orientation. The species of *Chilodus* are handsomely marked fish less than 7.5 centimeters (3 inches) long and are popular in aquariums.

FRESHWATER HATCHETFISHES

The freshwater hatchetfishes (Gasteropelecidae) are insectivorous, surface-dwelling species with deep, compressed pectoral fin supports and modified pectoral fins. These fins are adapted to

PIRANHAS & THEIR ALLIES

Piranhas and their relatives belong to the characid subfamily Serrasalminae. The group consists of two parts: the first, a series of 7 genera and perhaps 60 species which are primarily plant-eaters. One of these, the Amazonian tambaqui *Colossoma macropomum* reaches 1 meter (a little over 3 feet) in length and 30 kilograms (about 66 pounds) in weight and is an excellent food fish; the second group consists of the piranhas and their relatives with 6 genera, some of which are flesh-eaters and some herbivorous. In addition to piranhas, this group includes the silver dollar fishes, genus *Metynnis,* well known to aquarium keepers.

The piranhas (or caribes) proper include 4 genera, *Pygopristis, Pygocentrus, Pristobrycon,* and *Serrasalmus,* totaling about 50 species, many of which are still to be described. Most people think of piranhas as traveling in schools and eating any living creature in their path. However, many of the species eat aquatic plants, especially seeds, and fishes. It is true that certain species can sometimes be dangerous to humans or livestock. For example, the common red-bellied piranha *Pygocentrus nattereri* of the Amazon basin may undergo a feeding frenzy when it comes across a bleeding animal, reducing it to a skeleton in a few minutes. The species seems not ordinarily dangerous and collectors have swum in waters occupied by this species without trouble. However, care and caution are advised as there is some evidence that individual fishes protecting their spawning sites may attack intruders, causing bleeding which may then trigger a feeding frenzy.

The subfamily is confined to the Orinoco, Amazon, Paraguay, and San Francisco river basins and the streams of the Guyanas.

▼ *Best known of all South American fishes, the danger of attack from piranhas such as the red-bellied piranha* Pygocentrus nattereri *is usually greatly exaggerated. When incited into a feeding frenzy, however, their short, powerful jaws and sharp interlocking teeth enable them to remove flesh in clean bites from the injured prey.*

increase the fishes' ability to leap from the water to escape predators. The mouths of freshwater hatchetfishes allow them to easily catch insects on the water's surface. There are seven or eight species in three genera—*Thoracocharax, Gasteropelecus,* and *Carnegiella.* The largest species reaches about 10 centimeters (4 inches) long, and lives in large streams or lakes. The smaller species, to about 3 centimeters (about 1⅕ inch), live in small streams. This family occurs from Panama south into the Paraguay basin. The relationships are unclear but it seems most likely that they are derived from some group of the American characins.

STANLEY H. WEITZMAN & RICHARD P. VARI

► *Reaching a maximum of about
10 centimeters (4 inches), the banded
corydoras* Corydoras barbatus *is among
the largest members of its genus.
Because they are attractive, hardy, and
do not grow too large, these fishes are
popular with aquarium hobbyists.*

CATFISHES & KNIFEFISHES

At first glance, there is little to suggest that catfishes and knifefishes are even remotely related. The often squat bodies of catfishes contrast sharply with the elongated tapering forms for which electric knifefishes are so aptly named. The whisker-like barbels that surround the catfish mouth and the pungent, sometimes venomous, protective fin spines for which catfishes are so well known are not to be found in any knifefishes. But, by looking beyond these external characteristics, it is possible to find a whole host of anatomical features that are shared by the knifefishes and catfishes and are found in no other group. These anatomical similarities indicate that these two remarkable and strikingly different groups of fishes started their evolutionary paths from a single ancestral species.

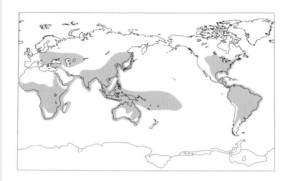

CATFISH CHARACTERISTICS
Catfishes may be one of the most widely recognized groups of fishes worldwide. In large part, this can be attributed to their ubiquitous distribution and the high regard given them as food fishes throughout much of the world.

Catfishes are most readily recognized by the thread-like barbels around the mouth. At least one pair of barbels, called the maxillary barbels, is invariably present at the two corners of the mouth. Usually two additional pairs are found on the lower jaw and sometimes a fourth pair on the snout, associated with the nostrils. When more than one pair of barbels is present, the maxillary barbels are usually the largest, sometimes exceeding the length of the entire fish. They are used in a number of ways, but are best known as sense organs. Barbels are covered with taste buds

Max Gibbs/Oxford Scientific Films

◄ *A head-on view of the toothless catfish* Anodontiglanis dahli *displays the "vacuum-cleaner" mouth and the arrangement of fleshy, sensitive barbels around the mouth that are typical of catfish: a pair at the gape and two pairs along the lower jaw.*

that allow a catfish to detect food in front, below, and alongside it.

Also characteristic of catfishes are the retractable thorny spines that are found at the front edge of the dorsal and pectoral fins. These spines are enlarged, stiffened rays, otherwise similar in form to the remaining rays that support the fin. The presence of spines provides the fish with a formidable weapon against predators. Erected spines can be locked into place, thereby enlarging the cross-sectional size of a fish and making it harder for a predator to swallow. Spines can be sharply pointed at the tip, further protecting the fish from attack. In a few species, such as the walking catfish *Clarias batrachus*, the pectoral spines are used like a pair of crutches with which a fish can "walk" out of water. The spines of some catfishes have an associated venom gland which can deliver a painful toxin into a wounded victim. While spines are an obvious characteristic of catfishes, not all catfishes have them. Dorsal and/or pectoral spines have been

lost in the evolution of various groups of catfishes.

The final conspicuous characteristic used to recognize catfishes is the absence of scales covering the body. Most species possess a tough, leathery skin that lacks any type of bony covering. In contrast, several catfish lineages, collectively referred to as armored catfishes, have developed bony plates that cover part or all of the body surface. These plates are different from typical fish scales, both in structure and in the way they are deeply embedded in the skin, and are not easily removed. Thus, it appears that some catfishes have evolved a protective coat of armor to replace the scales that were lost early in catfish evolution.

THE DIVERSITY OF CATFISHES

More than 2,200 species of catfishes are known at present and, at the rate with which new species are being discovered, the total number may well exceed 3,000. The large number of catfish species is reflected in the diversity of body forms and sizes found in this group. Many of the world's major

and the Congo and Nile rivers of Africa are all home to catfishes that exceed 50 kilograms (110 pounds). At the other end of the spectrum, catfishes that grow no larger than 25 millimeters (1 inch) are known from South America and Asia. The smallest species, which matures at 12 millimeters (½ inch), is a member of the South American catfish family Scoloplacidae, in which all four species grow to no more than 30 millimeters (1⅕ inches).

HABITAT

Catfishes have been found on all continents. Although no catfishes inhabit Antarctica at present, at least one species is known from fossil remains dating to the late Eocene or early Oligocene (about 37 million years ago). On the remaining continents, catfishes are currently found in almost all freshwater environments and they are often the dominant group of fishes. In regions where catfishes are well represented, such as South America and Asia, one species of catfish or another can be found in all parts of the riverine environment—from torrential mountain streams to slow-moving, broad river channels.

While primarily inhabitants of fresh water, some catfish species are found in estuarine and coastal marine environments. Most notable among these are species of the Indo-West Pacific family Plotosidae, in which about one third of the approximately 30 species are found either in coral reefs or brackish water estuaries. Most species of the circumtropical family Ariidae, of which there are somewhat over 100 forms, inhabit marine environments and only rarely venture into fresh water. A few of the species of this family, however, appear to be restricted to freshwater environments.

▲ *The wels* Silurus glanis *is the largest catfish in the world and the only one native to western Europe.*

▼ *Emitting and receiving impulses for self-defense and navigation, the electric organs of the electric catfish* Malapterurus electricus *(top) consist of modified muscle tissue under the thick skin of its belly and sides. The boldly marked young of the striped sea catfish* Plotosus lineatus *(bottom) form tightly packed schools to confuse and discourage predators.*

river systems are home to at least one catfish species which often ranks among the largest fishes of the river. Among these giant catfishes is the European wels *Silurus glanis*, the largest known catfish, which has been reported to grow to 5 meters (16 feet) in length and weigh more than 300 kilograms (660 pounds). Nearly as large as the wels is the royal catfish of Thailand *Pangasius gigas*, called *pla bŭk* in Thai, which is found in the Mekong River system, and can weigh more than 200 kilograms (440 pounds). Similarly, the Amazon, Orinoco, and Paraná river systems of South America, the Mississippi of North America,

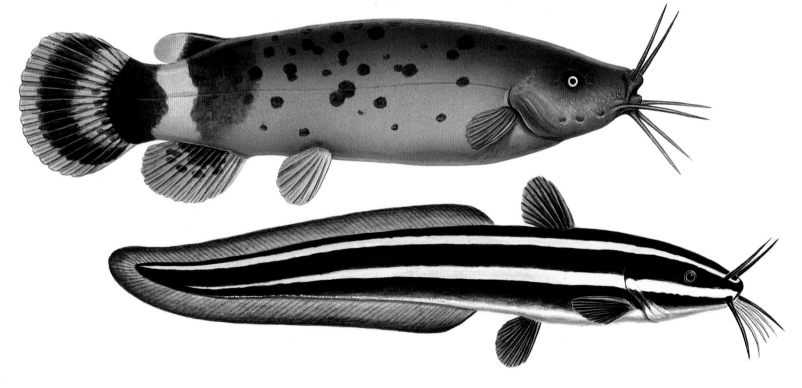

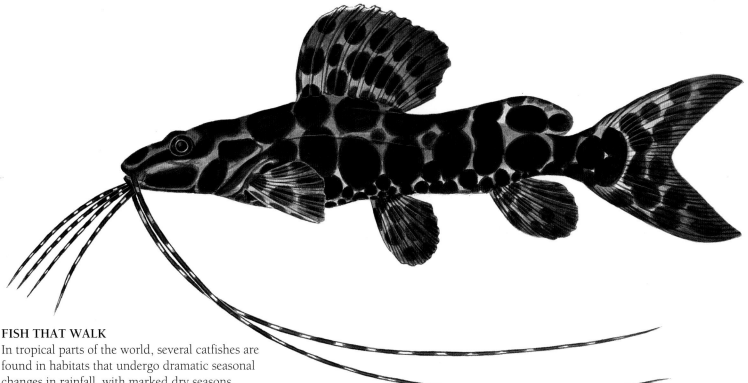

FISH THAT WALK

In tropical parts of the world, several catfishes are found in habitats that undergo dramatic seasonal changes in rainfall, with marked dry seasons. During these dry periods, the river channels and lakes dry up and become crowded with trapped fishes. Some catfishes that inhabit these regions have accessory air-breathing structures that allow them to take oxygen directly from the atmosphere.

The most famous of these air-breathing catfishes are the so-called walking catfishes of the Afro-Asian family Clariidae, especially an albino form of the walking catfish *Clarias batrachus*. Introduced into Florida fishponds by the aquarium fish industry, the walking catfish left their overcrowded ponds and walked, by means of their sturdy pectoral spines, into the natural waterways of southern Florida where they have succeeded in establishing many thriving colonies throughout the state.

FOOD GATHERING

While catfishes have gained a nearly universal, but nevertheless undeserved, reputation for being "mud-eaters", they do, in fact, display an enormous diversity of food-gathering activities. A large number of species live closely associated with the bottom and feed primarily on the aquatic invertebrates that live there. Many are piscivorous (fish-eating). Among the piscivores at least two groups of catfishes show unusual forms of prey capture. In members of one small family from southern Asia, the Chacidae, the catfish lure their prey toward them in a manner that resembles the better known anglerfishes. These catfishes possess minute maxillary barbels that are directed vertically and wiggled in a manner that resembles two aquatic worms. Small fishes that are attracted to the barbel tip of these otherwise motionless catfishes are readily inhaled into their extraordinarily capacious mouth.

Electric catfishes of the African family Malapteruridae stun prey that strays too close with strong electric shocks. The electrical discharge, which may exceed 350 volts, is generated by the massive electric organ found just beneath the skin of these sausage-shaped fishes. Among the catfishes that do not feed on other fishes or benthic (bottom-dwelling) invertebrates, there are several species that actively swim in the water column and feed exclusively on zooplankton. Members of the largest catfish family, the

▲ *The sailfin pimelodid, genus* Leiarius, *displays the features that characterize the Pimelodidae: wide mouth, large adipose fin, and no barbels on the upper part of the snout.*

▼ *An albino catfish genus* Clarias. *Members of this group are often called "walking catfish" because they can travel on land using their stiff pectoral fins rather like crutches.*

Jane Burton/Bruce Coleman Limited

neotropical Loricariidae, are almost exclusively herbivorous, taking advantage of the wide variety of plant items found in the aquatic environment. Among these botanic food sources are aquatic algae, fallen leaves, and even the woody parts of fallen trees, which at least one group of South American catfishes appears able to digest with the aid of symbiotic micro-organisms.

Parasitism as a form of nutrition gathering occurs in a few groups of catfishes and is best known in the family Trichomycteridae of South America. Two forms of parasitism are practiced by a small but notorious species group within the family. In one form, called lepidophagy, scales, mucus, and even bits of flesh are rasped off the skin of prey. In the second form, called hematophagy, a catfish swims into the gill cavity of a host fish (often a larger catfish), lodges itself in place by means of patches of opercular spines, bites off the tips of gill filaments and swallows the released blood. Several hematophagous catfishes of South America, called *candirú*, are notorious for their occasional mistakes in choice of a place to feed. Instead of a gill cavity, *candirús* sometimes swim into the urethra of a urinating mammal (including bathing humans), presumably to feed there. The opercular spine patches makes the removal of the misplaced catfish most difficult and painful. Except for the activities of the candirú, both gill- and blood-feeding catfishes move from host to host without irreparably injuring the host fish in most cases.

SOCIAL ACTIVITY

Catfishes are generally thought of as solitary nocturnal fishes that live primarily on or near the bottom of lakes and rivers. While a great many species fit this description, there are exceptions. Many members of the neotropical families

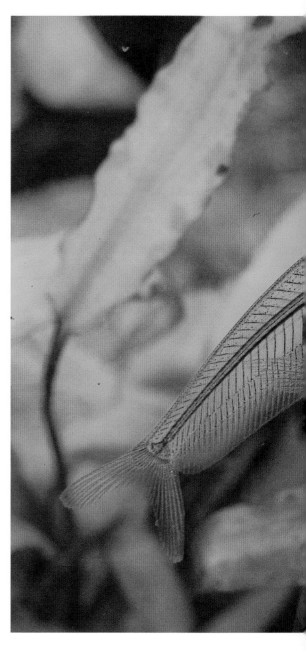

▼ *The needle catfish Farlowella amazonica, a member of the catfish family Loricariidae, is confined to northern South America. The mouth is provided with broad, fleshy rough lips to help skim algal films from rocks and similar surfaces.*

Callichthyidae and Loricariidae (the armored catfishes) actively feed throughout the day. While tight schooling behavior, like that found in herrings, is rare in catfishes, many species move in loose aggregations of several to several hundred individuals. The well known *Corydoras* catfishes of South America can often be seen in large shoals over sandy or muddy river bottoms. Some shoaling catfishes are found well up in the water column, usually feeding on planktonic organisms. Among these free-swimming planktivores are the glass catfishes of Asia (*Kryptopterus* species), the Debauwi catfishes (*Eutropiellus* species) of western Africa, and the South American loweye catfishes (*Hypophthalmus* species).

REPRODUCTION

The range of reproductive biology exhibited by catfishes does credit to their diversity of forms. Some amount of parental care can be found in

Max Gibbs/Oxford Scientific Films

almost all catfish species, and some go to extraordinary lengths to care for either the eggs or both eggs and young. While most species of catfishes build and guard some form of nest, the more remarkable examples of parental care are those in which one parent carries the developing embryos around during the period of incubation. For example, males of the marine catfish family Ariidae incubate their marble-sized eggs in their mouth until the eggs hatch and the fully formed juveniles are able to fend for themselves. The spotted African squeaker, *Synodontis multipunctatus,* from Lake Tanganyika (family Mochokidae) actually begins its life in the mouth of another species of fish. The squeaker eggs are strategically tossed into the water column by its parents during the spawning activity of one of the many oral-brooding cichlids (family Cichlidae) that also inhabit the lake. The female of the pair of spawning cichlids picks up the catfish eggs and

broods them as if they were her own. Several species of the armored suckermouth catfish (family Loricariidae) have males that carry their clutch of developing eggs around with them, held in place by a greatly expanded lower lip.

Species of the South American family Aspredinidae that inhabit muddy coastal estuaries provide not only shelter for their developing young, but nutrition as well. Females of the whip-tailed banjo catfishes of the genera *Aspredo* and *Aspredinichthys* somehow place their newly fertilized eggs on a patch of spongy abdominal skin that holds the eggs in place. Microscopic blood vessels from the abdominal wall surround the yolk sac and provide the developing embryo with oxygen and nutrients.

ELECTRIC KNIFEFISHES

In contrast to the ubiquitous catfishes, electric knifefishes (or gymnotoids) form a relatively small

▲ *Entirely translucent except for the intestines, the phantom glass catfish* Kryptopterus bicirrhus *from Southeast Asia is another popular aquarium fish because of its small size, easy care, and remarkable appearance.*

group of about 85 species known only from the warm inland waters of South America and southern Central America. Most are less than 30 centimeters (1 foot) in length. The term "knifefishes" is quite appropriate for these fishes, as they have long slender compressed bodies that usually taper toward the tail and are held straight while swimming. The swimming motion is accomplished by the elongated anal fin that extends for nearly the entire length of the body, from just behind the head to near the tip of the tail. The anal fin rays are moved laterally in coordinated patterns, causing continuous series of waves that ripple along the length of the fin. The generated waves can be initiated at either the front or rear end of the fin, thereby allowing the fish to move forward and backward, with equal ease, while maintaining a practically rigid body.

The most striking characteristic of gymnotoids is an organ system that allows these fishes to produce and receive electrical impulses. This additional sensory mode, called the electric sense, is used by knifefishes to navigate in their environment, to identify and locate food, and to communicate with members of the species. In nearly all species of electric knifefishes, the voltages generated by electric organ discharges (EOD) are in the order of thousandths of a volt, or millivolts, which are too weak to be felt by scientists. This led these fishes to be classified as "weak" electric fishes. In contrast, many of the better known electric fishes, such as the electric eel, the electric catfish, and the torpedo, are called "strong" electric fishes as they generate potent electrical discharges in the order of hundreds of volts, which can be felt by humans touching them.

In much of South America, electric knifefishes are only encountered infrequently in nature, giving the impression that they are but a minor component of the fish fauna of South America. This perception derives in part from the fact that gymnotoids are nocturnal animals that hide in dense vegetation or lie buried in sand during the day. It was discovered recently that gymnotoids are, in fact, quite abundant in the deep channels of some of the larger South American rivers and may well be among the dominant components of the food web there.

Little is known about the biology of most species of electric knifefishes. Of those that have been studied, most feed on a variety of invertebrates that live on the river bottom. Reproduction in gymnotoids is equally poorly known. From laboratory observations, some species have been observed to spawn repeatedly, over a period of several months, during the rainy season. In a couple of species, the end of the spawning season was followed by the disappearance of nearly the whole adult population. It has been suggested that this may indicate a post-spawning mortality similar to that of sea-run salmon.

CARL J. FERRARIS, JR

ELECTRIC EELS

The best known, and yet the most atypical, of the electric knifefishes is the electric eel *Electrophorus electricus,* the sole member of the Electrophoridae. In appearance, the electric eel little resembles the knife-blade form that is otherwise the mark of a gymnotoid. Instead, it is nearly round in cross-section from head to blunt tail. The feature that most readily distinguishes the electric eel from its close relatives is the nature of the electrical discharge it produces. In addition to the weak electrical discharges that are characteristic of gymnotiform fishes, the electric eel is able to unleash a burst of electricity that is measured in hundreds of volts. The intensity of these strong electrical discharges is related to the size of the individual fish, with maximum voltages of 550 volts at 1 amp, recorded for fish of 2 meters (6 ½ feet) or more in length. Discharges consist of a rapid series of brief electrical pulses, each of which lasts for only a few milliseconds. Electric eels use these powerful electrical discharges primarily to stun prey fishes, which are immobilized by the shock and then quickly engulfed by the toothless eel. The discharges can also be used to repel would-be threats, including both predators and other organisms with which they come into accidental contact (including people). The discharge from a large electric eel is strong enough to knock a person down, but statements that an electric eel can kill a human are untrue. Moreover, it is not possible for an electric eel to light up an ordinary household light bulb, immersed directly in the water, because the discharge produced by one of these fishes is too brief to heat the filament.

Hans Reinhard/Bruce Coleman Limited

▲ *Highly modified and evolved from muscle tissue, the electricity generating organs of the electric eel are arranged like cells in a battery along each flank.*

SALMONS & THEIR ALLIES

The allies of salmons include trouts, chars, pikes, smelts, and whitefishes. The salmons and trouts certainly rank as most important fishes. They are highly prized as food and angling fishes, and are also of great interest to biologists because of their amazing migratory habits.

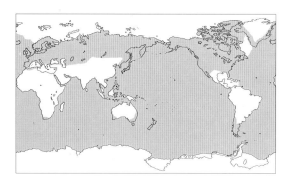

CHANGING RELATIONSHIPS

Scientific opinions are in a state of flux about the relationships of the 14 families that make up salmons and their allies. All of the families are rather primitive teleosts, sometimes regarded as closely related, though recent studies suggest that they may not be as close as was formerly believed. The families can be arranged into three groups—salmoniforms, esociforms, and argentiniforms.

The salmoniforms include a number of families. The salmonoids (family Salmonidae) comprise about 70 species of trouts, chars, salmons, and whitefishes, which occur widely across the northern temperate lands of Europe, Asia, and North America. The osmeroids include about 10 species of northern smelts (family Osmeridae), which also occur widely across the cool temperate regions of the northern hemisphere. There are about 12 species of so-called icefishes (family Salangidae), present in lands bordering the cool northern Pacific Ocean. The 6 species of southern smelts and graylings (family Retropinnidae) and 40 species of whitebaits, galaxiids, and pelladillos (family Galaxiidae) occur in the southern cool temperate lands of Australia, New Zealand, Patagonian South America, and the southern tip of southern Africa. The salmoniforms also include the single species in the family Lepidogalaxiidae—the highly peculiar Western Australian salamanderfish (see box, page 118).

The esociforms are made up of the pikes and their relatives—five species in the family Esocidae and four species in Umbridae. These fishes are characteristic of cool northern temperate lands.

Several oceanic or deepsea families make up

J. M. Labat/AUSCAPE International

◀ Hatched in the upper reaches of a freshwater stream, these Atlantic salmon Salmo salar fry are still thriving on the nutrients in their yolk sacs. When they begin foraging they will gradually move downstream as they grow, spending their adult lives in the ocean until returning to the stream to breed.

▲ *Like other Pacific salmons, the sockeye salmon* Oncorhynchus nerka *(top) undergoes a huge physical transformation as breeding approaches. This transformation is particularly spectacular in the male, shown here, which starts as a sleek, silvery, trout-like fish before going upriver, then develops large, hooked jaws, a humped back, and bright red and green coloration. A member of the family Umbridae, the Alaska blackfish* Dallia pectoralis *(above, right) is found in Arctic and subarctic pools and lakes in Alaska and Siberia. Like its close relatives the pikes, the blackfish is a lurking predator and has its dorsal and anal fins placed far back on its body for maximum thrust when darting at prey. Arthurs' paragalaxias* Paragalaxias mesotes *(above, left) of the family Galaxiidae is another small lurking predator. Although this species has a limited distribution in Tasmania, the family as a whole is widespread across the southern hemisphere, with representatives in Australia, New Zealand, New Caledonia, South America, and southern Africa.*

the argentiniforms. They include 12 species of herring smelts (family Argentinidae), about 30 species of slickheads (Alepocephalidae), and about 100 species in other families.

Salmoniformes, esociforms, and argentiniforms are primitive fishes of ancient origins with few overall defining features (which in part explains why our understanding of the relationships is changing). They are all characterized by having two bones forming the margin of the mouth in the upper jaw—the premaxilla, the chief tooth-bearing bone, which nearly always bears teeth; and the maxilla, which is behind and below the premaxilla and bears teeth in some groups but not in others.

An adipose fin is a well-known feature in salmons and trouts and in osmerid and retropinnid smelts, but it also occurs in other groups. In salmons and their allies, the pelvic fins occur at mid-abdomen, and usually there is a single, often quite small, dorsal fin (in addition to the adipose fin, when present). This dorsal fin may be high on the back (as in salmonids and osmerids), or may be well back above the anal fin (as in some retropinnids, galaxiids, esocids, and umbrids). All the fins are supported only by fin rays, which are

segmented and usually branched; there are no spines in the fins of any of these groups.

SALMONS AND TROUTS

Salmons and trouts probably best represent the basic form and habits of the salmoniform group as a whole. They are slender, spindle-shaped fishes, often somewhat laterally compressed, usually with smallish fins and a well-forked tail, and are well-adapted to a free-swimming life in midwater. They have quite small rounded scales. The mouth is armed with single rows of sharp, recurved teeth, and the digestive canal is simple, with a single loop at the stomach and a short intestine that runs directly back to the anus.

Salmon migration

Although many salmoniform families occur in fresh waters, a distinctive feature of the salmons (and some other families) is that in many species a regular part of the life cycle is spent in the sea, and there are thus regular migrations to and from the sea. Reproduction occurs in fresh water, but the young, either immediately following hatching, or after a few days to one or more years feeding and growing in fresh water, move to sea where most of

◄ Adult pink, or humpback salmon *Oncorhynchus gorbuscha* make their way up a British Columbia river in preparation for spawning. Having lost their oceanic metallic blue-green sheen, they have begun the metamorphosis that will see the development of long, hooked jaws and greatly humped napes in the males. Pink salmon spawn at two years of age and die shortly afterwards, so in any one area they can be divided into non-interbreeding, odd, and even year races.

G.E.Schmida

▲ *The Australian smelt* Retropinna semoni *is a member of the small family of southern smelts, the Retropinnidae. Southern smelts are close relatives of the northern smelts and share, amongst other features, a peculiar cucumber-like odor.*

their life is spent. The adults typically return to fresh water to spawn after a few months to several years at sea; in most species, feeding ends at the time the adult fish begin that return migration. This life-history pattern, described as anadromous, is best known in the salmons, both Atlantic (*Salmo*) and Pacific (*Oncorhynchus*) species.

Feeding

Members of the family Salmonidae are predators. In some species the prey is plankton which may be taken by filtering large volumes of water across rows of fine, slender gill-rakers (as, for example, in the sockeye salmon *Oncorhynchus nerka* of the northern Pacific); many species live off small invertebrates found in their habitats—often crustaceans in the marine plankton or aquatic insects in streams and rivers; and many salmoniforms, especially the larger species like the lake char *Salvelinus namaycush* of northern North America, are voracious fish predators.

Food value

Many salmonids are of great importance as food fishes—important for aboriginal, commercial, and recreational fisheries, and also in aquaculture. In historical terms, the Atlantic salmon *Salmo salar* is probably the most significant, having been valued by the peoples of both Europe (eastern Atlantic) and North America (western Atlantic) for as long as there are records. Drawings of salmon feature in the rock art of early human occupants of France; salmon were mentioned in the Magna Carta of thirteenth century Britain; and salmon were

represented on the medieval coats of arms of the nobility of several countries in Europe. They were also a valued food fish of the North American Indian peoples of eastern North America, from the New England states north to the Inuit peoples of Arctic Canada.

The salmons of the northern Pacific (genus *Oncorhynchus*) were equally important to the peoples who lived close to the ocean there—to the North American Indian and Inuit peoples, as in the east, and along the Asian coasts of the northern Pacific, in Siberia, Kamchatka, and Japan. These salmons were initially taken by the native peoples, but in more modern times fishing fleets have exploited the vast shoals of salmon that occupy the cool and cold waters of the northern Atlantic and Pacific oceans, where they form the foundations for large industries, especially in the Pacific. Nearly all the 70 or so species of salmonids have been important food species. Trouts have been introduced very widely into other lands, including Australia, New Zealand, South America, and southern Africa.

SMELTS

The northern smelts (family Osmeridae) closely resemble the salmons and trouts, but are much smaller. Many of them move to and from the sea, although a few, like the Pacific surf smelt *Hypomesus pretiosus*, actually complete their life cycles at sea. Because they are small they are of less importance as food fishes, although some are valued, especially the eulachon *Thaleichthys pacificus*. This species was caught by the North

▶ *Life cycle of an anadromous salmon, the sockeye salmon* Oncorhynchus nerka. *The eggs are laid in gravel nests in freshwater streams. The hatchlings, or alevin, remain in the nest until they absorb the nutrients in their yolk sac, then emerge as feeding, free-swimming fry. The fry leave the stream and live as parr in rivers or small lakes before turning silvery and entering the sea as smolt. After a period of life in saltwater the adult undergoes dramatic changes in shape and coloration and returns to the stream in which it hatched to spawn and die, beginning the cycle again.*

THE LIFE CYCLE OF A SOCKEYE SALMON

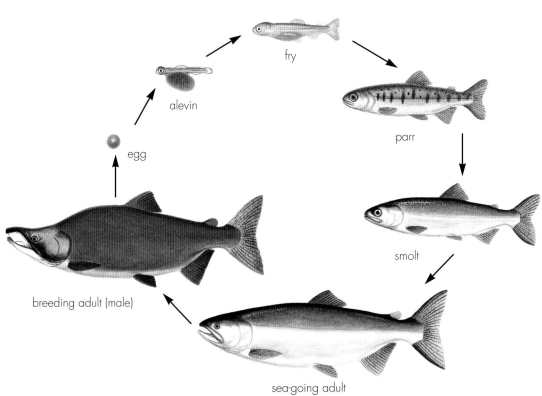

fry

alevin

egg

parr

smolt

breeding adult (male)

sea-going adult

◀ *One of large numbers of Australian species of galaxiid, the western galaxias* Galaxias occidentalis *shows the smooth, scaleless skin and tubular shape of this group. Galaxiids differ from most salmoniform fishes that migrate to and from salt water in that they return to fresh water while still immature, long before they are able to reproduce.*

G. E. Schmida

American Indian peoples, both for food and for its oil, which they burnt to produce light.

Southern smelts (family Retropinnidae) resemble osmerids and both families have a strong cucumber-like odor, the function of which is unknown. Japan's ayu (family Plecoglossidae) is a specialized osmerid, distinctive among the northern families in feeding on river algae; it has specialized teeth to permit this. The southern graylings (species of *Prototroctes* in Australia and New Zealand) are also herbivorous, although this feeding characteristic seems to have evolved separately in the two families. In both families the intestine is longer than in others of this group, an adaptation to help them digest plant material. The ayu is one of Japan's most favoured food fishes, in spite of its small size (25 centimeters or 10 inches in length). Even the tiny oriental icefishes (family Salangidae, to about 16 centimeters or 6 ½ inches) are taken as food in Japan, Korea, and China.

GALAXIIDS

The southern family Galaxiidae differs from the other families mentioned in many features. They tend to be rather stocky, tubular fishes that live close to the substrate rather than swimming in midwater. They have their fins back toward the tail, and the tails are often square. All these features adapt them to a life within cover. These species have no scales. They also differ in their life history. Like the salmons, they spawn in fresh water and the newly hatched larvae move to the sea. There they feed and grow for a time, but as small juveniles return to fresh water and feed and grow there for several months to years before maturing and reproducing (a life cycle described as amphidromous).

The southern freshwater families have less importance to fisheries than the northern ones, although the immense, swarming shoals of juvenile galaxiids (less than 5 centimeters or 2 inches long) that enter New Zealand's rivers during the spring were traditionally regarded as important food fishes by the indigenous Polynesian Maori people. Their example was followed by European settlers in New Zealand during the second half of the nineteenth century. They called the little fish "whitebait", after similar small fishes taken in British seas, although the resemblance is only superficial. Small fisheries for the galaxiid juveniles also exist in Tasmania, Australia, and in some parts of Chile. The Tasmanian whitebait *Lovettia sealii* was also once a valued food fish, although this is no longer true.

PIKES, PICKERELS, AND MUSKELLUNGES

The pikes, pickerels, and muskellunges of the family Esocidae are large, strongly toothed predators that, unlike most of the "freshwater" species in the group, spend their whole lives in fresh water. They resemble the galaxiids in having their fins concentrated back near the tail, an adaptation for being ambush predators. They have attracted attention from anglers throughout the northern temperate lands, especially the muskellunge and northern pike which grow to a very large size and are powerful fighters.

MARINE FAMILIES

It is when we turn to some of the marine or oceanic salmoniform families that more substantial and peculiar specializations are seen. The opistho-proctids, as their common name "barreleyes" indicates, have tubular and upward- or forward-looking eyes; some also have photophores (light organs) that are probably important for species recognition in the dimly lit environment of the deep sea. But not all the marine families are specialized, and some species of argentines, or herring smelts, closely resemble the freshwater osmerid smelts, being slender, elongate, and very silvery.

ROBERT M. MCDOWALL

▼ *With its large, tooth-studded mouth allowing it to take food one-third to one-half its own size, the northern pike* Esox lucius *has been known to feed on anything from other fishes to muskrats and ducklings. Easily enticed into taking artificial lures, this voracious predator is a popular freshwater game fish in North America and Europe.*

H. Berthoule/Jacana/AUSCAPE International

WESTERN AUSTRALIA'S SALAMANDERFISH

In 1961 a tiny, insignificant fish was first described as *Lepidogalaxias salamandroides* from the peat-swamps of southwestern Australia. The name *Lepidogalaxias* ("the scaled galaxias") was something of a contradiction, since no previously recognized galaxiid had scales, and ichthyologists soon removed *Lepidogalaxias* into its own family, Lepidogalaxiidae. No one really knew what it was related to, and there were suggestions that it was the only southern hemisphere representative of the pikes (family Esocidae). However, this view was subsequently discarded, and it is now thought to be closest to the other southern salmoniform families, Galaxiidae and Retropinnidae. Apart from its classification, there was little interest in this fish for many years, but recent studies have shown that it is one of the most peculiarly specialized fishes known.

Under observation in captivity, *Lepidogalaxias* was seen to be able to bend its body behind its head, as if it had a neck—something that is most unusual in other fishes. It looked rather like the Concorde aircraft on the ground with its cockpit deflected downwards. This strange attitude prompted questions about why a fish would develop this adaptation, and the explanation was soon obvious. *Lepidogalaxias* has its eyes fixed in the orbits: they cannot rotate to follow prey or predators because the skin is continuous from the surface of the head and over the eyes. There is no circumorbital sulcus (groove of loose skin) around the eyes, as is found in virtually all vertebrates, to make rotation of the eyeball possible; there are no external eye muscles, either. If the eyes cannot move, why have muscles to move them? No definite reason for this continuous skin covering is yet available, but it is probably related to the habitats and habits of the fish.

Salamanderfish live in temporary pools filled with vegetation and twigs and it can be imagined that debris would find its way into the sulcus around the eye, if there was one. Loss of the sulcus could be attributed to the need to avoid such problems. But, lack of movement would restrict vision and the salamanderfish has developed the ability to search its habitats by moving the head, rather than the eyes. It does this by the special adaptation of bending its neck, which most other fish cannot do. In the salamanderfish there are large gaps between the first few vertebrae behind the head, and these allow bending. This would probably not be practicable in larger fishes but the salamanderfish grows to only about 60 millimeters (2½ inches) and so the strength and interlocking nature of the vertebral column are not as important as they would be in a larger and heavier fish.

These are not the only specializations in the salamanderfish. It has morphological and physiological features that allow it to survive in pools that dry up for months at a time; and it has reproductive specializations, including internal fertilization of the eggs, that also help the species to thrive in these small, temporary and inhospitable habitats.

▲ A salamanderfish Lepidogalaxias salamandroides *in its aestivation burrow. Living in small temporary pools and ditches, this unusual fish possesses a number of features that allow it to survive extensive dry spells by burrowing in the substrate.*

▼ *This close-up of the salamanderfish shows clearly the presence of scales, one of the features that sets it apart from the galaxiids.*

DRAGONFISHES & THEIR ALLIES

KEY FACTS

ORDER STOMIIFORMES
• 4 families • 51 genera
• c. 250 species

SMALLEST & LARGEST

Cyclothone pygmaea
Location: Mediterranean Sea
Body length: 1.5 cm (⅗ in)

Opostomias micripnis
Location: Atlantic, Pacific, and
Indian oceans
Body length: up to 50 cm (20 in)

CONSERVATION WATCH
■ Very little is known about
population structure, numbers of
individuals, or breeding. The
species appear to face no threats
to their survival, other than the
general danger posed by massive
pollution of the oceans.

D ragonfishes and their relatives (order Stomiiformes) make up a significant part of the ocean's fishes. The group consists mostly of mid- or deepwater fishes that are found in the open waters of all major temperate and subtropical oceans, with some species extending into subarctic and Antarctic waters. There are two major lineages, one comprising the families Gonostomatidae and Sternoptychidae, and the other made up of the families Phosichthyidae and Stomiidae. Various members of the order are known as lightfishes, hatchetfishes, bristlemouths, loosejaws, snaggletooths, scaly dragonfishes, black dragonfishes, and viperfishes, which says something about their diversity and suggests that some are rather impressive predators.

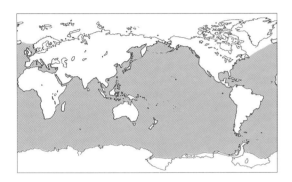

several bones of the gill basket. Depending on the species, adult stomiiforms range from less than 2 centimeters (about ⅕ inch) to over 50 centimeters (20 inches) from the tip of the snout to the base of the tail. In an evolutionary sense, these fishes are more advanced than smelts, salmons, minnows, and catfishes, but more primitive than perches, codfishes, and lanternfishes.

PREDATORS WITH LIGHT ORGANS

As with many deepwater fishes, there are no common names for most stomiiforms. All members have features that show they are descended from a unique common ancestor; these include light organs (photophores) of particular construction, hinged teeth that have an anterior axis of rotation, and specializations of the jaw muscles and ligaments as well as of

TWO LINEAGES

Primitive stomiiforms are slender-bodied, usually silvery and brown, with rows of egg-shaped light organs along the lower body, and a mouth armed with numerous teeth. The two major evolutionary lineages in this group lead to two widely divergent body plans and associated ecology. The families Gonostomatidae and Sternoptychidae comprise one lineage, which includes the bristlemouths and marine hatchetfishes. Bristlemouths are among the most primitive members of the stomiiforms,

▼ *Special lighting reveals the numerous photophores on the underside of the viperfish* Chauliodus sloani, *one of the most fearsome of deepsea predators, despite its small size of around 30 centimeters (12 inches).*

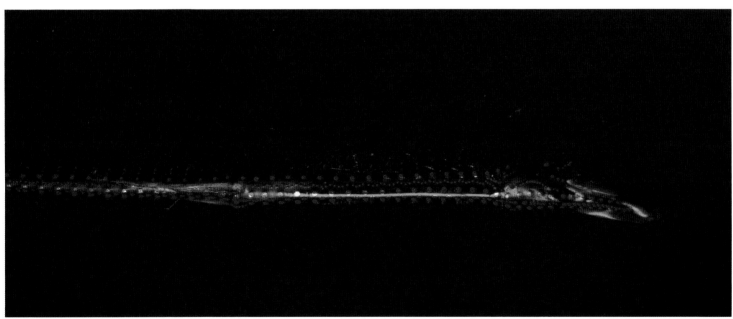

while marine hatchetfishes represent one of the evolutionary extremes. A group within the Sternoptychidae, the hatchetfishes are disk-shaped, with thin, deep, bright silvery bodies, small mouths, and relatively small teeth; apparently they feed primarily on plankton. The other evolutionary group consists of the families Phosichthyidae and Stomiidae. The Phosichthyidae shares the name lightfishes with some gonostomatids; many of its members have the primitive body plan already described. The Stomiidae, which includes the dragonfishes, represents the second extreme in divergence; many of these fishes have elongate, snake-like black bodies, large mouths, and extremely large teeth. Most stomiids feed on large prey, including other fishes and crustaceans.

FEEDING MIGRATIONS

Many stomiiforms migrate nightly from resting areas in deep waters up to the more productive areas near the surface, where food in the form of small fishes and invertebrates is plentiful. Migrators maintain their place in the depths by measuring the downwelling light visually and swimming within an area of a particular intensity; as the sun lowers to the horizon, the fishes follow the fading light to the surface. After feeding through the night, the stomiiforms then swim back to the depths during the day. This phenomenon of a vertically migrating fauna is well documented and is seen in other kinds of fishes as well as in many invertebrates. Some of the larger predatory stomiiforms remain at great

depths, feeding on the migrators as they return from the surface after dawn.

REPRODUCTION

Stomiiforms usually breed near their deep resting areas, but the eggs are buoyant and float to the surface where they hatch and the young form part of the plankton. As the larvae begin to transform into juveniles and attain something like their adult body shape, they move to greater depths. Like many other deepwater fishes, some stomiiforms can change sex. Both *Gonostoma* and *Cyclothone* include species known to start life as males and then change into females as they age.

LIGHTING THE WAY

Regardless of their shape or size, all but a single species of stomiiforms have light organs. These range from tiny luminous dots scattered over the skin to quite elaborate arrangements of luminous material, lenses, and reflectors. The most common arrangement is to have one or two rows along the lower sides of the body, from head to tail base. The light produced can be a steady glow or intense flashes. The color of the photophores varies from white and pale-yellow to bright-red and violet. Members of the family Stomiidae typically have light organs on a stalk, or barbel, at the front of the lower jaw. The barbel can simply have a luminous end or it can have multiple filaments and other ornamentations. In addition, in more phylogenetically advanced stomiids, there are luminous tissues, some quite elaborate, associated with the pectoral fins. In other such stomiids there are no lights on the pectoral fins; in some cases there are not even fins, but there are tiny bony fin remnants under the skin, indicating that their ancestors once had fins and perhaps lights as well. Both barbels and pectoral fin lights may be used for prey attraction or for communication among members of a species. Most stomiids also have one or more light organs in the cheek region. The cheek organ is sexually dimorphic in many cases, that of males being larger than that of females. In some species the large cheek organ is attached to a muscle in the face that can rotate the organ, so that it can either shine outwardly, or rotate behind black pigment. This essentially turns the light off and on, so that it functions much like a headlight. This light organ is of a color that prey organisms cannot see because visual pigments in their eyes do not react to those particular light wavelengths. So, the stomiid can see the prey, but the prey remains "in the dark".

Another visual specialization of many stomiiforms is a yellow eye lens that enhances biologically produced light and filters out downwelling sunlight.

BRISTLEMOUTHS AND THEIR ALLIES
Bristlemouths and lightfishes are names applied to the relatively primitive stomiiforms in the family Gonostomatidae. These fishes range from rather large (up to nearly 30 centimeters, or 12 inches), elongate fishes such as members of the genus *Gonostoma* to small (2 to 8 centimeters, or ¾ inch to 3 inches), delicate inhabitants of the marine plankton such as genus *Cyclothone*. The latter genus is considered by some scientists to include more individuals than any other genus of vertebrates, since literally billions of these small fishes are present in ocean plankton.

Marine hatchetfishes (family Sternoptychidae)

resemble silvery coins floating vertically in the water. This striking silver color is enhanced by the ventrally oriented light organs, with their reflectors scattering light laterally as well as ventrally. The combination of pigment and light organs serves to camouflage these fishes, making them nearly invisible in the water. The back is dark so that it matches the dark water below them, and a predator above them would see only the dark of the abyss. The photophores produce light that is like that of the downwelling light, so that a predator searching from below would see only light coming down. Some sternoptychids have tubular eyes that concentrate dim light; in addition, these eyes permanently look upward.

▲ *Sternoptyx diaphana*, a deepsea hatchetfish. Approximating a silver dollar in size, shape, and general appearance, hatchetfishes have upward-pointing mouths and rather large, bulging eyes. More than 25 species are known.

◄ A snaggletooth *Astronesthes gemmifer*. These small active predators resemble dragonfishes in having photophores along the belly and a long, luminescent barbel on the lower jaw, but are generally somewhat less snake-like in body shape. All inhabit the deep sea, some occurring at depths as great as 3,000 meters (about 9,800 feet).

Peter Parks/Oxford Scientific Films

▲ One of the deepsea viperfishes (genus Chauliodus) in a close-up display of the extraordinary teeth and jaws that typify this group of fearsome pocket predators.

A significant feature of stomiid evolution has been the association of light organs with the pectoral fins. The function of such lights is also thought to be prey attraction and communication among members of a species. The skeleton of these pectoral fins is modified to be especially mobile, suggesting that the lights are waved about. In some cases, for example members of the genus *Trigonolampa*, the lights are near the fin base, while in others, such as the genus *Thysanactis*, the lights are at the end of long, flexible rays.

Chauliodus and its closest relative *Stomias* (scaly dragonfishes) are the only stomiids with scales. They also both have a jelly-like covering over the body. The black dragonfish *Idiacanthus fasciola* is one of the most extreme of the dragonfishes, in terms of life history and body size. The female larvae get quite large, up to 7 centimeters (2¾ inches), but their most distinctive feature is the presence of elongate stalked eyes that extend from the head a distance of nearly one-third of the body length. As the larva transforms into a juvenile and begins to approach adult body shape, the eye rods are absorbed and the eyes move into sockets in the skull. Male *Idiacanthus*, whose larvae are small, remain small after transformation, and retain many larval features. Adult males have no teeth and never feed. Females, on the other hand, are elongate and snake-like, with one of the longest bodies in the order, exceeding 40 centimeters (15¾ inches).

DRAGONFISHES AND ALLIES

Stomiids (family Stomiidae) make up the large evolutionary group called dragonfishes, scaly dragonfishes, loosejaws, snaggletooths, and viperfishes. Some workers recognise six separate families for these. For the most part these fishes are elongate and very darkly pigmented. With few exceptions, there are no scales, and the skin resembles dark velvet, often with small light organs scattered over it. Nearly all dragonfishes have a barbel extending from the lower jaw. The barbel is controlled by muscles behind the jaw bones and is movable; its function can only be estimated, but it likely is used as a lure to attract prey and as a signaling device for communication among individuals. Many stomiids have extremely large teeth, often elaborately sculptured; in the viperfish *Chauliodus sloani* the lower jaw teeth actually extend up over the head. Many stomiids can eat large prey and have a number of specializations for such a diet, such as an ability to lower the internal skeleton of the pectoral fins to allow prey to pass into the gullet, and extremely muscular stomachs. These are lined with black tissue so light organs of prey will not shine through their bodies, exposing them to predators.

Another group of dragonfishes is sometimes called loosejaws, in reference to the lack of skin between the lower jaws—there is no floor to the mouth in these fishes. There is instead a slender column of muscle and tissue between the gill basket and lower jaw; contraction of the muscle pulls the mouth shut.

Eustomias is a very large genus of dragonfishes. One of the most striking is the modification of the front of the vertebral column into a spring-like structure which might act to counteract forces generated when the fish strikes its prey.

WILLIAM L. FINK

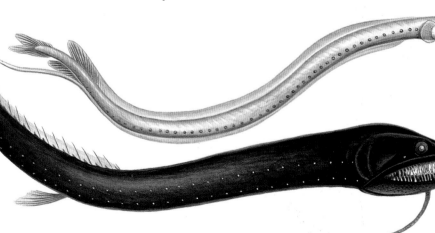

▶ Larval forms of the widespread black dragonfish Idiacanthus fasciola have elongated stalks supporting their eyes (top). These stalks are absorbed as the fish grow, until the eyes retreat into sockets. The adult female is illustrated here (bottom).

LIZARDFISHES & THEIR ALLIES

The order Aulopiformes is very diverse and includes estuarine and coastal species as well as deepsea fishes of midwaters and the bottom. They are well known to scientists because of peculiar modifications of the eyes in some groups, their frequently extensive metamorphosis (changes from larva to juvenile), and hermaphroditism. Lizardfishes and their allies share some features of the order Myctophiformes (lanternfishes and their allies), but are placed together in the order Aulopiformes because of a specialized feature in the gill arches.

PRIMITIVE AND ADVANCED

Lizardfishes and their allies have both primitive and advanced features. The primitive features include abdominal pelvic fins, an adipose (fatty) fin, absence of fin spines, and scales that are usually rounded (cycloid) when present. Advanced features are a swim bladder without a duct, and a maxillary bone excluded from the gape (not on the edge of the jaw). Many, especially the deepsea species, are bisexual synchronous hermaphrodites, that is, they may have the ability to self-fertilize. They are known from the Upper Cretaceous (135 million years ago).

The order contains 14 families, 43 genera, and about 220 species. Two genera, both deepsea benthic, *Bathysauropsis* (2 species, allied with the Ipnopidae and Notosudidae) and *Bathysauroides* (1 species, allied with the Bathysauridae and Giganturidae), remain unassigned to family and may warrant family-level recognition.

LIZARDFISHES

The family Synodontidae contains 4 genera with about 55 species. It is divided into two

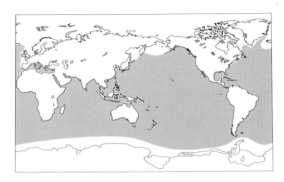

subfamilies: Synodontinae and Harpadontinae. The Bathysauridae was placed with this group, but has recently been allied with the Giganturidae.

The shallow-living lizardfishes (Synodontinae) comprise 34 species of cigar-shaped fishes with large toothy mouths. When viewed from the side, the head is very lizard-like, and the eye is situated well before the end of the mouth. The largest species are about 60 centimeters (24 inches) long, but most are smaller. The dorsal fin is at the midbody, and they have a small adipose fin before the tail. They

◀ *Lizardfishes (family Synodontidae) generally have camouflage coloration and blend well with the bottom. This species, the variegated lizardfish* Synodus variegatus, *is widespread in the Indo-Pacific region from the Red Sea to Hawaii and south to Lord Howe Island.*

KEY FACTS

ORDER AULOPIFORMES
- 14 families • 43 genera
- *c.* 220 species

SMALLEST & LARGEST

Stephens pearlfish *Scopelarchus stephensi*
Total length: 7 cm (2¾ in)

Lancetfish *Alepisaurus ferox*
Total length: 2 m (7 ft)

CONSERVATION WATCH
■ None of these fishes are listed as threatened.

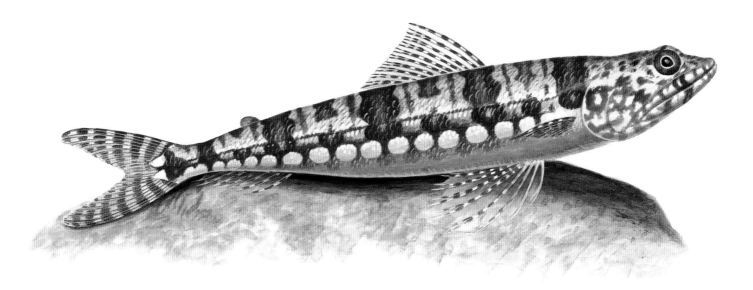

▲ *Deriving its name from its habit of sitting on the bottom with its head raised in a lizard-like pose, the variegated lizardfish* Synodus variegatus *will also sometimes bury itself in sand, leaving only its eyes exposed, then dart out after prey.*

occur in all warm seas, mostly inshore, with the deepest at 400 meters (1,300 feet). Lizardfishes typically sit on the bottom in sandy areas, with their front end propped up by their ventral fins, and they can also bury themselves, leaving just their eyes exposed. Their cryptic coloration enables them to ambush prey, mostly fishes. Like some other members of the order, they have very distinctive pelagic (free-swimming) young that are scaleless and transparent, but with large black blotches in the skin on their lower sides; these "pigment shields" are actually internal and associated with the gut, but their function is unknown.

BOMBAY DUCK AND ALLIES

The mostly inshore Bombay ducks (Harpadontinae) contains about 21 species in two genera. *Saurida*, with 15 species, is predominantly shallow water, and in appearance, distribution, and behavior, the species are similar to lizardfishes. Six species (one undescribed) comprise the genus *Harpadon*. They resemble lizardfishes, but have long pectoral and pelvic fins, a short rounded snout, and a median

tail-fin lobe of prominent lateral line scales. The genus is divided into two groups, with three offshore and pelagic or benthopelagic species, and the other three coastal and estuarine. The genus is widespread in the Indo-Pacific from east Africa to Japan, Korea, and Australia.

The family Aulopidae, with one genus of at least ten species, occurs nearly worldwide, but is absent from the eastern Pacific. The aulopids resemble lizardfishes and occur on the bottom from nearshore to about 1,000 meters (3,300 feet), usually over sand. The dorsal fin is large and adult males have an extended first dorsal fin ray. The species of this family are unique in having two supramaxillae (small bones lying along the upper edge of the jawbone).

The family Pseudotrichonotidae includes a single species known from Japan and the southeastern Pacific. *Pseudotrichonotus* superficially resembles the sandfish, genus *Trichonotus* (a member of the highly evolved order of perches), and is similar in the habit of sand diving. It is apparently closely related to lizardfishes, Bombay ducks, and aulopids.

▶ *In profile the slender lizardfish* Saurida gracilis *(family Harpadontinae) shows the distinctively reptilian appearance that gives this group of fishes their common name. Growing to a length of about 28 centimeters (1 foot), this species is widespread in the Indo-Pacific; other members of the group are about equally divided between the Indo-Pacific and Atlantic regions.*

Kevin Deacon

Rudie H. Kuiter/Oxford Scientific Films

GREENEYES AND IPNOPIDS

The family Chlorophthalmidae contains at least 20 species in 2 genera known as greeneyes. They are also known as cucumberfishes because some smell like cucumbers. Most have iridescent eyes, the lens greenish to yellowish in color. Some have large eyes with keyhole-shaped pupils, others have tiny eyes. Greeneyes occur worldwide in tropical to warm-temperate seas on the outer shelf and slope. Maximum size is about 30 centimeters (12 inches), but most are smaller. They live on the bottom and prop themselves up like lizardfishes.

The remaining benthic (bottom-dwelling) family of the order contains the 26 or so species of ipnopids (Ipnopidae), sometimes included as a subfamily in the greeneye family. They live very deep in all oceans, even at abyssal depths. Most have a flattened head and pencil-shaped body, with tiny or rudimentary eyes, or lenseless plates, that cover the entire head but cannot form an image. Most have thickened elongate fin rays. Some, known as tripod fishes, have very extended pelvic fin rays; they have been photographed on a mud bottom propped up on their pelvic fins, with the tail fin forming a tripod. The maximum length of these species is 36 centimeters (14 inches).

In recent classifications the waryfishes (family Notosudidae) and the unassigned genus

Bathysauropsis (2 species) have been allied with the chlorophthalmids and ipnopids. The 19 species (in 3 genera) of paperbones or waryfishes (Notosudidae) are scaled but otherwise similar in shape and fin placement to other midwater aulopiforms. They too occur at mesopelagic and bathypelagic depths as adults, although some may be pseudo-oceanic or benthopelagic; their larvae are found near the surface and are unique in having teeth on the maxillae (upper jawbones).

LANCETFISHES AND DAGGERTOOTHS

Among the midwater families there is tremendous diversity in body shape and size, and marked transformations from larvae to adults. Two small families contain the largest members of the order: lancetfishes (Alepisauridae, 2 species) and the daggertooth (Anotopteridae, 1 species). Lancetfishes have a high sail-like dorsal fin that starts behind the head and continues nearly to a small adipose fin before the tail. There is a fleshy keel on the side of the tail, and the body is scaleless and flaccid. The mouth is huge, with large dagger-like teeth. Lancetfishes are some of the largest predators in the deep sea (they grow more than 208 centimeters, nearly 7 feet) and have destroyed undersea cables. Their digestion is evidently slow, and their stomach contents have

▲ *The Sergeant Baker* Aulopus purpurissatus *(family Aulopidae) is endemic to the southern coast of Australia. It favors shallow waters to about 80 meters (about 260 feet).*

▼ *A ferocious predator, the hammerjaw* Omosudis lowei *(family Omosudidae) inhabits deep seas beyond 1,000 meters (3,300 feet) almost worldwide.*

Peter David/Planet Earth Pictures

▲ *In the depths of the abyss the bizarre tripodfish genus* Bathypterois *(family Ipnopidae), waits patiently for prey, using its extraordinarily elongated fins to hold itself clear of the bottom.*

provided some unusual specimens. They range widely in the midwaters of the Atlantic and Pacific, sometimes found near the surface and occasionally seen floundering in the surf.

The single hammerjaw (Omosudidae) is a widespread bathypelagic (deepsea-dwelling) relative, living mostly below 1,000 meters (3,300 feet). It has enlarged fangs on each side of the lower jaw. Like lancetfishes, the hammerjaw lacks scales.

The daggertooth (Anotopteridae) resembles a lancetfish in that it has the same large mouth and dagger-like teeth, but it lacks the sail-like dorsal fin. It is also a large predator, growing to about 147 centimeters (58 inches). The presumed single species occurs nearly worldwide but is rare in the tropics; it too ranges from deep to shallow depths.

The barracudinas (Paralepididae), with 50 to 60 species, make up the largest family in the order. Interrelationships of the 12 recognized genera remain poorly understood, but the daggertooth (*Anotopterus*) should possibly also be included in the family. Barracudinas are mesopelagic, although some may occur at greater depths, and they are found nearly worldwide. They have a slender, elongate body with the dorsal fin typically placed at the middle of the back, abdominal pelvic fins, and a long, pointed snout with fang-like teeth. Superficially they resemble barracudas, hence their name. Scaled species are usually silvery; unscaled ones are transparent. The largest members of this family grow to 1 meter (39 inches). They have been observed swimming vertically, with head up. Whales and oceanic predators, such as tunas, feed on them.

The pearleyes (family Scopelarchidae) form another group with unique visual properties. Their eyes are somewhat telescopic and directed upward, or forward and upward; this is not unusual, as telescopic eyes are known in a number of deepsea fish families. Besides increasing light admission

they also allow binocular upward vision, which cannot be achieved with eyes on the side of the head. The pearleyes also have a glistening white spot on the eye (called the pearl organ) and a second (accessory) retina in each eye. The pearl organ (or lens pad) is transparent in life and may serve as a light guide, picking up light from a wide lateral field and guiding it to the lens. It is thought that the pearl organ and accessory retina may allow the fish to see (although poorly) a prey item to its side and then turn towards it to get a better view with its main retina. There are about 18 species of pearleyes, and they have typical features of the order, including an adipose fin, dorsal fin at the midback, cycloid (rounded) scales, forked tail, and a large toothy mouth. Most live very deep at 500 to 1,000 meters (1,625 to 3,300 feet), but some are found at shallower depths and may possibly approach the surface at night. They sometimes turn up in the stomachs of sport fish.

The eight species of sabertooth fishes (family Evermannellidae) resemble the pearleyes in body shape and other features of the order. Some have tubular eyes and their teeth are long and fang-like. They are mesopelagic in tropical and temperate waters to depths of about 1,000 meters (3,300 feet).

TELESCOPEFISHES AND THEIR ALLIES

Members of the remaining midwater families are smaller and unlikely to be encountered except by specialists. Toothy mouths, vision adaptations, hermaphroditism, and modified larvae continue to feature in these groups. A rare but well-known family is the Giganturidae, or telescopefishes, discussed in ichthyology textbooks. Adults have tubular eyes that point forward—in dim light they are telescopic in the way they maximize the light reaching the retina by having an enlarged pupil and lens. It is likely that they hover vertically in the water column with the eyes pointed straight up toward the surface. The two species have large mouths, with the jaw extending way past the eye. They have numerous teeth that can be depressed backwards to engulf and then hold large prey. Their pectoral fins are high on the sides above the gill opening; the lower lobe of the tail fin is elongate. The larvae do not resemble the adults— they have different teeth, fin placement, and body shape, the eyes are laterally directed, and they have pelvic fins which adults lack—and until recently, when individuals were discovered in the bizarre process of transformation, the larvae were placed in a separate family.

Related to the giganturids, the two species of bathysaurids (family Bathysauridae) resemble lizardfishes, but they have a flatter head, with curved and barbed teeth; they live mostly below 1,600 meters (5,250 feet).

The poorly known deepwater *Bathysauroides* is related to both families, but remains unassigned.

ROBERT K. JOHNSON & WILLIAM N. ESCHMEYER

LANTERNFISHES

L anternfishes (family Myctophidae) are the most widely distributed, the most species-diverse, and the most abundant of all fishes in the deep open seas. They therefore play an important role in the cycling of energy in the oceanic food-web, and represent a potential resource for commercial exploitation. Their allies, the neoscopelids (family Neoscopelidae), are less abundant. They occur over the continental and island shelf regions in tropical and subtropical waters. Together, these two families make up the order Myctophiformes.

LIGHT PRODUCTION

Lanternfishes are so-called because of the light organs (photophores) that occur on their head and body, mostly on the ventral surface. Each photophore is overlain by a modified scale that acts as a lens to focus the light. They are small fishes with large eyes, a single dorsal fin followed by a small adipose fin, and with pectoral fins situated well in front of the ventral fins. A rudimentary spine precedes the dorsal, anal, and ventral fin rays. The mouth is terminal (subterminal in some genera) and the jaws extend behind the posterior margin of the eye. The jaws are armed with bands of small, closely set teeth, the inner series of which may be enlarged in some species (for example *Diaphus fragilis*). Rounded (cycloid) or comb-shaped (ctenoid) scales cover the body, and freshly caught specimens of heavily scaled species may be iridescent blue, green, or

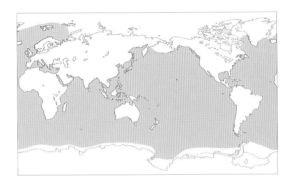

silver in color. Deeper-living species tend to be gray or dark brown.

Photophores are present in all species of lanternfishes except *Taaningichthys paurolychnus* and are arranged in distinct groups. This not only provides scientists with a reliable character for species identification, but suggests that the unique

KEY FACTS

ORDER MYCTOPHIFORMES
• 2 families • 35 genera
• *c.* 250 species

SMALLEST & LARGEST

Diogenichthys panurgus
Location: tropical and subtropical waters of the Indian Ocean
Total length: 3 cm (1⅕ in)
Weight: 0.4 g (⅟₁₀₀ oz)

Gymnoscopelus bolini
Location: Antarctic and subantarctic waters of the Southern Ocean
Total length: 35 cm (14 in)
Weight: 250 g (8¾ oz)

CONSERVATION WATCH
■ None of these fishes is listed as threatened.

▼ *The combination of an unremarkable fishy shape and numerous light organs arranged in groups rather than long rows characterizes the myctophids or lanternfishes, among the most numerous and widespread of deepsea fishes. The species of* Gonichthys *are some of the shallowest dwelling in the family.*

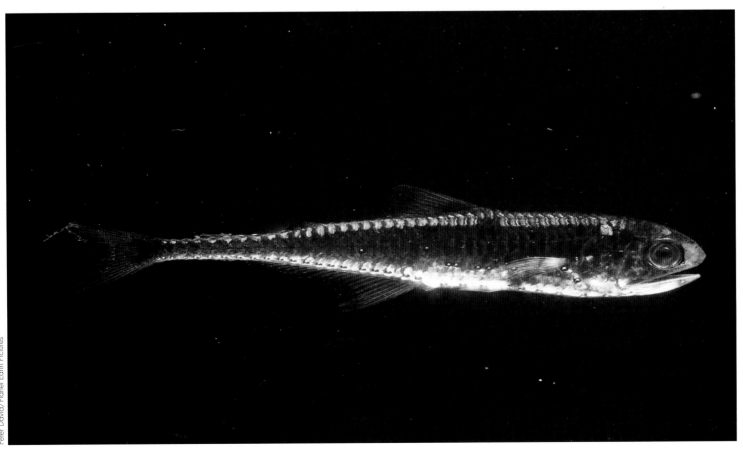

Peter David/Planet Earth Pictures

pattern in each species serves as an identifying signal in shoaling behavior. However, the arrangement of an almost continuous series of photophore groups on the ventral surface also suggests that they play a role in camouflaging the fish's silhouette against light from above. In the genus *Diaphus*, two pairs of light organs situated immediately in front of the eyes are well developed and presumably act as headlights for illuminating prey.

Other light-producing structures may include minute "secondary" photophores that cover the head and body, and more diffuse patches of luminous tissue at the bases of the dorsal, anal, and ventral fins. In addition, luminous glands may be found on the dorsal and ventral surfaces of the tail stalk. The structure of these glands varies in complexity, ranging from a single (sometimes black-bordered) organ to a series of overlapping luminous scales. Typically males and females have different light organs near the tail, suggesting they may be used for sex recognition, although this is not true for all species (*Lampanyctus* and *Lampadena* species are exceptions). It has been postulated that the flashing of these tail glands is an escape strategy which may confuse a predator.

Blue, green, or yellow light is emitted from the light organs and photophores, following nervous stimulation of a chemical reaction within the light-producing cells (photocytes) of these structures.

FOOD AND PREDATORS

Planktonic crustaceans are the main food items of lanternfishes. In temperate latitudes where there is seasonal variation in food abundance, they lay down energy reserves in the form of fats in the body tissue and swim bladder. The reserves may be used for egg production during the late winter to early spring spawning period, and may provide additional buoyancy to the swim bladder.

Lanternfishes are preyed on by tuna, bonitos, albacore, dolphin fishes, hakes, horse mackerel, kingklip, snoek, and large squids. Many species of seabirds (particularly penguins), seals, and cetaceans (whales and dolphins) take myctophids in addition to their staple diet.

DISTRIBUTION

Lanternfishes live in the open oceans, occurring from Arctic to Antarctic waters. Species distributions are related to both ocean currents and the physical and biological characteristics of the water. By day, lanternfishes live in depths between 150 and 2,000 meters (490 to 6,600 feet) depending on their size and sexual maturity. Mesopelagic species migrate vertically at night to feed in the top 50 meters (165 feet) of water; some may be netted at the surface of the sea on moonless nights. However, this daily migration is not undertaken by all fishes. Even within a single species, the behavior may vary with latitude,

season, sex, and stage in the life cycle. Deeper-living species apparently do not migrate vertically.

As adults, some larger species (for example, *Notoscopelus kroeyerii, Gymnoscopelus piabilis,* and *Gymnoscopelus bolini*) live a few meters above the bottom, especially over continental slope regions. Here their shoreward penetration appears to be limited by both water depth and water temperature, although their planktonic larvae can be found further inshore.

A few species, like *Lampanyctodes hectoris*, that are descended from oceanic ancestors, occur above the continental shelf itself and are termed pseudoceanic species. Their close proximity to the coast, coupled with their shoaling behavior, makes them suitable targets for commercial fishing.

Unquestioned myctophid fossils, some with photophores still visible, are known from as far back as the Oligocene, 37 to 24 million years ago. More than 100 fossil species in some 20 genera have been described, many from otoliths (ear bones) only.

COMMERCIAL POTENTIAL

Lanternfishes make up about 65 percent of all mesopelagic fishes, and their global mass has been estimated at 600 million tonnes (660 million tons). Acoustic surveys using echosounders near Shag Rocks (western South Atlantic) suggest a stock of 160,000 tonnes (176,000 tons) of *Electrona carlsbergi*, while a value of 100 million tonnes (110 million tons) has been reported for *Benthosema pterotum* in the Arabian Gulf. Recent acoustic investigations off the west coast of South Africa have demonstrated the existence of large shoals of subantarctic species along the edge of the continental shelf break, with biomass values up to 33 tonnes per square kilometer (94 tons per square mile) for *Symbolophorus boops* and 7 tonnes per square kilometer (20 tons per square mile) for *Diaphus hudsoni*. A single haul off Argentina yielded 30 tonnes (33 tons) of *Diaphus dumerilii* in an hour.

Limited commercial exploitation occurs off South Africa, where annual purse-seine landings (mainly of *Lampanyctodes hectoris*) have fluctuated between 100 and 42,400 tonnes (110 to 46,725 tons). The lanternfishes are reduced to fish meal and fish oil. Because of lanternfishes' high oil content, processing plants are forced to mix them with other species to prevent clogging the machinery. Around South Georgia and Shag Rocks, experimental fishing on *Electrona carlsbergi* (mainly juveniles) averaged about 20,000 tonnes (22,000 tons) per year between 1988 and 1990, but increased dramatically to 78,488 tonnes (86,494 tons) in 1991. The Commission for the Conservation of Antarctic Marine Living Resources therefore introduced a 200,000 tonne (220,400 ton) TAC (total allowable catch) for this species for the 1992 season.

P. ALEXANDER HULLEY

TROUTPERCHES & THEIR ALLIES

KEY FACTS

ORDER PERCOPSIFORMES
• 3 families • 6 genera • 9 species

SMALLEST & LARGEST

Chologaster cornuta
Total length: 5 cm (2 in)

Percopsis transmontana
Total length: 20 cm (8 in)

CONSERVATION WATCH
!!! The Alabama cavefish *Speoplatyrhinus poulsoni* is listed as critically endangered.
! The Ozark cavefish *Amblyopsis rosae*; northern cavefish *Amblyopsis spelaea*; and *Typhlichthys subterraneus* are listed as vulnerable.

These nine species of small, North American freshwater fishes are classified together in the order Percopsiformes for technical anatomical reasons, which some ichthyologists question. They are considered intermediate in structure between soft-rayed, smooth-scaled fishes such as trouts and herrings, and strong-spined, rough-scaled fishes such as snappers and groupers. They apparently comprise the remnants of a formerly more abundant and widely distributed group. The marine fossil genus *Sphenocephalus* from the Upper Cretaceous period (about 65 million years ago) of Europe was thought to be anatomically similar to both Percopsiformes and Gadiformes (codfishes and their allies), but its relationship to the percopsiforms is now in doubt. However, there are three fossil genera of freshwater percopsids from the North American Eocene epoch (57 to 37 million years ago). The nine species of living percopsiforms are grouped into three families: Percopsidae (troutperches), Aphredoderidae (pirate perches), and Amblyopsidae (cavefishes and swampfishes). Although the living percopsiform species are few, the three families are so different from each other that they must have been evolving for a long time.

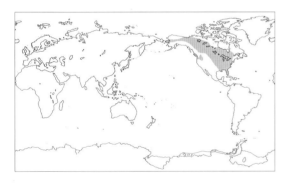

TROUTPERCHES
The two species of troutperches belong to the family Percopsidae. They have weakly toothed scales, a few weak fin spines, and an adipose (fatty) fin resembling that of the trouts and smelts. Troutperches are northern North American in distribution, and are apparently more active at night than in the daytime.

PIRATE PERCHES
The pirate perch family, Aphredoderidae, consists of only a single species, living in quiet waters of the eastern half of the United States. It has weakly spiny scales and a few fin spines, but lacks an adipose fin. A unique character of this species is the far-forward position of the anus in adults, directly below the head.

CAVEFISHES AND SWAMPFISHES
The cavefishes and swampfishes of the family Amblyopsidae number six species classified in four genera. They were thought to be related to the killifishes, which they superficially resemble in having all soft rays, a paddle-shaped tail fin, a flattened head, and the dorsal and anal fins nearly opposite each other. In internal anatomy, however, they are more similar to troutperches and the pirate perch. Like the pirate perch, the amblyopsids have the anus placed far forward. One species lives in swampy regions on the mid-Atlantic plain of the United States. The other five are restricted to limestone caves and springs in the midwestern and southeastern United States. Although the two species in the genus *Chologaster* have tiny eyes, the other four species in the family, inhabitants of a dark subterranean world, have none at all. Rows of sensory papillae (small fleshy projections) compensate for the absence of eyes.

DANIEL M. COHEN

▼ *The only member of its family, the pirate perch Aphredoderus sayanus has a number of unusual features, particularly the placement of its anus at its throat. In young specimens the anus is situated farther back on the body, migrating forward as the fish matures to its final position directly below the head.*

KEY FACTS

ORDER GADIFORMES
• *c.* 9 families • *c.* 66 genera
• *c.* 500 species

SMALLEST & LARGEST

Codlet (genus *Bregmaceros*)
Total length: 7–10 cm (3–4 in)

Atlantic cod *Gadus morhua* and
ling *Molva molva*
Total length: 1–2 m (3¼–6½ ft) +

CONSERVATION WATCH
!!! The morid cod *Physiculus
helenaensis* is critically endangered.
! The Atlantic cod *Gadus morhua*
and haddock *Melanogrammus
aeglefinus* are listed as vulnerable.

▼ *A school of bib* Trisopterus luscus
*(family Gadidae). Sometimes known as
pout, this fish is common in coastal
waters of the eastern Atlantic, supporting
a small commercial fishery in Spain.*

CODFISHES & THEIR ALLIES

Some of the codfishes in the order Gadiformes are so anatomically different from each other that ichthyologists do not agree either on why they are classified together or on the details of their family tree. Although many species resemble each other in having a distinctive tail-fin skeleton, even more have a long tapering tail with a reduced fin at the end and virtually no skeleton. The group is widely distributed, found around the world inhabiting deep to shallow seas, and ranging through the tropics from north to south polar regions; a few species are even found in fresh waters.

TASTE RECEPTORS
The presence or absence on the chin of a barbel bearing microscopic taste receptors is for many gadiform species an indication of the habitat in which they most often feed. Bottom feeders use their barbels (and sometimes elongate pelvic fin rays) to explore and taste the substrate before ingesting it. Midwater feeders do not need a barbel to sense their prey. The pelagic species, as well as the blue whitings (genus *Micromesistius*), the pollock *Pollachius pollachius* of the northeast

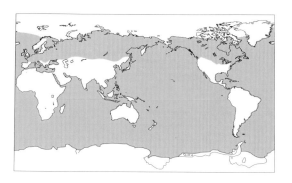

Atlantic, and the morid (genus *Halargyreus*), are examples of gadiforms that lack a barbel.

FOOD VALUE

Gadiform fishes are of particular importance to humans as a food resource. The family Gadidae, which is found chiefly, although not exclusively, in cool waters of the northern hemisphere, is exceeded in catch only by the sardine and herring family. In 1989 the total world catch of gadids was 11,413,127 tonnes (12,577,265 tons), nearly 11.5 percent of the total world fish catch. The highest catch of all fishes was the Alaska pollock *Theragra chalcogramma* of the North Pacific Ocean and Bering Sea; in 1989, 6.3 million tonnes (6.9 million tons) were fished. They are trawled mainly during daylight hours when vast densely packed schools hover close to the bottom, where they are easily swept into a towed net. Most of the catch of this species was once used for animal food, but now it is consumed directly by humans in a variety of products. The North Atlantic cod *Gadus morhua* is another abundant gadid, with nearly 1.8 million tonnes (1.9 million tons) caught in 1989. One reason there are such huge populations of gadiform fishes is that they produce an enormous number of eggs, among the most of any fishes. A large Alaska pollock may spawn 15 million eggs in a year and the record for a large Atlantic cod is 9.1 million.

RATTAILS

The largest gadiform family, containing well over half the known species in the order, is the Macrouridae, commonly known as rattails or grenadiers. These long-tailed fishes are found in virtually every ocean, where they are most abundant between 200 and 2,000 meters (660 to 6,600 feet), although they have been caught as deep as 6,000 meters (19,680 feet). They are the most characteristic of all deepsea fishes, living on or near the bottom where they hover, often at a

Labat-Lanceau/AUSCAPE International

nose-down angle, making an occasional foray to the substrate for feeding. A few of these bottom-dwelling species have occasionally been caught a thousand meters or more (around 3,500 feet) up in the water, and a very few species spend their entire lives in midwater far above the bottom. Other midwater gadiforms include the silvery pout genus *Gadiculus* of the northwest Atlantic and the rattail genus *Cynomacrurus* of far southern seas. Two other macrourids that probably spend most of their time in midwater are the bizarre bubble-headed genera *Squalogadus* and *Macrouroides*. Most species of rattails live beneath tropical seas, especially in areas with high productivity, such as off the mouths of large rivers. Although rattails are caught far below the limits of surface light penetration, most species have large eyes, although their function remains a puzzle. Are rattails attracted to luminescent animals for food, or do they in fact spend more time in the surface layers of the ocean than we realize?

Most rattails are small or have watery tasteless flesh; however, a few of the larger species support

▲ *A cool-water species, the Atlantic cod* Gadus morhua *(family Gadidae) is one of the most important European food fishes, supporting a large commercial fishery, especially in the North Sea. The source of cod-liver oil, the annual catch is estimated at around 2.5 million tonnes. It congregates in huge shoals in midwaters, especially when gathering to spawn in March and April over shallow banks and reefs just east of Scotland.*

Peter David/Planet Earth Pictures

◄ *The roundhead grenadier or rattail* Odontomacrurus murrayi *(family Macrouridae) is a deep sea fish that is abundant on continental slopes in the Atlantic and Indian oceans, but nothing is known of its reproductive behavior and very little of its diet. Juveniles like this one can be strikingly colored, while most adults are drab gray or brown.*

Rudie H. Kuiter/Oxford Scientific Films

▲ *The bearded rock cod* Lotella rhacinus *(family Moridae), endemic to southern Australia and New Zealand, favors caves and overhanging ledges in shallow waters along rocky shores and reefs. It grows to a length of about 50 centimeters (20 inches) and a weight of about 2.5 kilograms (5½ pounds).*

minor deepwater fisheries (about 48,000 tonnes or 52,900 tons in 1989) in the Bering Sea and North Pacific and in the North Atlantic.

HAKES
The hakes (family Merlucciidae) of the genera *Merluccius* and *Macruronus* supported a fishery in 1989 of about 2 million tonnes (2.2 million tons). These fishes are widely distributed along continental shelves of the world oceans. *Merluccius* is inexplicably absent from the Indian Ocean and western Pacific shores except for New Zealand and a single record from Japan; however, the three species of *Macruronus*, which resemble rattails because of their long tapering tails, are abundant off southern South America and New Zealand, less common along southern Australian shores, and are caught only occasionally at South Africa. In keeping with a preference for cool water, hakes living in tropical regions are found in deeper water than species found to the north or south.

MORID CODS
The family Moridae, the morid cods or moras, is represented by about 100 species which are characterized by the structure of the tail-fin skeleton, a connection between the swim bladder and the back of the skull, and the shape of the otolith (ear bone). They are found around the world at depths ranging from the intertidal to more than 2,500 meters (8,200 feet) but are most abundant in southern waters at intermediate depths. Even so, the blue hake genus *Antimora*, which is characterized by a flattened V-shaped snout, is quite common in many regions of the deep sea. Many morid species so closely parallel the appearance of gadids in body shape and fin pattern that they cannot be identified to family by external appearances alone.

BLACK MELANONIDAE
The family Melanonidae, which lacks a common name, is represented by only two species that live in midwaters of the deep sea far from shore in the Pacific, Atlantic, and Indian oceans. These relatively rare small black fishes grow to 15 centimeters (6 inches) at most.

CODLETS
The codlets of the family Bregmacerotidae, represented by more than a dozen species, are remarkable in being the only gadiform family that is limited to warm water. These small fishes are pelagic in the upper layers of most tropical seas, sometimes close to shore but also in mid-ocean. They are distinguished by their fin pattern; the first dorsal fin is a single strong ray growing out of the top of the head, and the long pelvic fin rays extend back from under the head.

FRESHWATER CODS
The burbot *Lota lota* is the only gadiform restricted to fresh waters, where it is found in north temperate and boreal regions throughout the northern hemisphere. The Atlantic tomcod *Microgadus tomcod* also enters fresh waters along the northeast coast of North America, and the Arctic cod genus *Boreogadus* frequents very low salinity habitats.

POLAR CODS
The most northerly dwelling gadiforms are the two species of polar cods of the genus *Arctogadus*, which live in association with polar ice, both near and at great distances from shore. The most southerly gadiforms are the four species of the far southern and Antarctic family Muraenolepididae.

LUMINESCENCE
Many gadiform species in the families Macrouridae, Moridae, Steindachneriidae (a single species in the tropical western North Atlantic), and Euclichthyidae (a single species around temperate Australia, New Zealand, and New Caledonia) have organs that house light-producing bacteria. Although there is no definite proof of the function of luminescence in these fishes, the organ in most is associated with the urogenital area. In many its exact location and size is unique to a particular species, suggesting that perhaps there is some connection with species recognition and breeding. If so, this might also help to explain the large eyes of some rattails, although many large-eyed species are not luminescent.

DANIEL M. COHEN

▼ *The circumpolar burbot* Lota lota *is the only member of the codfish order restricted to fresh water and is also unusual in that is spawns in mid-winter, under the ice, in the middle of the night.*

CUSKEELS & THEIR ALLIES

KEY FACTS

ORDER OPHIDIIFORMES
• 4 families • 90 genera
• 380 species

SMALLEST & LARGEST

Grammonoides opisthodon
Total length: 5 cm (2 in)

Lamprogrammus shcherbachevi
Total length: 2 m (6½ ft)

CONSERVATION WATCH
! There are 7 species listed as
vulnerable: *Lucifuga simile;* New
Providence cuskeel *Lucifuga
spelaeotes; Lucifuga subterranea;
Lucifuga teresianarum; Saccogaster
melanomycter; Stygicola dentata;*
dama ciega blanca *Ogilbia pearsei.*

T he species in the order Ophidiiformes were classified for a long time in the spiny-rayed perciform order, but more recently they have been considered more closely related to the codfishes by most authors. A typical ophidiiform fish has a small head, a tapering body, and smooth scales. It has a long dorsal fin, and a long anal fin that is most often united with the caudal fin. The ventral fins may have up to two rays each and are placed either below the gill cover or farther forward.

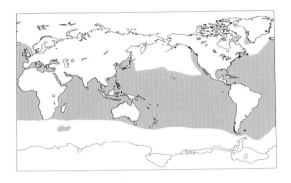

CORAL REEFS TO OCEAN DEPTHS

The ophidiiform fishes comprise a very successful group with at least 380 species in 90 genera, and there are many species yet to be described. Fossil evidence, especially otoliths (ear bones), shows that the group was one of the most common in the Lower Tertiary period. Ophidiiform fishes are known from all oceans, with the northernmost record off West Greenland and the southernmost from the Weddell Abyssal Plain, but the majority of species occur below subtropical and tropical surface waters. Neither deep bottom-dwelling nor deep pelagic species show any obvious adaptations to life in the deep, so it seems probable that ophidiiform fishes descended from shallow depths relatively late. When comparing a deep-living species and one from shallow water there is little to distinguish them.

The ophidiiform fishes live in a variety of habitats and depths. The deepest recorded fish, caught at 8,370 meters, is an ophidiiform; there are also numerous ophidiiforms inhabiting the shallow waters of coral reefs. In addition some species are found in fresh water, such as in limestone caves in the West Indies.

The reproductive biology varies from oviparous (egg-laying) species which lay several hundred thousand eggs to viviparous (live-bearing) species giving birth to about 100 larvae. Most ophidiiform fishes live near the bottom, some occur pelagically, and a few live inside various invertebrates. Larvae of only a few species are known. They are most often caught in the upper 100 meters (330 feet), rarely near the bottom.

PEARLFISHES

The pearlfish family (Carapidae) includes about 30 species in 4 genera, all of which are long and slender, with the anus placed far forward. In these egg-laying, scaleless fishes, the dorsal fin rays are shorter than the corresponding anal fin rays. Their biology is quite special, as the adults of most of the species typically live inside invertebrates. Those that live in holothurians (sea cucumbers) are known to eat the gonads of their host, while pearlfishes found in ascidians (sea squirts), asteroids (starfishes), and bivalves (clams) normally just hide within the host.

The larvae of all species go through a planktonic stage that is recognized by the presence of a long, thin filament placed anterior to the dorsal fin and often provided with fleshy appendages. Pearlfishes are widely distributed in all oceans. They rarely exceed 50 centimeters (20 inches) in length.

Jen & Des Bartlett/Bruce Coleman Limited

◀ *The pearlfish* Carapus bermudensis *is
typical of its family in living in the shells
of mollusks or the body cavities of sea-
cucumbers, emerging to feed at night.
It is widespread in the Caribbean.*

Ken Lucas/Planet Earth Pictures

▲ One of the largest members of its family, the red brotula Brosmophycis marginata (family Bythitidae) has a separate caudal fin and a filamentous pelvic fin on the breast. A secretive fish, it inhabits rocky shores from Alaska to Baja California.

▶ A juvenile cuskeel, probably of some abyssal species. Many forms inhabit bathyal and abyssal waters, and the deepest fish ever found was a cusk eel. Cusk eel eggs float to the surface and hatch to join plankton swarms, and the young fishes gradually move to greater depths as they grow older.

CUSKEELS

The cuskeels (family Ophidiidae) are the most speciose of the ophidiiform families, with about 240 species in 50 genera. The elongate body is covered by smooth scales; the dorsal fin rays are equal to or longer than the corresponding anal fin rays; and there are usually more than seven long rakers on the anterior gill arch. All cuskeels are oviparous (egg-laying), with the larvae living in the open sea in the upper layers of water. Many species occur at bathyal and abyssal depths (from 2,000 to 6,000 meters or 6,600 to 19,800 feet) living on or close to the bottom; *Abyssobrotula galatheae*, reaching a length of 20 centimeters (8 inches), holds the depth record among fishes at 8,370 meters (27,455 feet) in the Puerto Rico Trench. A few are pelagic at several hundred meters and the rest are found on the continental slopes and shelves. Two genera, each with a few species, are commercially important—the kingklip, *Genypterus* species, growing to 1.5 meters (5 feet) on temperate coasts of the southern hemisphere, and the bearded brotula, *Brotula* species, which reaches 1 meter (3¼ feet) and occurs on the continental shelf in most tropical oceans.

BROTULAS

Viviparous (live-bearing) brotulas (family Bythitidae) are represented by about 85 species in about 25 genera. They are generally less elongate than the cuskeels and there are rarely more than seven long rakers on the first gill arch. In front of the anal fin the males have a copulatory organ

which varies among the genera. The viviparous brotulas are divided into two groups: one in which the tail fin is united to the dorsal and anal fins, and another with a separate tail fin. Most species occur at shallow depths, often concealed in corals; many of these species are only 5 to 10 centimeters (2 to 4 inches) long. The others live on the continental shelf and slope, with a few in the abyss. Species of the genus *Lucifuga*, however, are found in the caves and sinkholes of Cuba and the Bahamas in waters where the salinity ranges from fresh to highly saline. Species of a related genus live in freshwater caves in Yucatan. There are many undescribed species in this family.

APHYONIDS

The Aphyonidae is a small viviparous family with about 25 species divided into 6 genera. Most of them are about 10 centimeters (4 inches) long when adult. They have no scales and the loose, transparent skin is gelatinous. The long-based dorsal and anal fins are united to the tail fin. Except for the genus *Barathronus*, adult aphyonids have retained a number of larval features: they have cylindrically shaped vertebral centra, slightly ossified bones, and poorly developed musculature, gill rakers, and gill filaments. The eyes, the lateral line, and the olfactory (nasal) organ are much reduced, and there is no swim bladder.

The great majority of aphyonids live near the bottom between 2,000 and 6,000 meters (6,600 to 19,800 feet), where food is scarce; consequently the populations are scattered and it is difficult for the sexes to find each other. However, aphyonids (and viviparous brotulas) produce bundles of sperm encapsuled in thin sacks, called spermatophores. The fact that spermatophores are found in completely unripe ovaries shows that mating can take place even before the eggs are ready to be fertilized—the spermatophores are stored in the ovaries until the eggs are ripe, when the capsule is dissolved. This ability to store spermatophores means that ripe males can mate with unripe females. In this way it may be possible for aphyonids to compensate for the difficulties in locating a mate that result from low-density populations and poorly developed sense organs.

JØRGEN G. NIELSEN

Peter David/Planet Earth Pictures

TOADFISHES

Toadfishes are truly amazing fishes. These shy, sluggish creatures have evolved a method of raising and caring for their young which is second to none. What they lack in efficiency of form and motion has been made up for in the parental care given to each new generation. Producing only a relatively small number of eggs, this nurturing is obviously an important adaptation that has helped to ensure their survival.

KEY FACTS

ORDER BATRACHOIDIFORMES
FAMILY BATRACHOIDIDAE
• 3 subfamilies • 22 genera
• 69 species

SMALLEST & LARGEST

Thalassophryne megalops (a venomous toadfish species)
Total length: 7.5 cm (3 in)

Pacuma toadfish *Batrachoides surinamensis*
Total length: 57 cm (22 in)
Weight: 1.5 kg (3 lb)

CONSERVATION WATCH
! 5 species are listed as vulnerable: Cotuero toadfish *Batrachoides manglae*; whitespotted toadfish *Sanopus astrifer*; whitelined toadfish *Sanopus greenfieldorum*; reticulated toadfish *Sanopus reticulatus*; splendid toadfish *Sanopus splendidus*.

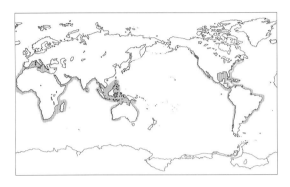

BROAD HEAD AND BIG EYES

The toadfish has been described as "having a face that only its mother could love". Members of the toadfish family are easily recognized by their broad, flattened heads and fleshy flaps and barbels, especially around the mouth. The trunk is rather cylindrical, tapering to a rounded caudal fin. The soft dorsal and anal fins are relatively long. In front of the soft dorsal fin are two or three short, sharp spines which form the small spiny dorsal fin.

As the name implies, toadfishes resemble toads and frogs. Obvious similarities include the prominent forward-facing eyes that are located high on the head, the large down-turned mouth, and the smooth, slimy skin. Members of the subfamily Porichthyinae, however, differ most noticeably from other toadfishes in having lines of bioluminescent (light-producing) organs along the sides of their heads and bodies. As these organs resemble the gleaming buttons of a naval uniform, these fishes are more commonly referred to as "midshipmen".

The majority of species of toadfishes are found in the Americas, followed by Africa, and then the Australian/Indonesian region. They are bottom dwellers, and generally inhabit shallow to moderately deep coastal waters, although some species prefer brackish water. At least one species is found in fresh water.

Toadfishes can survive for considerable periods of time out of water; some individuals have remained alive for more than 24 hours after capture. One species has been reported to move across exposed mudflats at low tide, using its ventral fins as legs.

Toadfishes are often mistaken for members of the venomous scorpionfish family Scorpaenidae, in particular the extremely dangerous stonefishes. Scorpionfishes have numerous sharp spines on the head and venomous spines in the dorsal and anal fins. Toadfishes, on the other hand, possess one to four strong sharp spines on the gill cover and another two to three in the spiny dorsal fin (being rather short, these spines are are often hidden in thick skin tissue). However, they are venomous only in members of the subfamily Thalassophryninae, which occurs in Central and South America. The spines of these species are hollow, with a venom gland connected to the base of each spine. On puncturing a victim's flesh, a jet of venom is expelled from the spine's tip, in an action similar to that of a hypodermic needle. Although not deadly, these spines are capable of inflicting painful wounds.

CAMOUFLAGED PREDATORS

Coloration within the toadfish family tends to be drab, consisting usually of browns, greens, and

▼ The starry toadfish *Sanopus astrifer* of the Gulf of Mexico and the Caribbean. Females lay their eggs under stones or shells on the sea bottom, then leave them for the male to guard until they hatch after about three weeks.

Marty Snyderman

Ken Lucas/Planet Earth Pictures

▲ *With its broad mouth and bulging eyes, the plainfin midshipman* Porichthys notatus *has a toad-like appearance. Toadfishes are highly vocal, and make sounds that might be mistaken for those of a toad or frog. The rows of silvery spots on the body are light organs and the light produced is involved in courtship displays.*

grays which are arranged in patterns that help to camouflage the fish. The exception to this, however, occurs in the coral reef-dwelling members of the genus *Sanopus* from Central America. These fishes have quite striking color patterns, a feature that is commonly asssociated with animals inhabiting coral reefs.

Toadfishes enjoy a varied diet, including crabs, prawns, octopuses, shellfish, sand dollars, and fishes. They sit motionless on the bottom, often at the entrance to their lairs, blending perfectly with the background. Some hunt away from their holes, stealthily tracking their prey. When close enough, the unsuspecting animal is swallowed whole in one swift movement. As toadfishes have a very elastic stomach, they are capable of swallowing quite large prey. Their teeth are mostly large, conical, and blunt, although some species have smaller and sharper teeth. Many toadfish species are very aggressive, particularly during the mating season, and will bite if handled.

REPRODUCTION

The most interesting aspects of toadfishes center on their breeding behavior, in particular the calls made during courtship and the nests in which the young are raised. It is not known if the following applies to all members of the family, but it is certainly relevant for some species from North America and Australia.

Toadfishes are shy creatures, normally inhabiting muddy holes or rocky crevices. The holes are dug next to rocks so that at least one side of the burrow is made of solid rock. Some occupy dead bivalve shells, while others take over refuse discarded by humans, such as glass jars and cans. The presence of toadfishes in these areas is often advertised by grunts, growls, croaks, or a sound similar to that of a steam whistle. Particularly during courtship, the noise may be so loud that people living nearby may find it intolerable. On the east coast of North America, for example, people on houseboats have been kept awake by the bellowing of the oyster toadfish *Opsanus tau*.

During the mating season—from late spring to summer—males attract gravid females to their

holes by calling. The female releases her eggs after entering the nest; they are then fertilized by the male. The eggs of most species are relatively large, and number between 20 and about 100. There is a sticky disk on the ventral surface of each egg by which it is attached to the side of the nest. Other females in turn may also be attracted to the same hole, laying their eggs in clusters adjacent to those of the previous visitors. A single nest, therefore, may contain various groups of eggs, each at a different stage of development.

The females, however, never return to the nest; the developing eggs become the responsibility of the male. During the brooding period, the male never leaves the nest, as the unguarded eggs would be quickly consumed by other predators. Most brooding males appear to reduce their intake of food, probably relying on what can be captured at the entrance to the nest. They become noticeably thinner than those individuals not involved in nest-guarding activities.

In the nest, the eggs develop quickly. A small embryo forms on top of each large yolk sac and soon resembles a tadpole, its tail continually lashing from side to side. Markings presently appear on the fish, and they start to look like miniature versions of the adult, the yolk sacs becoming smaller and finally disappearing. At this stage (three to four weeks after fertilization), the young fish are still attached at the breast to the substrate, but they eventually break free. However, they do not immediately leave the nest. The male continues to provide care until the young fish are big enough to fend for themselves.

Aquarium observations have shown that the free-swimming young crowd around the male, and even sit on his head or take refuge beneath his body. Some reports suggest that the male trains the young fish to catch their own food during this period. Finally the young leave the nest to seek out homes of their own. As toadfishes are cannibalistic, the newly departed juveniles are vulnerable at this time to larger members of their own species, as well as to other predators. Obviously their first priority would be to seek a safe shelter.

J. BARRY HUTCHINS

▼ *A coral reef-dweller of Central America, the splendid toadfish* Sanopus splendidus *is the most colorful member of a generally drab family.*

ANGLERFISHES

T|he highly specialized anglerfishes (order Lophiiformes) occur in all marine temperate and tropical environments from the shallow littoral zone to the abyssal depths of the oceans. Throughout history their unique appearance and peculiar way of life have attracted attention. More than 2,000 years ago Aristotle described how the common Mediterranean species, the "fishing frog", hides on the sea bottom and, by means of the filaments on its head, lures other fishes within reach of its huge, greedy mouth. Much later, when rationalism tried to remove fiction from scientific fact, the tale about the "angling fish" was regarded as a product of the well-known and lively imagination of sailors and fishermen. More recently, however, studies have not only confirmed (and added to) the ancient story, but revealed biological specializations that surpass the most daring fantasy.

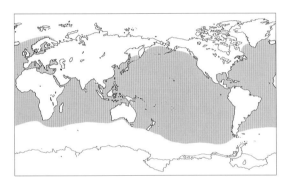

SHARED SPECIALIZATIONS
Anglerfishes form a well-defined but relatively isolated order of advanced teleost fishes. Their high degree of functional adaptation makes it difficult to trace their evolution, but it is assumed that they are related to toadfishes and more distantly to codfishes.

Their efficient and energy-saving feeding methods have enabled anglerfishes to find niches in widely different environments, resulting in a great diversity within the order. There are nearly 300 species, grouped in 65 genera and 18 families that are separated into 5 suborders. They differ mainly in their adaptations to different living conditions. Anglerfishes differ most significantly from other fishes in the position and development of the first two or three spines of the dorsal fin. In all members of the order these spines are separated from the rest of the fin and placed on top of the head. The foremost is the lure they use to attract their prey. It consists of a bony rod with a fleshy bait at the tip. At the base it is freely movable on

Rudie H. Kuiter

◄ *A characteristic unique to the Lophiiformes is demonstrated by this antennariid anglerfish* Rhycherus filamentosus *guarding her egg mass— in some species, up to several million eggs may be enveloped and embedded in the large gelatinous mass.*

▶ *Common in European waters, the goosefish or monkfish* Lophius piscatorius, *known since ancient times as the "fishing frog", has three widely spaced spines on the head. The foremost (called the illicium, and shown tilted over the left eye in this photograph) can be swivelled in any direction and carries a small fleshy bait at the tip used to coax prey within range.*

G.I. Bernard/Oxford Scientific Films

the tip of a fin-spine carrier (the basal bone) that is partly or completely embedded beneath the skin of the head.

The characteristic appearance of most anglerfishes is closely related to their feeding methods, which allow them to remain stationary instead of having to hunt for food. Without need for the more usual streamlined shape of other fishes, anglerfishes have short bodies and huge heads. The mouth is enormous and more or less protrusible and the gill covers are greatly enlarged, while the gill openings are narrow spouts placed below or behind the bases of the pectoral fins. Combined with strong bony support and well-developed musculature, this creates a most efficient sucking device. By extreme expansion of the oral and gill cavities, followed by a sudden opening of the mouth, attracted prey are easily captured.

Matching the huge gape, the throat and stomach are very expansible, enabling anglerfishes to engulf extremely large prey animals. Some eat other fishes, while others have been found with prey longer than themselves coiled in their stomach. When full, the bulging belly adds to their grotesque appearance. The bottom-dwelling species have especially large pectoral fins that are placed low and have limb-like elongated bases. The order was once named Pediculati, which means "small legs". Anglerfishes have no scales, but the skin may be amply supplied with spines, warts, and filaments.

Unique to this order is the provision made for the eggs. These are enveloped in a secretion which at spawning swells into a large, free-floating, gelatinous mass. In the large goosefishes this so-called "egg-raft" may reach a length of at least 12 meters (39 feet) and contain more than 3 million eggs.

GOOSEFISHES

For many centuries the "fishing frogs" of the genus *Lophius* were the only anglerfishes known. Characteristic features of most goosefishes (suborder Lophioidei) are their extremely broad, flattened, spiny head with upturned eyes, and wide mouth with large, pointed teeth. The lure is a slender fin-spine with one or more simple flaps of skin as bait. It is followed by one or two similarly elongated spines on the head and, unlike other anglerfishes, one to three additional spines just behind this.

The goosefish is practically invisible lying flattened against the sea bottom, with its darkly marbled skin matching the bottom color and the outline of its body obliterated by a fringe of branched skin flaps. Bottom-feeding fishes, attracted by the sight and vibrations of the swinging bait, are swallowed with lightning speed. Remarkably large fishes and even diving birds are sometimes found in the stomachs of goosefishes, which may reach lengths of more than 1.5 meters (5 feet).

The suborder consists of a single family (Lophiidae) divided into four genera with a total of 25 species. Together they cover a depth range from a few to more than 1,000 meters (3,280 feet), but most species prefer depths of more than 200 meters (656 feet). The family is represented on the continental shelves and slopes of all oceans except in polar regions.

The largest and most abundant species are subject to commercial fishing in Mediterranean and North Atlantic waters.

FROGFISHES

Frogfishes (suborder Antennarioidei) differ most distinctly from the goosefishes in having a high, slightly compressed head and body, and in lacking the series of spines just behind the head. In the typical frogfish the lure is thin and slender with a fleshy bait, while the second and third spine of the head are more or less covered by skin; the third spine looks somewhat like a pixie-cap riding on top of the head.

Frogfishes are masters of ingenious camouflage—an essential requirement for "lie-in-wait" predators living in clear, shallow water. With the lumpy body variously provided with warts and filaments, combined with an extreme ability to change color, they can simulate their background almost exactly (usually coral with brightly colored patches of algae, sponges, and other sessile invertebrates).

The behavior of frogfishes has been studied extensively because of the ease with which they can be kept in aquaria. Their angling behavior appears most deliberate: the potential prey is sighted and, carefully followed by moving eyes, the lure is directed towards the prey. The shape, color, and movements of the bait may be a remarkably life-like imitation of a wriggling worm, a hopping crustacean, or an undulating little fish.

Being too swift to follow by eye, the "gape and suck" method by which the prey is caught has been examined in high-speed cinematography. This has shown the extreme efficiency and velocity by which an unsuspecting would-be predator is changed into an engulfed food item within less than one-hundredth of a second!

Observations of the use of the limb-like paired fins show two kinds of gaits on the sea bottom: one similar to the normal tetrapod walk, in which the matching limbs are moved alternately, and another in which the pectoral fins are used in parallel movements like a pair of crutches.

The total of 15 genera of frogfishes are divided into 4 families, of which the Antennariidae contains the great majority of the 50 or so known species. In most species the known maximum length is less than 10 cm (4 inches), but some may reach 30 to 45 centimeters (14 to 18 inches). They are most common at depths between 20 and 100 meters ($65\frac{1}{2}$ to 328 feet) and are represented in all tropical and subtropical coastal waters of the world, except the Mediterranean Sea. Some species occur in temperate waters, for instance off Japan and southern Australia.

BATFISHES

In the batfishes (family Ogcocephalidae) the head and trunk are markedly broad and flattened, disk-shaped, or with a triangular outline, and have a solid armor of broad-based spines. Compared to goose- and frogfishes, the fin-spines of the head are greatly reduced. The lure consists of a very short bone completely embedded in the bait. The simple or tri-lobed bait is placed in the opening of a groove from which it can be pushed forward by the basal bone of the lure. The second spine is

rudimentary and the third is lost. The shape and position of the pelvic and long, angular pectoral fins are well adapted for walking on the sea bottom.

The biology of batfishes is poorly known. Limited examination of their stomachs has revealed remains of fishes, crustaceans, and polychaete worms. In the dim light, where the majority of species live, attraction of prey may be based mainly on vibration of the lure. However, the lobes of the bait contain glandular structures that have no similarity to light organs, but might produce secretions of odor attractive to prey.

▼ *Seldom more than 25 centimeters (10 inches) in length, frogfishes are capable of swallowing fishes nearly as large as themselves, engulfing them in suction resulting from the sudden opening of their cavernous jaws, in one of the fastest movements known among animals.*

Marty Snyderman

The single family of this suborder (Ogcocephaloidei) contains 9 genera and about 60 species, of which the largest may reach a length of 50 centimeters (20 inches). Some species occur in shallow water, but the majority prefer a depth between about 200 and 1,000 meters (656 and 3,200 feet). The family is distributed along the coasts of all tropical and subtropical regions of the world, except the Mediterranean Sea.

SEATOADS

Being in some ways intermediate between frog and batfishes, the seatoads (family Chaunacidae) are placed in a separate suborder, Chaunacoidei. They have a flabby balloon-shaped, pink body, densely covered with minute spines. The very short lure has a flattened, fleshy bait fringed with short filaments and can be laid back in a shallow depression on the snout. Two genera and 12 species are recognized, the largest reaching about 30 centimeters (12 inches). They occur in all but polar seas at or near the bottom in depths from 90 to over 2,000 meters (300 to over 6,500 feet).

DEEPSEA ANGLERFISHES

The energy-saving way of life, achieved by luring instead of hunting, is especially advantageous in environments such as the free water masses of the deep sea, where food is scarce and populations very thinly spread. Two adaptations have enabled the deepsea anglerfishes (suborder Ceratioidei) to invade this dark and inhospitable regime: the development of a luminous bait and the achievement of a uniquely specialized sexual difference.

With few exceptions, the lure of the females has a bulb-shaped bait with an internal, globular gland containing luminous bacteria. The bait carries a specific characteristic and often very spectacular pattern of appendages, which in many species contains light-guiding, silvery mirrors and tubes. The length of the lure varies between less than the diameter of the bait in some species to more than three times the length of the fish in others. Similarly, the basal bone of the lure varies considerably among species, from short, completely covered by the skin of the head, and nearly immobile, to long, greater than the length of the head when extended forward beyond the snout and protruding as a long tentacle on the back when fully retracted.

Transformed into passively floating food traps, the females are highly adapted for survival and growth in their barren environment. However, this specialization has left them no ability to locate a sexual partner within their meager and widely spaced population. Through a number of adaptations this task has become the sole responsibility of the males, which after transformation from larvae differ from their females in nearly all features. They lack a lure and have feeble jaws without teeth. Unable to feed, they become dwarfed, reaching in most families a length of only 15 to 45 millimeters ($\frac{1}{2}$ to $1\frac{3}{4}$ inches), and less than 5 to 10 percent of the largest females of their kind. Based on their slender, streamlined shape the males are able swimmers and nearly all are equipped with huge olfactory organs, which indicate that they are led by a track of a characteristic-smelling substance secreted by the slowly drifting females of their species. In all males the snout and chin carry hooked denticles by which those who succeed in their search can attach themselves to the skin of the female while waiting for spawning. In some families the tissues of the pair become fused in such a way that the male is nourished through the blood of the female and able to continue growth and development of the testes, being in this way reduced to a mere sexual organ on its large mate.

▼ A deepsea anglerfish Linophryne polypogon. Ceratioid anglerfishes have a number of remarkable adaptations to a deepsea life. Females have a bulbous bait (called an esca), containing luminescent bacteria, whereas the comparatively minute males have no baits and feeble, toothless jaws. They do not feed but follow scent trails to locate females. Males of a few families permanently attach themselves to females, and live as parasites.

Peter David/Planet Earth Pictures

◄ Very little is known about this species, Phrynichthys wedli (family Diceratiidae), a small anglerfish inhabiting the waters off the continental shelf on both sides of the Atlantic. No male specimens have ever been found. This juvenile was caught in the Caribbean by the slurp gun of a research submarine at 1,000 meters (3,300 feet) and was photographed alive, untouched by any net.

The morphological diversity of the suborder is remarkably large, with divergent elaborations of the luring and jaw mechanisms, which include several exceptions from the general description given above. The 150 or so known species are divided into 34 genera and 11 families. Many of the species are known from single or very few specimens, indicating that many more species are waiting to be discovered. In the great majority of species the known specimens are less than 10 centimeters (4 inches) in length, but in some families females of 30 to 50 centimeters (12 to 19½ inches) have been caught. The northern seadevil Ceratias holboelli may reach a total length of about 1 meter (3¾ feet). They are distributed in depths from a few hundred to at least 2,000 meters (6,560 feet) in all oceans and adjacent seas, except for the Mediterranean Sea and the polar regions, but occurrence of larvae is restricted to regions with a surface temperature of at least 20° C (68° F) or within approximately 40° North and South.

E. BERTELSEN &
THEODORE W. PIETSCH

▼ Paxton's whipnose angler Gigantactis paxtoni, named after one of the consultant editors of this book, is one of a group of deepsea anglerfishes whose females are characterized by, among other things, a greatly elongated illicium, or "fishing rod".

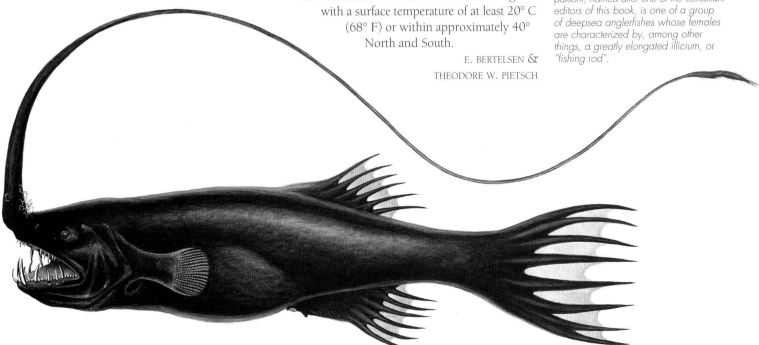

CLINGFISHES & THEIR ALLIES

The order Gobiesociformes contains mostly small inshore fishes belonging to three families: the family Gobiesocidae (the clingfishes) and the families Callionymidae and Draconettidae (the dragonets).

CLINGFISHES AND DRAGONETS

The species in the order Gobiesociformes lack scales and generally have large flattened heads and tapered bodies, so they look a little like colorful tadpoles. They occur in all tropical and temperate oceans, where they mostly sit on the bottom. Clingfishes have a single dorsal fin and a unique adhesive disk on the breast, although one subfamily, the Cheilobranchinae (shore-eels or singleslits) has either lost the disk or possesses only a rudiment. Dragonets on the other hand have two dorsal fins and always lack the adhesive disk on the breast.

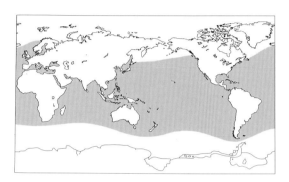

COLORFUL CLINGERS

The small subfamily Cheilobranchinae, with one genus and four species, is found only in southern Australia, Tasmania, and Norfolk Island. These are small, eel-like fishes without paired fins. There is only a superficial resemblance to other clingfishes, but certain skeletal characteristics seem to indicate a distant relationship. The gill opening is a single slit on the ventral surface just behind the head.

The clingfishes in the subfamily Gobiesocinae belong to 8 tribes comprising 43 genera and about 150 species. All are marine except for six members of the genus *Gobiesox* that inhabit New World freshwater streams. In the temperate zone, many species may be found in shallow water where there are strong currents and wave actions. They protect themselves by clinging to stones, seaweed, or seagrass. Others live in deeper water attached to invertebrates such as sponges and ascidians. In the tropics, many species are commensal with sea urchins and crinoids—some eat the tube feet of their host. In southern Australia, two species act as cleaner-fish, removing parasites from larger fishes.

Clingfishes are difficult to detect since they have a cryptic coloration. Many have the ability to change color rapidly, especially when moving from one habitat to another.

▶ *Small and inconspicuous, Alabes parvulus (subfamily Cheilobranchinae) is restricted to coastal southern Australia. It is not uncommon in the intertidal zone, especially in shallow rockpools where it hides under stones or among seaweed.*

Rudie H. Kuiter

◄ Aspasmogaster occidentalis (subfamily Gobiesocinae) inhabits tidal rock pools, like many other clingfishes. It grows to a length of about 6 centimeters (2½ inches). Clingfishes rely on their ventral sucker to hold them secure in wave surges; the suction is strong enough that a hooked clingfish can often be lifted complete with the rock it clings to. Clingfishes have also been observed moving smoothly over a rock surface without releasing the sucker's grip.

B. Hutchins

Most clingfishes are small, generally less than 5.5 centimeters (2 inches), with a variety of body shapes. The head is almost always depressed. A lateral line is always present with the sensory pores well developed, but small, on the head and often missing posteriorly. There are no scales but the body is protected by a heavy coat of mucus. They have one dorsal and one anal fin. The ventral fins are incorporated into a prominent adhesive disk located on the underside of the clingfish's body.

FOSSIL RECORD

No fossil clingfishes have ever been found so their evolutionary history must be deduced from living species. The most primitive, or generalized, tribe is the Trachelochismini. All but two of the species are southern hemisphere relicts, existing in the temperate waters of Australia and New Zealand. The related tribe Haplocylicini has a single species, confined to New Zealand. The Lepadogastrini occurs in the eastern Atlantic and Mediterranean, with one species in the deep waters of southern Africa. The monotypic Chorisochismini is confined to the warm temperate waters of southern Africa. The tribe Diplocrepini is an Indo-West Pacific group except for one species found in the deep waters of the Caribbean Sea. With the exception of this species, all New World clingfishes belong to the tribe Gobiesocini. One monotypic genus has reached southern Africa. The most advanced, or specialized, tribe is the Diademichthyini, found in the tropical Indo-West Pacific.

The more specialized species occupy the Indo-West Pacific tropics; those that are intermediate in evolutionary terms are found in the New World and in the eastern Atlantic; and the most generalized are southern hemisphere relicts. This pattern supports the hypothesis that the central East Indies has been a center of evolutionary radiation.

▲ Although both sexes are striped to hide among sea urchin spines, the urchin clingfish Diademichthys lineatus (subfamily Gobiesocinae) is sexually dimorphic: the male (illustrated) possesses a wider snout, possibly reflecting different dietary habits.

DRAGONETS

The dragonets (suborder Callionymoidei) comprise 2 families (Callionymidae and Draconettidae) and 175 species of small benthic marine fishes. Most callionymoids are sexually distinct; the males usually have longer dorsal spines and tail-fin rays, and more brilliant, sometimes spectacular, coloration. Dragonets often have a distinctive spine embedded in tissue on the side of the head.

JOHN C. BRIGGS & J. BARRY HUTCHINS

◄ The brilliant coloration of the mandarin fish Synchiropus splendidus may serve as a warning to predators, for the skin of this species gives off a foul-tasting, odorous mucus. Like most dragonets this species is sexually dimorphic, the males (illustrated) bearing an elongated first dorsal fin for display.

ORDER BELONIFORMES
SUBORDER EXOCOETOIDEI
• 4 families • 32 genera
• c. 200 species

SMALLEST & LARGEST

Freshwater needlefish *Belonion apodion*
Body length: up to 3 cm (1⅕ in)

Agujon needlefish *Tylosurus acus*
Body length: up to 95 cm (37½ in)

CONSERVATION WATCH
! There are 3 freshwater species of halfbeaks from Sulawesi—*Dermogenys weberi; Nomorhamphus towoeti; Tondanichthys kottelati*—that are listed as vulnerable.

FLYINGFISHES & THEIR ALLIES

Flyingfishes and their allies have long been classified together, often as a separate order within the Atherinomorpha—a larger group that also contains the silversides, killifishes, ricefishes, and allies. In the classification system adopted by this book, the ricefishes (discussed on page 152) are grouped with the flyingfishes in the order Beloniformes. The flyingfish suborder Exocoetoidei includes four families—the sauries (family Scomberesocidae), the needlefishes (family Belonidae), the halfbeaks (family Hemiramphidae), and the flyingfishes (family Exocoetidae). Most members of the suborder are epipelagic (surface-dwelling) marine fishes, but several genera of needlefishes and halfbeaks are restricted to fresh waters, and a few other genera contain estuarine and freshwater as well as marine species. Features common to the flyingfishes and their allies include dorsal and anal fins on the rear half of the body, abdominal pelvic fins with six soft rays, no fin spines, a lateral line running along the ventral edge of body, lower pharyngeal bones fused into a triangular plate, and an open nasal pit.

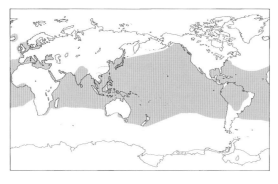

TWO GROUPS

Two groups have been recognized within this suborder under various names by a series of authors. Each group contains two families: the sauries and needlefishes in one group and the halfbeaks and flyingfishes in the other.

Most fishes of the suborder Exocoetoidei produce large spherical eggs with attaching filaments, a feature they share with other atherinomorph fishes. The primitive exocoetoid egg has long filaments distributed over the entire surface of the egg. Saury eggs and some flying fish eggs are pelagic and lack filaments.

SAURIES

Sauries (family Scomberesocidae) are oceanic fishes, found mostly near the surface, worldwide in subtropical and temperate waters. With two genera and four species, they are defined by one particular feature—dorsal and anal fins followed by a series of four to six small, flag-like fins or finlets. *Cololabis adocetus* is the smallest marine epipelagic fish in the world, with a total length of 7.5 centimeters (3 inches). The Pacific saury *C. saira* and the Atlantic saury *Scomberesox saurus* are schooling migratory fishes that are very abundant in subtropical and warm temperate waters. *C. saira* is an important commercial food fish and the fishery for it is based on its attraction to lights at night.

NEEDLEFISHES

Needlefishes (family Belonidae) are easily identified by their elongate upper and lower jaws, usually studded with relatively large sharp teeth, and their long body. They have one dorsal fin, nearly opposite the anal fin, and abdominal pelvic fins. Most needlefishes pass through a "halfbeak" stage in their development, in which the lower jaw, but not the upper jaw, is greatly extended. Transition from the halfbeak stage to the needlenose stage is associated with the change from eating plankton to eating fish. Juveniles of

▼ *Longtoms (genus Strongylura) are predatory fish that feed at night and favor shallow waters. They have caused fatalities among fisherman because of their habit of leaping violently in the direction of any strong light, their needle-snouts quite capable of skewering a limb or even a torso.*

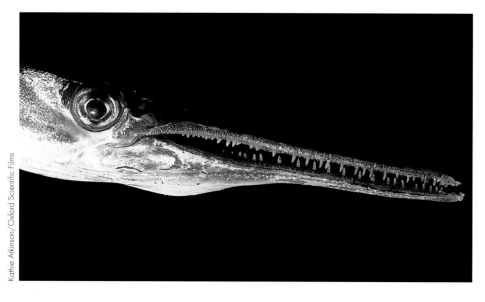

Kathie Atkinson/Oxford Scientific Films

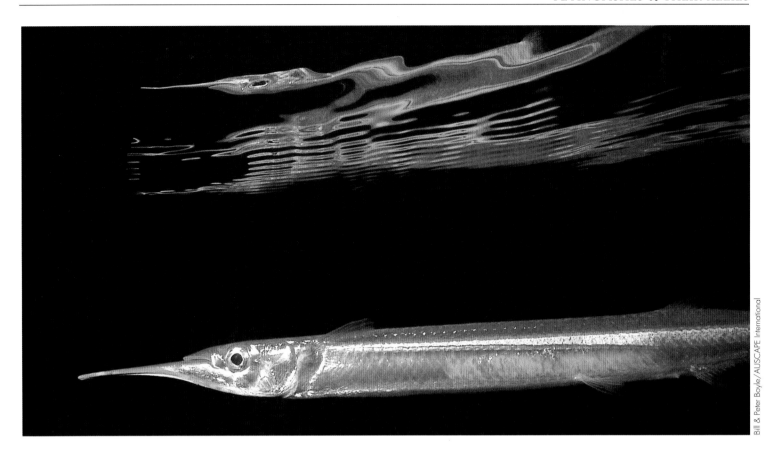

Bill & Peter Boyle/AUSCAPE International

the European needlefish *Belone belone* remain in the halfbeak stage until they are of a larger size than other needlefishes. The South American freshwater genus *Belonion*, which reaches a maximum size of up to 4 centimeters (1½ inches), matures while still in the halfbeak stage and is considered an example of paedomorphism, that is, it reaches sexual maturity while still appearing like a juvenile in other aspects.

Needlefishes are used for food in many countries, for example, in Australia, where they are marketed under the name "long tom". The flesh is good white meat, but it contains many fine rib bones that make the fish quite difficult to eat. Strangely, the bones are green.

Large needlefishes can be quite dangerous. At night they are attracted to light, like other beloniform fishes. A fatality occurred when a 1-meter (3-feet) long *Tylosurus* has leapt toward a light in a canoe and impaled a fisherman. There is even one reported fatality when a surfer in Hawaii ran into a *Platybelone* species whose beak went through the surfer's eye into his brain.

HALFBEAKS

Halfbeaks (family Hemiramphidae) are closely related to flyingfishes and are included in the same family by some authors. The lower jaw in most Hemiramphidae species is much longer than the upper jaw in juveniles of all genera, and adults of most genera, hence the name "halfbeak". They have short or moderately long pectoral fins. All genera are characterized by

particular combinations of lateral line characters too detailed to be explained here.

There are 12 genera and about 100 species of halfbeaks. Four genera have a very short beak or no beak at all. Two genera have long pectoral fins that allow these offshore fishes to remain in the air a little longer than other halfbeaks when they leap out of the water. This has led to the name "flying halfbeak" for both these genera.

Five genera (*Zenarchopterus*, *Dermogenys*, *Hemirhamphodon*, *Nomorhamphus*, and *Tondanichthys*), with 35 to 40 species between them, comprise a group of sexually dimorphic Indo-West Pacific estuarine or freshwater species. All of these species have a long nasal barbel. The anal fin of the males is modified into an andropodium.

Three genera are viviparous (that is, they bear living young). Internal fertilization appears to be accomplished by the enlarged genital papilla. The viviparous genera include some small species, and

▲ *Often known as the needle or splinter gar, the sea garfish* Hyporhamphus melanochir *(family Hemiramphidae) of southern Australian seas prefers shallow, sheltered waters such as large bays and estuaries. It is a shoaling species that feeds at night.*

▼ *The genus* Strongylura *has members in the Atlantic and Indo-Pacific regions but the needlefish* Strongylura krefftii *(family Belonidae) is the only freshwater species. It is restricted to New Guinea and northern and eastern Australia.*

G.E. Schmida

▲ *A number of halfbeak species occur in fresh water in Southeast Asia. The Malayan halfbeak* Dermogenys pusilla, *like other freshwater halfbeaks, is a livebearer, and males (illustrated) have the anal fin modified into a structure called an andropodium. The exact function of this organ is still unknown.*

therefore females face a space problem in carrying the developing embryos. One solution is superfetation—the female uses stored sperm to fertilize small batches of eggs every few days so that there may be five or six broods at different stages of development. As each brood is born, there is more space for the next broods to continue development.

The genus *Hemiramphus* (with 10 species) is a worldwide marine genus. Members of the genus *Rhynchorhamphus* (4 species) are confined to Indo-West Pacific marine waters. *Hyporhamphus* is the most speciose genus, with at least 34 species; some are marine, some estuarine, and some freshwater.

Most marine halfbeaks are largely herbivorous. They feed mainly on floating seagrasses, which they engulf by using the extended lower jaw, guiding the seagrasses down into the pharyngeal jaws. Others are carnivores, feeding on small fishes and crustaceans. Most species of the freshwater genera feed largely on floating insects. *Hemirhamphodon* species of Malaysia, Borneo, and the East Indies feed mostly on floating ants. The sensory pores of the head are very large, enabling them to detect the motion of a struggling ant. The teeth in these species extend out the length of the lower jaw and are directed forward, which helps

to entangle the legs of the ants.

Halfbeaks are important food fishes in some countries such as Australia, where they are sold as garfish; they are also important as baitfish for marlins in countries such as the United States.

The viviparous freshwater halfbeaks are used in the aquarium trade. Members of the genus *Dermogenys* have long been popular in Thailand as a fighting fish, although much less well known than *Betta splendens*, the anabantoid fighting fish. *Dermogenys* males are reported to grab each other by the beak and wrestle. Even marine halfbeaks of the genus *Hyporhamphus* sometimes wrestle with each other when two individuals meet after each has started to swallow opposite ends of a piece of floating seagrass.

FLYINGFISHES
Flyingfishes (family Exocoetidae) are easily identified by their exceptionally large, winglike pectoral fins that are sometimes used for gliding. The pelvic fins in some species are also very large. They have a very small mouth and no beak, except in the young of some species. Members of this family range in size from 14 centimeters (5½ feet) to 46 centimeters (1½ feet).

Flyingfishes inhabit the surface waters of all tropical oceans, often congregating around islands. They are well known for their habit of leaping out of the water when frightened and gliding above the sea surface. Their ability to glide depends on the size of their pectoral fins. *Fodiator* and

▶ *Belonging to the family Exocoetidae, flyingfishes somewhat resemble herrings in appearance, except for their (usually) much enlarged pectoral fins. They feed on plankton and live close to the surface of the open ocean, flying to evade attacks from predators below.*

Norbert Wu

◀ All flyingfish of the genus Exocoetus are "two-winged", but the genus Cypselurus has a number of "four-winged" members, like these in flight in the Red Sea—that is, their pelvic as well as their pectoral fins are modified for gliding.

Parexocoetus species, with the shortest pectoral fins, are rather poor gliders. In more advanced flyingfishes, the pectoral fins extend past the beginning of the anal fin. Two-wing flyingfishes (genus *Exocoetus*) build up swimming speed near the surface, set their pectoral fins in position, and launch themselves into the air for distances of 20 to 25 meters (65 to 80 feet). In members of the subfamily Cypselurinae (55 species) both pectoral and pelvic fins are enlarged, although the pectorals are much larger than the pelvics. They build up speed by taxiing like aircraft and continue to gather speed at the surface by sweeping the tail fin rapidly from side to side, keeping the lower lobe in the water. Towards the end of the taxi, they build up a speed of more than 60 kilometers (37 miles) per hour; the flight may last as long as 30 seconds and take the fish 200 meters (650 feet) or more from the starting point.

The largest species of flyingfish, the spotted flyingfish *Cheilopogon pinnatibarbatus*, inhabits the edge of the tropical zone with different subspecies off Japan, California, the Juan Fernandez Islands, Australia–New Zealand, South Africa, and the Azores. It reaches 50 centimeters (20 inches) in length and a weight of over 1 kilogram (2¼ pounds). The smallest flyingfish, *Parexocoetus mento*, reaches only 14 centimeters (5½ inches) and 25 grams (1 ounce). Despite these size differences, all flyingfishes appear to live for only a year or two and probably die after the first spawning season.

Flyingfishes feed mainly on plankton. They serve as food for many predators, including other fishes, especially dolphinfish, tunas, marlin, snake mackerel, as well as squids, birds, and porpoises.

There are fisheries for flyingfishes in many tropical countries. Several techniques have been used to catch them, including gill netting (Japan,

Carl Roessler

Vietnam, Barbados), dip-netting of spawning aggregations (India, Indonesia), and attraction by light and dip-netting at night (Pacific islands). One method of fishing is to anchor a canoe a little way offshore at night, fill the canoe with a foot or two of water, suspend a lantern over the middle of the canoe, and return in the morning. Flyingfishes are attracted by the light, leap toward it, and find themselves in water too shallow from which to escape. Upon return, the fisherman has a boat full of live, fresh fish. A method of fishing used in India is to float bundles of leaves near the surface at the end of a line during the spawning season. Flyingfishes approach the bundles of plants to deposit their eggs and, while so engaged, the fishers pull the bundles in close enough to their catamaran to be able to scoop the flyingfish out of the water. In addition to the fisheries for flyingfish meat, the eggs themselves are prized for use in Japanese sushi as *tobiko*.

BRUCE B. COLLETTE & NIKOLAY V. PARIN

▲ The track of a flyingfish, off Papua New Guinea. Aside from enlarged pectoral fins, flyingfishes also have asymmetrical tail lobes, the lower greatly enlarged. Functioning very like an outboard motor, the tail lobe vibrates in side-to-side motion at up to about 50 strokes per second, generating forward momentum to enable the fish to glide for up to 13 seconds, covering as much as 140 meters (460 feet), depending on wind and wave conditions.

▶ The guppy Poecilia reticulata *belongs to the family Poeciliidae, one of the few fish families whose members give birth to living young. Attractive and easy to rear and maintain, it is one of the most popular of all aquarium fishes.*

KILLIFISHES & RICEFISHES

K illifishes and ricefishes are among the most popular fishes in hobby aquariums and experimental biology. Not only are they easy to maintain in breeding aquariums but they have large eggs and a long time from fertilization to hatching, making them excellent specimens for scientific study. Killifishes, such as guppies and mollies (*Poecilia* species), platyfishes and swordtails (*Xiphophorus* species), and the mummichog (*Fundulus heteroclitus*), along with the ricefish or medaka *Oryzias latipes*, are among the most widely used fishes in laboratory studies of genetics, behavior, toxicology, and reproduction. Melanomas, or pigment tumors, develop in a regular, predictable fashion in some platyfish hybrids, indicating that some cancers may be inherited.

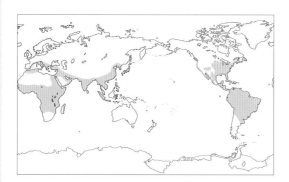

MEDIUM-SIZED SURFACE FEEDERS
Killifishes and ricefishes have until recently been classified together in the order Cyprinodontiformes, implying their close evolutionary relationship. Both groups are superficially similar, being small to medium-sized surface feeders that live in fresh or brackish water, often in coastal habitats but infrequently in strictly marine waters. They have small mouths, relatively large eyes, and single, usually soft-rayed, dorsal

Labat-Lanceau / AUSCAPE International

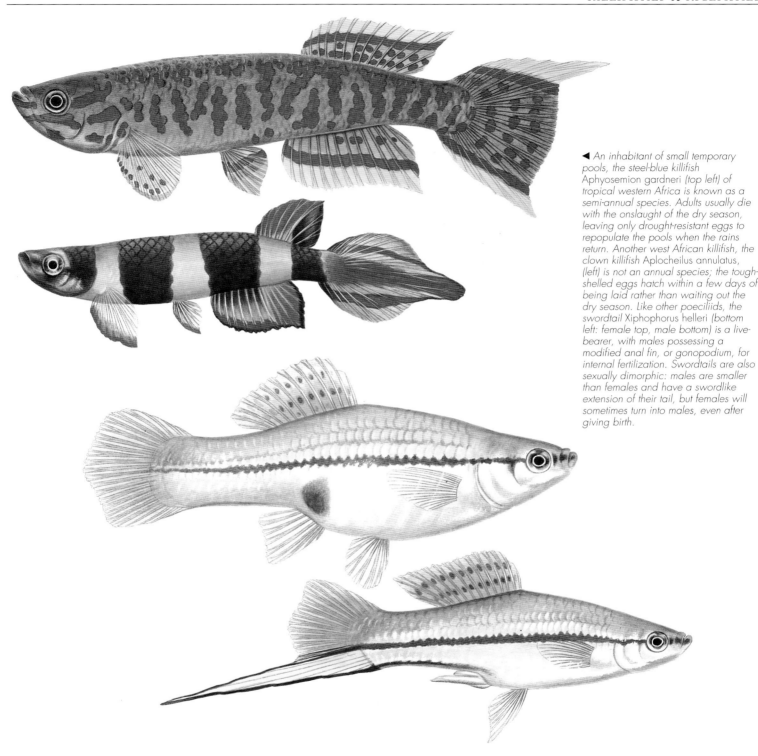

◄ An inhabitant of small temporary pools, the steel-blue killifish Aphyosemion gardneri (top left) of tropical western Africa is known as a semi-annual species. Adults usually die with the onslaught of the dry season, leaving only drought-resistant eggs to repopulate the pools when the rains return. Another west African killifish, the clown killifish Aplocheilus annulatus, (left) is not an annual species; the tough-shelled eggs hatch within a few days of being laid rather than waiting out the dry season. Like other poeciliids, the swordtail Xiphophorus helleri (bottom left: female top, male bottom) is a live-bearer, with males possessing a modified anal fin, or gonopodium, for internal fertilization. Swordtails are also sexually dimorphic: males are smaller than females and have a swordlike extension of their tail, but females will sometimes turn into males, even after giving birth.

fins. The pelvic fin may possess a spine. Many of these features are present in all fishes of the superorder Atherinomorpha, the larger group to which killifishes and ricefishes belong.

KILLIFISHES
Killifishes are placed in the Cyprinodontiformes, an order distributed widely in tropical and temperate fresh and brackish waters. They are readily distinguished from all other atherinomorph fishes by a number of anatomical features, including a rounded, symmetrical tail fin. Killifishes are diverse in form and structure and

exhibit a wide array of reproductive modes and behaviors. Annual killifishes of tropical South America (family Rivulidae) and Africa (family Aplocheilidae) are so-called because adults spawn at the end of the rainy season as their habitat is evaporating, leaving fertilized eggs in the drying, muddy substrate. The adults die and the eggs enter diapause (a resting stage) until the rains return, stimulating the eggs to develop and hatch. There is virtually no larval period, and individuals are sexually mature (able to reproduce) in just over a month after hatching. At least one rivulid, *Rivulus marmoratus*, has individuals that are self-fertilizing

▶ *A platy gives birth. A native of coastal wetlands of Mexico and Guatemala, the platy or southern platyfish* Xiphophorus maculatus *has become a popular aquarium fish. It is a live-bearer that delivers its young after a gestation period of four to six weeks.*

hermaphrodites. They produce both sperm and eggs and are able to fertilize themselves and produce populations of clones (genetically identical individuals).

The majority of killifish species are egg-laying, yet live-bearing is widespread, occurring in three families, as far as is known: the families Poeciliidae (mollies, guppies, and swordtails), Anablepidae (four-eyed fishes), and Goodeidae. There is internal fertilization of females by males, with sperm often being transferred in small, discrete packets or bundles. Sperm are deposited in females via the modified anal fin of the male, an elaborate structure that is highly variable from family to family. The modified anal fin of poeciliids and anablepids is called a gonopodium. The anal fin of male goodeids is less highly

modified and not usually called a gonopodium. Females give birth to live young. The three live-bearing families are not each other's closest evolutionary relatives, showing that this specialized reproductive mode has evolved more than once in killifishes. Superfetation (females carrying two or more broods in different stages of development) has evolved several times within the poeciliids.

The poeciliid fish *Poecilia formosa*, known commonly as the Amazon molly, has a special place in the history of reproductive biology as it is the first-discovered all-female vertebrate "species". Amazon mollies are not found in the Amazon River; they range naturally from the southern tip of Texas, USA, to Veracruz, Mexico. They are called Amazons because they occur naturally as

all-female populations, suggesting the lives of the legendary female Amazon warriors. Amazon mollies are believed to have arisen through hybridization between two *Poecilia* species. They breed in nature by a process called gynogenesis, a type of parthenogenesis (egg development without fertilization). Amazon mollies mate with males of one of the *Poecilia* species with which it co-occurs. The sperm of these males provides a stimulus for the egg's development, but does not contribute genetic material to the female's egg. Amazon mollies produce all-female clones; each daughter is genetically identical to her mother.

Species of the genus *Anableps*, ranging from southern Mexico to northern South America, have one of the more unusual adaptations among vertebrates to life at the water's surface. *Anableps* species are known commonly as cuatro ojos, Spanish for "four-eyes". Each eye, including the cornea and retina, is divided horizontally into an upper and lower section. Cuatro ojos are usually seen from above only by the tips of the upper portion of the eyes that protrude above the water's surface. The upper portion of the eye is used to see above the water, the lower portion is used to see below.

Discoveries of fossil killifish have been confirmed from North America, Mexico, and Europe. The oldest fossils are from freshwater deposits from the Oligocene epoch (37 to 24 million years ago) in Western Europe. The distribution of killifishes throughout all continents except Australia and Antarctica, however, indicates that killifishes as a group are probably older than the Oligocene and existed at least prior to the break-up of the ancient supercontinent Pangea over 180 million years ago. Fossil killifishes in western North America and Mexico are younger, no older than the Miocene epoch (24 to 5 million years ago).

The distribution and evolutionary relationships of the fossil and living killifish species help us understand what the environment was like in the Great Basin of the western United States before desert formation. Numerous killifishes survive here in small, relict populations, the most famous of which is the Devils Hole pupfish *Cyprinodon diabolis*, a small species that rarely grows to more than 2.5 centimeters (1 inch) and is further distinguished by the absence of pelvic fins. Its

habitat is Devils Hole, a small pool that is the head of a large underground water system in a portion of Death Valley National Monument, Nevada. As water was pumped out of the system for irrigation, the water-level in Devils Hole dropped, threatening the species with extinction. Now, by United States Supreme Court decree, pumping must maintain sufficient habitat to allow a stable population of pupfish.

Representatives of two other killifish genera, *Empetrichthys* and *Crenichthys*, survive also as Death Valley relics. They are primitive, egg-laying members of the family Goodeidae, the live-bearing members of which are found throughout the Mexican plateau. Before the extinction of the intervening species by desert formation, one can imagine a wet forest spanning the region from the northern extent of the Great Basin through to southern Mexico in which once lived the larger, ancestral populations of the modern-day goodeids.

The mosquitofish *Gambusia affinis*, a killifish native to North America, has been introduced into freshwater habitats around the world, including Australia, apparently to control mosquito populations, even though local, native species eat a greater proportion of mosquito larvae.

▲ *The four eyes or cuatro ojos* Anableps anableps *is so named because its eyes have divided pupils and retinas, enabling it to see above and below the water's surface. Consequently it can keep a look-out for predators as it searches for food.*

▼ *The genus* Gambusia *includes a number of species that are extremely similar in appearance but differ strongly in behavior. Several (especially* G. affinis*) have been widely introduced in tropical fresh waters in many parts of the world in attempts at mosquito control. Their effectiveness in this role is debatable but in some areas, such as Australia, they have become so well-established as to reach the status of pests.*

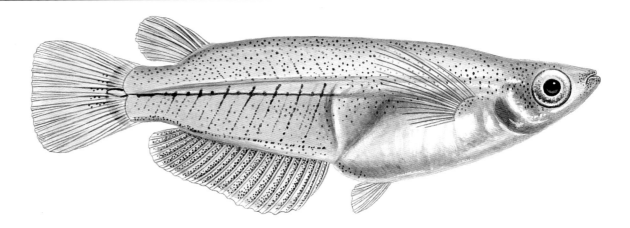

RICEFISHES

Ricefishes belong to the Adrianichthyidae, the sole family in the suborder Adrianichthyoidei, order Beloniformes. The order Beloniformes also includes the suborder Exocoetoidei, treated in the previous chapter (Flyingfishes & Their Allies). These fishes are called ricefishes in English, so-named nearly a century ago by American ichthyologists to indicate the occurrence of these fishes in Japanese rice paddies. The ricefishes' generic name *Oryzias* is based on the botanical name for the rice genus, *Oryza*. Other common names include buntingi in Indonesian Sulawesi, and medaka in Japan. Medaka is the Japanese word for killifish. Following the move of ricefishes from a close association with killifishes to the Beloniformes, the species known to many as the medaka *Oryzias latipes* is no longer considered a killifish. Despite the change in higher classification, *O. latipes* retains its status as perhaps the most widely used fish species in experimental embryology, genetics, and toxicology during the past century.

Ricefishes are found throughout Southeast Asia, and are widespread throughout the Japanese and Philippine archipelagos, but do not occur naturally on the island of Borneo. They occur on both sides of Wallace's Line, the imaginary boundary separating the Asian and Australian flora and fauna.

Ricefishes are distinguished by a number of specializations, such as a highly modified jaw and a jaw support that does not include numerous bones found generally in other fishes; a unique bone in the lower portion of the tail skeleton; and the dorsal portion of the gill arches with a greatly expanded supporting bone. Ricefishes are generally small to medium-sized fishes, ranging in standard length from 1.5 to 3 centimeters ($^3/_5$ to $1^1/_5$ inches) on average. Ricefish "giants" live in lakes of the Indonesian island group of Sulawesi (formerly the Celebes). Here, ricefishes, locally called buntingi, reach over 20 centimeters (8 inches) standard length and have elongate, duck-billed or beak-like upper and lower jaws. Species of three ricefish genera inhabit Sulawesi lakes; the endemic (only found there) genera are *Adrianichthys* and *Xenopoecilus,* and the more widespread *Oryzias*. All of the Sulawesi endemics are apparently threatened by the introduction of exotic species, such as the tilapia, for fisheries development.

Ricefishes are usually egg-laying; however, female medaka have been known to retain developing embryos, depositing fertilized eggs at or near the time of hatching. The carrying of fertilized eggs means that fertilization by males of females is internal. Females of the largest Sulawesi endemic species, *Xenopoecilus poptae*, have also been reported to expel fertilized eggs that hatch on impact with the water. In a closely related species, *Xenopoecilus oophorus*, females carry bunches of fertilized eggs in a hollow on their belly nearly to or until hatching. A functionally similar modification is present in the diminutive Indian ricefish *Horaichthys setnai*. This species reaches no more than 3 centimeters ($1^1/_5$ inches) standard length. Females are asymmetric; the right pelvic bone and fin rays are almost always absent. Absence of the fin provides a larger surface area to which the sperm bundle, passed from male to female via an elaborate gonopodium, can be attached. Other beloniform species, in addition to the killifishes already mentioned, exhibit internal fertilization and bear live young. Apparently, the ability to reproduce via internal fertilization is another characteristic found in most groups of atherinomorph fishes.

Lithopoecilus brouweri is a fossil species that has been placed in the Adrianichthyidae. It is from an undated, possibly freshwater, deposit in central Sulawesi. From the widespread distribution of ricefishes throughout coastal and fresh waters that border on a large portion of the central and eastern Indian and western Pacific oceans, however, we can infer that their evolutionary history has paralleled, or is somehow tied to, the evolution of the Indian and Pacific ocean basins. Ricefishes and other beloniforms can be estimated to be as old as their closest relatives, the killifishes. This means that recognizable, ancestral ricefishes had evolved at least by the time of the Pangean break-up 180 million years ago.

LYNNE R. PARENTI

SILVERSIDES & THEIR ALLIES

KEY FACTS

ORDER ATHERINIFORMES
• 7 families • *c.* 61 genera
• *c.* 285 species

SMALLEST & LARGEST

Bangkok minnow *Phenacostethus smithi*
Head–body length: 2 cm (¾ in)

Jacksmelt *Atherinopsis californiensis*
Head–body length: 44 cm (17½ in)

CONSERVATION WATCH
!!! 6 species are listed as critically endangered.
!! 5 species are endangered.
! 31 species are vulnerable.

T he order Atheriniformes is made up of seven families: Atherinidae (silversides), Dentatherinidae, Melanotaeniidae (rainbowfishes), Pseudomugilidae (blue-eyes), Telmatherinidae (sail-fin silversides), Phallostethidae (priapium fishes), and Notocheiridae (surf silversides). Many of these are small, elongate fishes and occur worldwide in tropical and temperate marine and fresh waters. They are generally schooling fishes and scatter their eggs—the eggs of marine species cling to the substrate or bottom, while those of freshwater species tend to cling to plants growing on the substrate. Typical distinguishing features include two separate dorsal fins (the first consisting of flexible spines), a single weak spine at the beginning of the anal fin, pelvic fins positioned on the abdomen, scales usually cycloid (they have a smooth margin), and no lateral line. The marine species are often a silvery color or at least have a bright silvery stripe down their sides. Some of the larger species are good for eating and smaller marine species are sometimes used as live bait in commercial tuna fishing operations. Several of the freshwater genera are popular aquarium fishes. The relationships of this group have been the subject of considerable controversy, but it is now generally accepted that it is allied to the Cyprinodontiformes (killifishes and their allies).

▼ *Silversides, also known as baitfishes, often form large dense schools near the surface over coral reefs.*

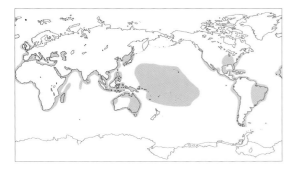

SILVERSIDES

The silverside family (Atherinidae) is the largest in the order, containing an estimated 35 genera and more than 160 species. Fossils of this group date back to the middle Eocene epoch, about 57 to 37 million years ago. The family currently has a wide distribution in shallow coastal seas around the world and also contains numerous freshwater species, particularly in the Americas and Australia.

The brook silverside *Labidesthes sicculus* is common in rivers and lakes of eastern North America. There are 18 species of *Chirostoma* on the western plateau of Mexico, and the family is also represented in South America, from high streams in the Andes to lowland waters of Brazil and Argentina.

The 19 species of the genus *Craterocephalus*, commonly known as hardyheads, are among the most numerous fishes in streams, lakes, and swamps of Australia and New Guinea. Five additional species in this genus are inhabitants of

G.E. Schmida

▲ *Restricted to a few river systems in Australia's top end, the strawman* Craterocephalus stramineus *is a member of the Atherinidae that shows considerable resemblances to the rainbowfishes, especially in its breeding habits.*

estuaries and coastal seas. The inland *Craterocephalus* species are believed to have evolved from marine ancestors that penetrated saline inland waters, eventually being isolated in the interior and gradually acclimatizing to fresh water. These species frequently have very limited distributions, being restricted to a localized watershed, a single river system, or even one small lake (for example, *Craterocephalus lacustris* of Lake Kutubu, Papua New Guinea).

Marine species in the genera *Atherion*, *Atherinomorus*, *Hypoatherina*, and *Stenatherina* of the tropical Indo-Pacific region often form dense schools containing several thousand individuals, which swim at or near the surface in the vicinity of

coral reefs. They offer a rich source of bait for tuna fishers, who catch them at night by attracting them into their nets with bright spotlights.

The Madagascar rainbowfishes of the genus *Bedotia* are currently classified as a very distinctive subfamily within the family Atherinidae, but future work may show they merit a separate family— Bedotiidae. Three species are presently known from fresh waters of eastern Madagascar, ranging from coastal lowlands to an elevation of about 500 meters (about 1,600 feet). However, it is highly probable that additional species may exist, mainly because freshwater habitats on this huge island are inadequately explored. These fishes bear a strong resemblance to the Australian rainbowfishes (family Melanotaeniidae). Although a few scientists have suggested a direct relationship between these families, it is more likely that the process of evolutionary convergence has molded their similar appearance and habits.

Interesting parallels exist between the fish faunas of Madagascar and Australia. True ("primary") freshwater species are largely absent in both areas; rather, inland fishes have evolved in relatively recent times (geologically speaking) from marine ancestors. Another small group of Madagascar freshwater fishes belong to the subfamily Rheoclesinae. These strange fishes have a goby-like head, but a body similar to *Bedotia*. The three known species in this group are presently classified in Atherinidae, but further study is necessary to determine their precise relationships.

THE DENTATHERINID
The family Dentatherinidae contains a single species, *Dentatherina merceri*, which inhabits

CALIFORNIA GRUNION

One of the best-known members of the family Atherinidae is the California grunion *Leuresthes tenuis*. This fish swims onto sandy beaches at night during spring and summer when the tides are highest. Males and females pair off to spawn; the female wriggles into the sand and lays her eggs while the male wraps himself around her and fertilizes the spawn. The parents then flip back into the water and struggle seaward into the breaking surf. The eggs remain buried in moist sand high on the beach until the next extreme high water period, about two weeks later. They then hatch within minutes of contact with water.

The spawning runs of the grunion relate to the phase of the moon; usually they occur two to six days after the new or full moon. During this period literally hundreds of fish ground themselves and are easily captured. Licensed recreational fishers are permitted to catch fish by hand at this time.

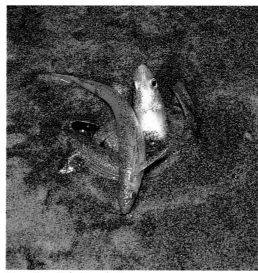

Waina Cheng/Oxford Scientific Films

◀ *A male grunion* Leuresthes tenuis *wraps himself around his mate at the moment of spawning on a moonlit beach at the peak of a high tide. This species occurs from about San Francisco south to the tip of Baja California; a similar species occurs in the Gulf of California.*

Max Gibbs/Oxford Scientific Films

shallow seas of the tropical western Pacific. This small, slender fish, to about 4 centimeters (1 1/2 inches) in length, has the general appearance of an atherinid, but possesses a number of distinctive skeletal characteristics.

RAINBOWFISHES

The colorful rainbowfishes of the family Melanotaeniidae are the second-largest family in the order Antheriniformes with 6 genera and 53 currently known species. They are entirely confined to the fresh waters of New Guinea and northern and/or eastern Australia. The past decade has seen a tremendous increase in the number of known species in this family as a direct result of scientific exploration of previously unsampled regions, particularly in New Guinea. Exciting new discoveries can be expected in the future, especially from the poorly known western half of the island (Irian Jaya). Most rainbowfishes reach a maximum length of under 12 centimeters (5 inches); a few dwarf species are less than half this size. Because of their small size, attractive appearance, and prolific breeding habits, rainbowfishes make excellent aquarium pets. The shape of the head, especially the jaws, is a useful character for separating genera.

The genera *Cairnsichthys* and *Rhadinocentrus*, each with only one single species, are endemic to eastern Australia. *Melanotaenia*, by far the largest genus with 32 species, is well represented in both New Guinea (20 species) and Australia (10 species). The unusual threadfin rainbow *Iriatherina werneri* is also shared between these two regions. The remaining genera, *Glossolepis*

▲ Inhabiting New Guinea and the tip of Cape York Peninsula, Australia, the threadfin rainbowfish Iriatherina werneri *is sexually dimorphic. This is a male; females are duller and lack the strikingly elongated plumes on the tail and fins.*

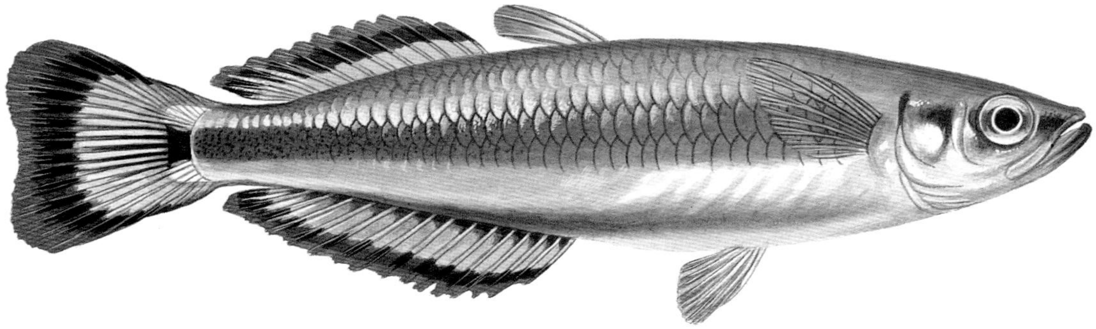

◀ Although the Australian rainbowfishes, such as Melanotaenia boesemani, *and Madagascar rainbowfishes, such as* Bedotia geayi, *bear some superficial resemblance, it is believed this similarity is the result of convergent evolution rather than reflecting a direct relationship. It is likely these two groups evolved from different marine ancestors rather than a single freshwater progenitor.*

G.E. Schmida

▲ *Among the most widespread of the blue-eyes, the Pacific or Cairn's blue-eye* Pseudomugil signifer *inhabits mangroves and coastal streams along the east coast of Australia, roughly from Sydney to Cooktown.*

and exhibit a similar pattern of spawning behavior. The genus *Kiunga* of the upper Fly River system in Papua New Guinea contains a single species. The genus *Scaturiginichthys* is a very recent addition to the family, having been discovered in 1990. Its sole representative, the diminutive S. *vermeilipinnis*, is known only from five spring-fed pools located on a cattle station in western Queensland, Australia. The other 13 species belong to the genus *Pseudomugil*.

SAIL-FIN SILVERSIDES

Sail-fin silversides of the family Telmatherinidae have a similar lifestyle and occupy much the same habitat as rainbowfishes and blue-eyes. The family contains 4 genera and 17 species, most of which were discovered quite recently. They are restricted to the Indonesian island of Sulawesi, except for the single species of the genus *Kalyptatherina* from brackish waters of Misool and Batanta, islands off the extreme western end of New Guinea. Most are inhabitants of the Malili Lakes in the central part of eastern Sulawesi. Geological evidence indicates that this half of the island was part of Australian Gondwana in pre-Jurassic times (before 200 million years ago). The sail-fin silversides are actually the first group of animals to provide confirmation of this former connection. Maximum size ranges from about 3 to 8 centimeters (about 1 to 3 inches). Males are very colorful and usually have an elevated, pennant-like first dorsal fin.

PRIAPIUM FISHES

The family Phallostethidae (including the former Neostethidae) contains about 20 species belonging to 9 genera. These tiny somewhat transparent fishes, less than 3 to 4 centimeters (1½ inches) long, inhabit fresh and brackish waters of Southeast Asia, including Thailand, the Malay Peninsula, Borneo, and the Philippines. Males of this family have a unique structure on the throat region, which is called the priapium. This muscular organ may possess tentacle-like projections. Together with the modified pelvic fins, it forms a clasping organ that is used to grasp the female during copulation. Fertilization is internal, unlike most fishes. The smallest known atheriniform fish, *Phenacostethus smithi*, belongs to this family.

and *Chilatherina*, are confined to northern New Guinea.

Rainbowfishes are very abundant in most freshwater habitats, including rivers, creeks, lakes, and swamps, below an elevation of 1,500 meters (about 5,000 feet). Spawning occurs all year round with local peaks of activity at the onset of the rainy season. A relatively small number of eggs are deposited each day among aquatic vegetation; hatching occurs within 7 to 18 days. Most species attain sexual maturity within the first year. Their diet consists of a variety of plant and animal items including algae, ants, aquatic insect larvae, and small crustaceans.

BLUE-EYES

Blue-eyes of the family Pseudomugilidae are close relatives of the rainbowfishes; in fact, until recently they were included in the same family. They have a very delicate appearance and most species are less than 4 to 5 centimeters (1½ to 2 inches) when fully grown. Blue-eyes inhabit fresh and brackish waters of Australia and New Guinea; they are usually found together with rainbowfishes

SURF SILVERSIDES

The family Notocheiridae contains six species, which are found in tropical and temperate seas. All belong to the genus *Iso*, except *Notocheirus hubbsi* from Chile. These small silver-striped fishes have a distinctive keel-like belly and are generally less than 10 centimeters (4 inches) long. They form huge schools in coastal waters, usually in the turbulent zone where breakers form—which is why they are sometimes called surf silversides.

GERALD R. ALLEN

OARFISHES & THEIR ALLIES

W ith common names as numerous and fanciful as unicornfishes, ribbonfishes, inkfishes, oarfishes, and tube-eyes, the order Lampridiformes is a curious assemblage of fishes. Perhaps the most colorful of all the fishes of the ocean, lampridiforms sport brilliant crimson fins and brightly colored bodies covered with white, brown, or red spots, or bands. Members of the order exhibit great extremes in form, from the ponderous and deep-bodied opah of the genus *Lampris*, to the longest of all bony fishes, the serpentine oarfish *Regalecus glesne*.

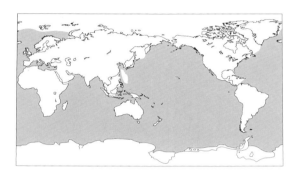

A NOVEL FEEDING MECHANISM

While bizarre and extremely divergent form and structure may belie their common ancestry, members of this order share several unique anatomical specializations, some of which are related to the evolution of a novel feeding mechanism. Unlike other fishes, the upper jaws (maxillae) of lampridiforms are not connected to the cheek bones (suspensorium), and the middle nasal cartilage is displaced at the back. These unusual modifications allow the upper jaw to be carried far forward when feeding. As a result, lampridiforms are capable of extreme jaw protrusion and expansion of the mouth cavity. This is as much as 40 times that of the closed mouth in one species!

The most generalized lampridiform family, the Veliferidae, contains two genera (*Velifer* and *Metavelifer*, each with a single species). They are moderate in size, up to 30 centimeters (12 inches), deep-bodied, and characterized by a scaly sheath at the base of the dorsal and anal fins into which anterior rays can be depressed. Opahs of the genus *Lampris* are strong swimmers, moving forward by

KEY FACTS

ORDER LAMPRIDIFORMES
• 7 families • 12 genera
• *c.* 21 species

SMALLEST & LARGEST

Tube-eye *Stylephorus chordatus*
Total length: 30 cm (12 in), not including the extremely long caudal fin rays

LONGEST Oarfish *Regalecus glesne*
Total length: average 5–8 m (16½–26¼ ft); record 17 m (55¾ ft)

HEAVIEST Opah *Lampris guttatus*
Total length: *c.* 1.25 m (4 ft)
Weight: 275 kg (606 lb)

ORDER ATELEOPODIFORMES
• 1 family • 4 genera
• *c.* 12 species

SMALLEST & LARGEST

Ateleopus natalensis
Total length: 60 cm (24 in)

Guentherus altivela
Total length: 2 m (6½ ft)
Weight: 4–6 kg (8–14 lb)

CONSERVATION WATCH
■ No species in these orders are listed as endangered, but most are considered rare.

Carlos Ivan Garces del Cid & Gerardo Garcia

◄ *Because of its bizarre appearance and extraordinary length, the oarfish Regalecus glesne is thought to be responsible for many sea serpent sightings. This small specimen (about 4.7 meters or 15 feet long) was struck by a boat's wheel near La Isla de San Marcos in Baja California, Mexico, on 28 December 1994. The island people reported that when they held their catch its color came off on their hands.*

Larry Madin/Planet Earth Pictures

▲ *Many lampridiform fishes, particularly ribbonfishes, have a highly distinctive larval phase quite unlike the adults in appearance. After metamorphosis, this larval scalloped ribbonfish Zu cristatus will display the bright red fins and silvery body typical of ribbonfishes.*

all lampridiform families, with approximately ten species arranged in three genera (*Zu, Desmodema,* and *Trachipterus*). The Trachipteridae and the Regalecidae (oarfishes) lack an anal fin, and have tiny, laterally projecting spines on each ray of the caudal and pelvic fins. In members of the family Regalecidae, each half of the pelvic fin has a single stout, elongate ray that is tipped with a blade-like swelling and sheathed in chemically sensitive skin. It is said that when oarfishes swim, their long pelvic rays are rotated like the oars of a row boat, hence their common name. However, folklore is not always based on fact; these appendages are thought to be used for taste perception, not locomotion. Oarfishes may generate electric currents—researchers in New Zealand have reported that a beach-cast *Agrostichthys parkeri* delivered repetitive pulsed shocks when touched.

One remaining family, the jellynose fishes (Ateleopodidae), has been allied with the Lampridiformes. Ateleopodids lack the anatomical traits that characterize the oarfishes, and are now believed to be close relatives of the dragonfishes and lizardfishes. Their exact classification is unknown, however. Ateleopodids are elongate, flabby fishes that may attain almost 2 meters (6½ feet) in length. Dark brown to black, they have a short dorsal fin near the head, a long anal fin continuous with the caudal fin, and a broad, gelatinous snout. These fishes are bottom dwellers in the deep ocean, but little else is known of their biology.

flapping large, muscular pectoral fins that are supported by a robust skeleton. Two species are recognized, both with fiery red fins, but one species does not have spots and is smaller. The deep-dwelling tube-eye *Stylephorus chordatus* (family Stylephoridae) has tubular eyes that are trained forward like binoculars. Extreme modifications of the front vertebrae allow the tube-eye to throw its skull up and back, and to protrude its tiny mouth forward while expanding a membranous pouch stretching between the skull and lower jaw. This remarkable increase in mouth volume creates powerful pipette-like suction, allowing *Stylephorus* to capture planktonic prey.

Members of the remaining lampridiform families are moderately to extremely long (all have more than 60 vertebrae), with many dorsal fin rays (about 100 to 400 in number). The inkfishes of the genera *Radiicephalus, Lophotus,* and *Eumecichthys* produce a black fluid from a tubular organ over the gut. Much like that of squids and octopods, the ink is presumably expelled as protection against predators. This does not occur in other bony fishes.

Unlike those of its relatives, the lateral line scales of the ribbonfishes (family Trachipteridae) bear stout spines, and the body surfaces are covered with a network of small lumps embedded in the skin. This family exhibits the greatest diversity of

REPRODUCTION

The large pelagic eggs of lampridiform fishes are about 2 to 6 millimeters (¹⁄₁₂ to ¼ inch) in diameter. While hardly rivaling the radiant hues of adults, the eggs are also brightly colored, usually in shades of amber, pink, or red. The eggs incubate for up to three weeks at the sea surface, so these colors may be a specialization to shield the embryo from the harmful rays of the tropical sun. Unlike the eggs of most bony fishes, which produce feeble larvae requiring a rich yolk sac for food, lampridiform embryos develop early, vigorously swimming and feeding on small planktonic crustaceans at hatching. Larvae are distinctive, with long, ornamented dorsal and pelvic fin rays, highly protrusible mouths, and dark pigment blotches arranged in species-specific patterns on the head and body. Little else is known about the young fishes, except that some species, especially ribbonfishes, undergo an abrupt metamorphosis before maturation.

DISTRIBUTION

The spinyfin and sailfin velifers (family Veliferidae) are only found in the Indian and Pacific Oceans, occurring in nearshore waters from 40 to 100 meters (130 to 330 feet) in depth from South Africa to Hawaii. These species are thought to be the closest living relatives of extinct, ancestral lampridiforms that probably lived in shallow seas during the Mesozoic era (245 to 65 million years ago). If true, then an important event in the evolution of the Lampridiformes was the exploitation of the deep ocean, for all other living examples are meso- and epipelagic fishes. Most lampridiforms are widely distributed in the temperate and tropical regions of all oceans except polar seas. One species, the rare southern opah *Lampris immaculatus,* occurs around the world in the subantarctic zone (below 45° S). Adults of some families, especially regalecids, lophotids, radiicephalids, and trachipterids, are rarely collected in scientific expeditions, and many distributional records result from strandings of dying individuals onto inhabited shorelines. Such occurrences are usually associated with strong storms at sea that probably injure, disorient, and radically displace individuals into inhospitable water masses. Some trachipterid species may exhibit somewhat restricted ranges. However, distributional information is sparse, and the family Trachipteridae, especially the genus *Trachipterus,* is in need of taxonomic revision.

FOOD, FEEDING, AND BEHAVIOR

Tales of ancient mariners, brief observations of captured or stressed individuals, and the food remains found in lampridiform specimens are the foundations for our scant knowledge of the general behavior and survival strategies of these extraordinary fishes. The elongate lampridiforms have extremely modified or reduced caudal fins, and swim slowly by continuous lateral undulations of the long dorsal fin. Sightings of giant oarfish lolling in the surface waters of the open ocean are probably the basis for unsettling accounts of sea monsters. This species normally inhabits depths of 20 to 200 meters (65 to 660 feet), where it feeds on invertebrates and fishes. Indeed, most lampridiforms consume krill, squid, and fishes as primary dietary choices. Some species well endowed with grasping jaw teeth, such as the crestfish *Lophotus lacepedei,* are probably more inclined toward other fishes as prey. The tube-eye *Stylephorus chordatus* resides in depths from 300 to 800 meters (985 to 2,625 feet),

and feeds exclusively on minute crustaceans, slurped through its cylindrical mouth. This diminutive species is known to undertake extensive daily vertical migrations of hundreds of meters, pursuing its prey. Although never observed in nature, the tube-eye probably assumes an oblique or vertical feeding position, with the head up and its long caudal filament serving as a sea anchor. Such conjecture is primarily based on observations of some other elongate lampridiform fishes (*Trachipterus* and *Zu*) trapped in large nets or placed in aquariums.

Because of their rarity, fragile nature, and large adult size, the secret lives of lampridiform fishes may never be fully revealed. While existing collections of preserved specimens may be sufficient to study their general form and evolutionary relationships, we can only speculate on the function and adaptive significance of their peculiarities. For the moment, we must wait for future observations of living lampridiforms in their natural environment and fresh examples for inspection of their delicate tissues.

JOHN E. OLNEY

▼ A popular food fish, worldwide in distribution, the unusual opah Lampris guttatus (below) bears little external resemblance to most members of the order Lampridiformes other than in the brilliant color of its fins. In parts of its range the ribbonfish Trachipterus altivelis (bottom) is known as king-of-the-salmon, from an native American legend claiming that these fishes lead Pacific salmon back to the rivers at the onset of the breeding season.

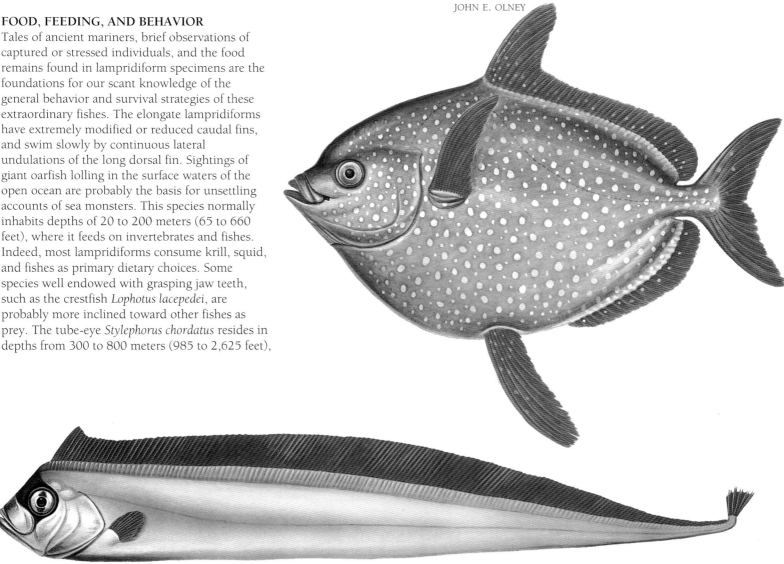

ORDER POLYMIXIIFORMES
• 1 family • 1 genus • 10 species

SMALLEST & LARGEST

Polymixia longispina
Total length: 18 cm (7 in)

Polymixia busakhini
Total length: 57 cm (23 in)

ORDER BERYCIFORMES
• 7 families • 29 genera
• *c.* 140 species

SMALLEST & LARGEST

Paratrachichthys pulsator
Total length: 8 cm (3 in)

Saber squirrelfish *Sargocentron
spiniferum*
Total length: 61 cm (24 in)

ORDER STEPHANOBERYCIFORMES
• 9 families • 27 genera
• *c.* 90 species

SMALLEST & LARGEST

Mosaic scale fishes (family
Megalomycteridae)
Total length: 4–7 cm (1½ to 2¾ in)

Rosenblatts whalefish
Gyrinomimus sp.
Total length: 45 cm (18 in)

CONSERVATION WATCH
■ None of these marine fishes are
listed as threatened.

► *The swallowtail nannygai Centroberyx
lineatus is widespread in the coastal
waters of southwestern Australia. Its tail is
long and forked, and it has the orange-
red color typical of its family, Berycidae.*

SQUIRRELFISHES & THEIR ALLIES

S quirrelfishes, whalefishes, and their allies are all marine dwellers, some living
in shallow waters but most in the deep sea. Almost all avoid bright sunlight,
either living permanently in the deep sea or sheltering in shallow-water caves or
deeper water in the daytime and becoming active at night. The classification of the
group has been much debated, with a recent split into three separate orders. The
order Polymixiiformes contains just one family of ten beardfish species. The seven
families of the order Beryciformes include squirrelfishes, alfoncinos and redfishes,
pineapple fishes, flashlight fishes, roughies, and fangtooths. Whalefishes, pricklefishes,
and their allies make up the order Stephanoberyciformes.

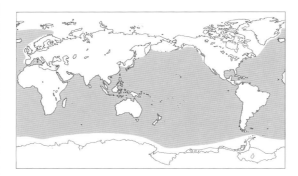

GENERAL CHARACTERISTICS
While strong fin spines are present in all of the
shallow-water members of this group, these have
been secondarily lost in some deepsea families.
The pelvic fins usually have more than five rays,
but in two deepsea families the entire fin has been
lost. Most of the shallow and slope species have
large eyes, while the bathypelagic whalefishes and
relatives have tiny eyes. The reproductive biology
has been studied in only a few species. The sexes

Kevin Deacon/AUSCAPE International

Marty Snyderman

are separate, sex reversal is unknown, and the eggs are fertilized outside the body.

POLYMIXIIFORMES

The ten species in the beardfish family Polymixiidae are placed in their own order, Polymixiiformes. They are now recognized as more generalized (with fewer specializations) than other fishes of the group. The family occurs in the tropical and subtropical waters of all major oceans. These silvery fishes, with a distinctive pair of barbels behind the chin (hence the common name), live on the outer continental shelf and upper slope at depths of between 20 and 800 meters (65 to 2,500 feet), usually on the bottom. With only one species longer than 40 centimeters (15¾ inches), beardfishes currently have little commercial importance and are typically seen only during exploratory fishing or as an accidental by-catch when fishing for some other target species.

BERYCIFORMES
Squirrelfishes

The squirrelfishes or soldierfishes, with some 75 species in 8 genera, belong to the order Beryciformes. They comprise the largest family (Holocentridae) in the group. Usually reddish in color, they have large eyes, strong fin spines, and heavy spinoid scales (with spines along their back edge), pelvic fins with one spine and seven rays, and a forked tail. Members of one of the two subfamilies have a long preopercular (cheek) spine.

Unlike the majority of species in the order, most squirrelfishes live in shallow water around tropical reefs. A few species are associated with rocky bottoms down to 200 meters (656 feet). The family is found in tropical waters around the world, with some species reaching subtropical waters. Squirrelfishes are active at night, sheltering in caves and under ledges during the day. They feed on small fishes and invertebrates. The maximum length of most species is 17 to 27 centimeters (6½ to 10½ inches), but one species exceeds 60 centimeters (23½ inches). While the adults are found in association with the bottom, the larvae have a long pelagic life and are often found far out to sea.

Alfoncinos and redfishes

The family Berycidae contains two genera with three species of alfoncinos and six species of redfishes or nannygais. These fishes are red in color and have a large mouth and eyes and spiny scales. They live on the continental shelf and slope in association with the bottom in temperate and tropical waters around the world. With the largest reaching 60 centimeters (23½ inches) in length, several species are commercially important.

Pineapple fishes

The three species of the family Monocentrididae are called pineapple, pinecone, knight, and coat-of-mail fishes, alluding to the yellow color and heavy, plate-like scales covering the body. They also have exceptionally stout spines in the dorsal and pelvic

▲ *The Hawaiian squirrelfish* Sargocentron xantherythrum *is widely distributed throughout the Hawaiian Islands and is most common at depths of 6 to15 meters (about 20 to 50 feet). When courting, male and female vocalize and hold their tails flat together, bending their heads away so that from above the pair somewhat resemble a capital letter Y.*

active at night when they use their light organs in feeding on large and small crustaceans and presumably for communication. In *Monocentris*, the light may be used to lure small prey, while *Cleidopus* uses the light to locate prey. The Australian pineapple fish can reach 30 centimeters (12 inches) total length, but the other species are smaller.

Flashlight fishes

The six genera and seven species of the flashlight fish family Anomalopidae all have a large light organ under the eye containing luminous bacteria. These fishes are gray or brown in color. They live on coral reefs and in deeper waters of the tropical Indian, Pacific, and western Atlantic oceans. All live in deeper water during the day. Some species move into the shallows on moonless nights, with at least one forming small schools. The light organs are used in feeding, communication among individuals, and to confuse predators.

As with all light organs utilizing luminous bacteria, some method of shading the continuously glowing organ is required to avoid predators. The

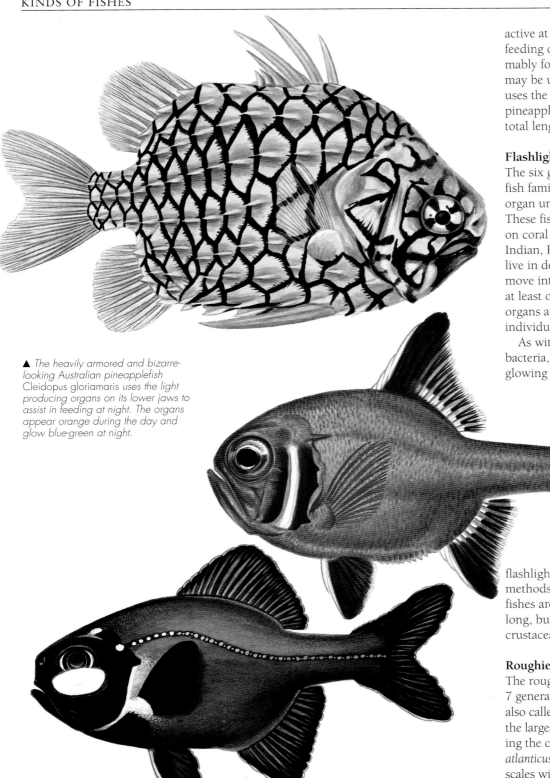

▲ *The heavily armored and bizarre-looking Australian pineapplefish* Cleidopus gloriamaris *uses the light producing organs on its lower jaws to assist in feeding at night. The organs appear orange during the day and glow blue-green at night.*

▲ *The large eyes of the southern roughy* Trachichthys australis *(above right) indicate that it is a noctural fish, hiding in caves during the day. The flashlight fish* Photoblepharon palpebratus *(above) turns its light organs on and off by covering them with flaps of skin. Situated beneath the fish's eyes, the bright light from these organs is produced by luminescent bacteria.*

flashlight fishes have evolved a surprising array of methods to shade the light. The largest flashlight fishes are 28 to 29 centimeters (about 11 inches) long, but most are only half that size. Small crustaceans are their primary food.

Roughies

The roughy family, Trachichthyidae, includes 7 genera and about 40 species, some of which are also called sawbellies or slimeheads. *Hoplostethus* is the largest genus, with more than 20 species, including the commercially important orange roughy *H. atlanticus*. Distinctive features are a row of enlarged scales with a posterior spine along the belly and mucous cavities separated by ridges of bone and covered with skin on the head. *Sorosichthys* and most species of *Paratrachichthys* have internal light organs. Roughies are bottom dwellers, most living on the outer shelf and slope down to at least 1,500 meters (4,920 feet). A few of the smaller Australian and New Zealand species (*Trachichthys* and *Optivus*) are shallow-water, inshore species.

Spinyfins

There are four species in three genera in the spinyfin family Diretmidae. These dark silvery

fins. Pineapple fishes are shallow-water fishes of inshore and continental shelf depths to 150 meters (490 feet) found in the tropical and temperate waters of the Indian and Pacific oceans. The Australian pineapple fish *Cleidopus gloriamaris* has a large light organ on either side of the lower jaw, while the two *Monocentris* species have a smaller light organ at the tip of the lower jaw. The light is produced by a colony of symbiotic luminous bacteria living in the organ. Pineapple fishes are fishes have a deep, compressed body, nearly vertical mouth, and spines both at the bases and along the axes of the finrays. Spinyfins are caught worldwide in both the midwaters and near the bottom of the deep sea down to 2,000 meters (6,560 feet). The maximum size is 37 centimeters (14½ inches).

Fangtooths

The fangtooth family Anoplogastridae includes one worldwide species living in the midwaters of the deep sea down to below 2,000 meters (6,560 feet). A second species in the same genus has recently been described. The adults are dark brown to black with a short, deep body, large mouth, and, as the common name suggests, very large teeth. The juveniles are very different, light gray in color with long head spines. The juveniles and adults were described as different species in the 1800s and even placed in different genera. Because juveniles do not transform to the adult stage until they are about 8 centimeters (3 inches) in length, only in 1955, as more deepsea specimens became available from exploration, was the true relationship discovered. Fangtooths eat crustaceans and small fishes. The maximum size is about 17 centimeters (6½ inches).

STEPHANOBERYCIFORMES
Pricklefishes and allies

Four families of obscure, small deepsea fishes, up to 20 centimeters (7 inches) in length, make up one part of the order Stephanoberyciformes. They differ from almost all members of the other two orders in the group in having small eyes and weak fin spines (or none in one species) and are similar to the roughies in having mucous cavities on the head. The pricklefishes (family Stephanoberycidae) include three genera of poorly known, bottom-dwelling fishes living at 1,000 meters (3,280 feet) and below. Two of the species have scales with fine spines. The bigscales (family Melamphaidae) are a large family, with 5 genera and about 35 species of midwater fishes living at 1,000 meters (3,280 feet) and below in almost all seas. As their name indicates, many have large, smooth scales and all have ridges of fine bone on the head. Gibberfishes (family Gibberichthyidae), with two species, have such distinctive larvae that they were originally described in a separate family. These midwater fishes live at depths to 1,000 meters (3,280 feet). The bristlyskin *Hispidoberyx ambagiosus* was described in a new family (Hispidoberycidae) in 1981 and is now known from only six specimens taken between 580 and 1,020 meters (1,900 to 3,350 feet) in the northeastern Indian Ocean and South China Sea. Similar to many of the fishes in the next grouping, the species has a very large mouth.

Whalefishes and allies

Five families of deepsea fishes make up this last group. All have small eyes, posteriorly placed dorsal and anal fins and no fin spines. All families have been taken in all major oceans. The flabby whalefishes (family Cetomimidae) is one of the largest families of deepsea fishes, with 9 genera and about 35 species. More than ten of the species have yet to be described. These whalefishes have a huge mouth, tiny eyes, and a very large lateral line; they lack external scales, pelvic fins, and ribs.

◄ *Until recently the bizarre and unmistakable fangtooth* Anoplogaster cornuta, *a worldwide deepsea species, was thought to be the sole member of its family, but a second species has lately been discovered. Despite its nightmare appearance, its maximum size is only about 17 centimeters (6½ inches).*

Bruce Robison

with small scales with a tiny central spine. This species appears to live in the midwater as a juvenile, but specimens 25 to 40 centimeters (10 to 16 inches) long are most often taken in bottom trawls. The two species of orangemouth whalefishes (family Rondeletiidae) are less than 15 centimeters (6 inches) long, orange-brown in color, and lack external scales. The last two families of this group were once placed in the oarfish order (Lampridiformes), but share a number of features with the whalefishes. The poorly known mosaic scale fishes of the family Megalomycteridae include perhaps three genera and six to seven species. They are also known as bignose fishes, referring to the very well-developed nasal organ typical of males of many deepsea fishes. All of the 20 bignose fishes tested are males, but they have not as yet been matched with their females, if they are known. The hairy fish *Mirapinna esau* is known only from one extraordinary 6-centimeter (2¼-inch) specimen with hairlike growths on the body, expansive pelvic and tail fins, and a small mouth. The tapetails, now placed in the same family Mirapinnidae, have a long caudal streamer when young that can be longer than the body of the fish. They feed on tiny crustaceans and none exceed 7 centimeters (2¾ inches) in length. None of the tapetails captured to date have been sexually mature. While it is thought that mature tapetails live in the deep sea, all the known specimens come from shallow waters.

JOHN R. PAXTON

▲ *The flabby whalefishes, family Cetomimidae, are characterized by huge mouths, tiny eyes, no external scales, no pelvic fins, no fin spines, very prominent lateral lines, and the fact that the dorsal and anal fins are situated very far back on the body.*

They are among the deepest dwelling of all midwater fishes, with centers of distribution at 1,500 meters (4,920 feet) and below. One species has been taken only between 2,500 and 3,500 meters (8,200 and 11,500 feet). Although one species exceeds 40 centimeters (15¾ inches) in length, and most reach 15 centimeters (6 inches), the few males that have been found are all less than 8 centimeters (3 inches) long. As with the deepsea anglerfishes, male flabby whalefishes appear to be dwarfs.

The only species of red velvet whalefish *Barbourisia rufa* is bright red in color and covered

ORANGE ROUGHY FISHERY

The orange roughy *Hoplostethus atlanticus* occurs worldwide in deepslope waters between 800 and 1,500 meters (2,625 and 4,920 feet). Exceptional commercial catches have been taken only around New Zealand and Australia, where over 50 million kilograms (110 million pounds) were caught annually for a number of years. Catches in the North Atlantic have been much smaller. In the south, huge schools aggregate for a few weeks every year in the same areas to spawn at depths from 800 to more than 1,000 meters (2,625 to 3,280 feet). The species is the largest in the roughy family, exceeding 55 centimeters in total length. It has proven very difficult to age, but most accept that the roughy is very long-lived—some results show the largest individuals reach 150 years in age. They are more than 30 years old when they reach sexual maturity and not all of the mature females breed every year. The very slow growth rates are presumably accompanied by low predation rates. These aspects of life history appear to be adaptations to the low level of food typical of most deepsea environments. Catch limits in Australia and New Zealand have been dramatically reduced to conserve the fishery.

▲ *A net full of orange roughy off New Zealand. The size of the catch indicates the density of the spawning schools.*

DORIES & THEIR ALLIES

Many of the genera of this interesting order (Zeiformes) include only a single species. Fossil records testify to the appearance of two families late in the Cretaceous period (145 to 65 million years ago). Their classificatory position between the Beryciformes (squirrelfishes and their allies) and Perciformes (perches and their allies) is still unclear, and there is still contention as to whether all the families of this order form a natural unit. About 30 percent of all zeiforms are considered to be rare and knowledge of their biology is very limited. The complete development from egg to the juvenile is known only for two species. Limited data exists for a few other species but even basic information on feeding, behavior, and distribution is scarce for most species.

ROUGH TO TOUCH

The bodies of zeiforms are mostly deep and compressed sideways; most have a large head, often with small spines or spiny ridges on the superficial bones, making them quite rough to touch. The eyes are located high up and close to the dorsal profile. The mouth cleft is always strongly oblique with characteristically highly protrusible jaws. Mouth size is variable, from mostly large to reasonably small. Small teeth are arranged in single series or rows. All zeiforms have long dorsal and anal fins preceded by spines. In the dorsal fins, spines and soft rays are separated from each other by a notch. The rounded pectoral fins are placed high on the body and the pelvic fins have one spine and five to seven soft rays, or seven to ten soft rays only. All soft rays are unbranched in the dorsal, anal, and pectoral fins. All genera except *Zenopsis* possess scales. (In the John dory *Zeus faber* the skin appears naked but there are minute scales). Scales are often ctenoid and rarely cycloid.

Zeiforms are usually small to medium-sized, occur in tropical and temperate areas of all oceans, and live in midwater or deepwater, generally above shelves, slopes, or seamounts down to depths of about 600 meters (2,000 feet). The Oreosomatidae is the only true deep-sea family, with some species living at 1,000 meters (3,300 feet).

DORIES

There are 7 genera and 13 species in the family Zeidae, or dories, of which the John dory *Zeus faber* and the mirror dory *Zenopsis conchifera* are the best known. Both have extremely compressed bodies with long and nearly straight fin spines. The membranes between the dorsal spines are extended to form filaments. Spiny scutes or large bony plates are found along the bases of the dorsal and anal fins, and on the breast and belly. Most are silvery, with juveniles having dark spots.

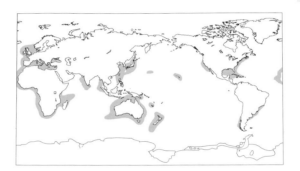

The John dory occurs in the eastern Atlantic, Mediterranean, and the western Pacific. It has been appreciated as an excellent food fish since Roman times and continues to be of commercial importance, although it is only caught in relatively small numbers.

▼ *Widespread in the tropical and temperate Atlantic, the common boarfish Capros aper lives near the bottom at about 40 to 600 meters (130 to 2,000 feet). It grows to about 16 centimeters (6 inches) in length and feeds mainly on small crustaceans.*

Kevin Deacon/AUSCAPE International

▲ The greatly compressed John dory Zeus faber offers a narrow head-on profile as it drifts up to unsuspecting prey and slurps them up with its highly protrusile mouth. A popular food fish itself, the origin of its common name is uncertain, but legend has it that the dark spots on its sides are the thumbprints of St Peter.

The John dory is a golden-greenish-gray color, with a dark blotch surrounded by a whitish ring on the center of each side. Legend has it that this mark is Saint Peter's thumbprint. It lives alone or in very small groups from close inshore to depths of 200 meters (660 feet)—rarely 400 meters (1,300 feet)—above sandy bottoms or weed-covered rocks. Although it is a poor swimmer the extreme mobility of the jaws makes it an efficient predator, feeding on herrings, sand eels, anchovies, pilchards, and occasionally crustaceans. In British waters males are seldom larger than 45 centimeters (18 inches); females attain on average a length of 66 centimeters (26 inches). They spawn inshore at depths of less than 100 meters (330 feet), usually during summer at higher latitudes, but during winter or spring in warmer seas. The large free-floating eggs measure about 2 millimeters (1/12 inch) in diameter. The newly hatched larva measures 4 millimeters (1/6 inch), and gradually attains the adult characters; by the time it is 19 millimeters (4/5 inch) in length it becomes recognizable as a young John dory. However, in another member of the Zeidae, the look-down dory Cyttus traversi, unlike adults, the juveniles

have some extremely extended, flexible fin spines and rays with leaf-like appendages.

OREOS

The family Oreosomatidae includes four genera and nine species. In 1829 the French zoologist G. Cuvier described a curious little fish and named it Oreosoma atlanticum because of the conical "hills" or cones on its body. (The genus name comes from the Greek: óros, "mountains", and soma, "body".) The second description of a member of the family Oreosomatidae, Cyttosoma boops, appeared as late as 1904—and it took another 70 years to discover that Cyttosoma is nothing but the adult form of Oreosoma!

The body shape of oreos varies from oval to rhomboidal, sometimes with a strongly concave predorsal profile. The broad head is dorsally flattened and the eye very large; in the oxeye dory Oreosoma atlanticum it is more than half of the head length. The scales are usually strongly ctenoid and firmly fixed, with a texture like that of a grater. Oreos attain sizes of 30 to 50 centimeters (12 to 20 inches). The warty dory Allocyttus verrucosus of South Africa, has a maximum length of 40 centimeters (15¾ inches).

Eggs have been described only for one species (*A. verrucosus*), but no larvae are known for any of the species. Oreosomatidae have so-called "prejuveniles" that look very different from the adults but are much older than any likely larval stage. In *Oreosoma* this prejuvenile grows to 10 centimeters (4 inches), and in another species even to 15 centimeters (6 inches). Following this stage an abrupt transformation occurs. The greatly expanded belly shrinks, the cones and leathery skin disappear, and scales develop. The eyes become huge and the colors (ranging from gray or greenish, to violet, but always shiny) change to the dark coloration of adults, consistent with a change from mid- to deepwater life.

Members of the family Oreosomatidae inhabit mainly the continental slope to about 1,000 meters (3,300 feet) depth, and feed on shrimps, mollusks, and fishes. Five species live along the southern coasts of the southern hemisphere, and two of these are common enough near New Zealand to support a deepwater fishery.

The prejuveniles of *Pseudocythus maculatus* are exceptional in distribution in that they have also been caught far southwards, between 63° and 68°S, in association with Antarctic krill.

FAMILY MACRUROCYTTIDAE
The family Macrurocyttidae includes three genera and five species. *Macrurocyttus acanthopodus* was described in 1934 from a single specimen deposited in the National Museum of Natural History in Washington D.C., USA. Much later, body parts of another five or six specimens of the same size were encountered in the mud at the bottom of the very same jar! Among these fragments were also found ovaries with few eggs of about 3 millimeters (⅛ inch) in diameter, huge compared to the size of the species of 43 millimeters (1⅔ inches).

BOARFISHES AND TINSELFISHES
The family Caproidae, or boarfishes, comprises 2 genera and 11 species and the family Grammicolepididae, or tinselfishes, has 2 genera and 2 species. Members of these two families prefer tropical regions. The species are generally small and deep-bodied, but are variable in shape, even within one species. Boarfishes are a bright orange to red in color. Tinselfishes are silvery, with linear, vertically elongated scales, a small head, and very weak fin spines.

FAMILY PARAZENIDAE
The orange-red *Parazen pacificus* is the only member of its family. It grows to a length of 25 centimeters (10 inches), and differs from all other zeiforms by having abdominal pelvic fins of seven rays which are soft. The species lives above the upper continental slope off Japan, but there are also records from all other oceans.

CHRISTINE KARRER & HANS-CHRISTIAN JOHN

▼ *A spotted oreo* Pseudocyttus maculatus *(family Oreosomatidae). Oreos look very like dories but have an extraordinary 'prejuvenile' stage differing from adults in appearance to such an extent that some were once described as a distinct species. Oreos are cosmopolitan but the greatest number of species occurs in the southern hemisphere.*

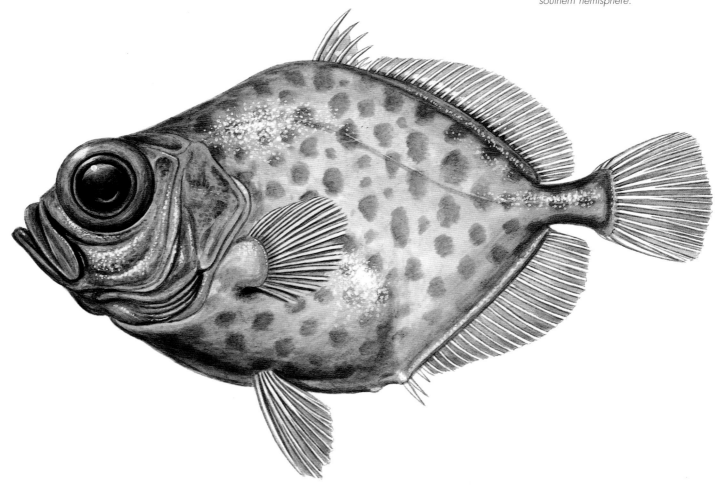

PIPEFISHES & THEIR ALLIES

Pipefishes and seahorses, ghost pipefishes, snipefishes, shrimpfishes, trumpetfishes, cornetfishes, seamoths, sticklebacks, tubesnouts, sand eels, and the paradox fish make up this diverse and often bizarre group of fishes. As most of their common names imply, they all have an elongate snout, a characteristic that makes them readily recognizable and gives them broad popular appeal. Because of this feature and a few other anatomical similarities, two quite distinct evolutionary lineages—suborders Syngnathoidei (pipefishes) and Gasterosteoidei (sticklebacks)—as well as a single puzzling species (*Indostomus paradoxus*, family Indostomidae), have been historically linked.

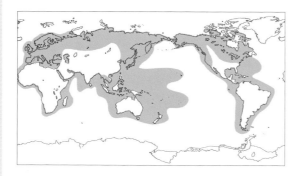

PIPEFISHES

The suborder Syngnathoidei is the most specialized and diverse in the order Gasterosteiformes and contains eight well-defined families.

Instantly identified by their elongate snouts and bodies encased in relatively thick bony armor, the pipefishes (family Syngnathidae) are found throughout warm and temperate marine waters of the world, having their greatest diversity in the Indo-West Pacific. The family is composed of about 55 genera and 200 species, exceeding by far the number of species in any other family of the order Gasterosteiformes. Pipefishes typically inhabit shallow waters near the shore, which have complex substrata such as eelgrass beds and coral reefs. However, members of this family are also known from such diverse habitats as the open ocean, in association with sargassum seaweed, and the inland streams of Central and South America.

Seahorses (contained in a single genus, *Hippocampus*) possess a prehensile tail and a head that, unlike typical pipefishes, is positioned more or less at a right angle to the body. These curled

▶ The harlequin ghost pipefish *Solenostomus paradoxus*. Both pipefishes and seahorses swim, not by flexing the body like most other fishes, but by rapid fanning movements of their fins. Their range and speed over distance is therefore restricted, but on the other hand they have exquisite control over their precise position in the water.

Kevin Deacon/AUSCAPE International

pipefishes have been the subject of mythology and folklore since ancient times when the creature was described as half fish and half horse. Aristotle (384 to 322 BC) was the first natural historian to describe the seahorse, which he said "breaks apart during the spawning season" to release the young. He was referring to the highly unusual reproductive strategy of all syngnathids, which entails an elaborate entwining courtship dance during which the female passes her eggs to the male who then fertilizes and broods them. In seahorses, the eggs are incubated inside a closed pouch on the male's belly; after hatching, the young are pumped out through a small hole at the front end of the pouch. Several other species of pipefishes also possess a fully enclosed brood pouch, but in some the eggs are placed in a partially closed pouch or they are simply attached beneath the tail.

In contrast, in the small but fantastic family of ghost pipefishes (Solenostomidae), the female broods the eggs in her greatly expanded pelvic fins. The family contains only a single genus (*Solenostomus*) and three species which are found in warm tropical water from Japan to South Africa. Although thought to be closely related to the family Syngnathidae, there are several structural

differences between the two families. Ghost pipefishes have pelvic fins, which are absent in syngnathids, star–shaped skin plates covering the body, and a prominent dorsal fin made up of numerous spines.

▲ *Closely related to seahorses and much resembling them, seadragons differ in their somewhat more conventional breeding habits. This is the leafy seadragon* Phycodorus eques.

◄ *Underwater male delivery. Seahorses are notable for their highly unusual breeding habits: the female lays her eggs in the male's pouch, where they are incubated and reared, and in due course expelled one at a time. Here a male* Hippocampus breviceps *is portrayed at the moment of "giving birth" in this way.*

The shrimpfishes or razorfishes (family Centriscidae) and the snipefishes (family Macrorhamphosidae) share detailed similarities in their anatomy. They also exhibit some striking differences. Two distinctive features characterize the shrimpfishes: a sharply pointed, horizontally oriented dorsal covering (in one genus topped with a small articulating spine), which looks and feels like a shrimp's rostrum or beak (hence their common name), and a sharp razor-like belly formed by the smooth lateral body plates (hence the alternative common name "razorfishes"). Shrimpfishes often orientate the razor's edge toward potential predators or competitors of their own species. They are subtly colored with spots and a horizontal stripe, a cryptic pattern useful for blending in with sea urchins with which they often associate. In contrast to the sleek shrimpfishes, snipefishes are deep-bodied and oval in cross-section, topped with a massive, erect dorsal spine. They are usually brilliantly banded in hues of red and violet and are found worldwide in deep, temperate to warm waters. Shrimpfishes are restricted in distribution to the shallow, tropical waters of the Indo-Pacific.

Trumpetfishes and cornetfishes make up two small families (Aulostomidae and Fistulariidae), each with a single genus. Members of these two families have extremely elongate bodies and snouts and are among the largest of the entire group, with trumpetfishes reaching up to 80 centimeters (32 inches) and cornetfishes up to about 2 meters (6 feet). They are active predators of fishes in tropical and subtropical waters around the world. Trumpetfishes hunt by ambush, hiding behind coral outcroppings or even behind the bodies of larger fishes in order to creep up on unsuspecting fishes. Cornetfishes are most often found in open areas, such as sand flats, where they attack passing

Carl Roessler

▲ Shrimpfish (genus Aeoliscus) are restricted to the Indo-Pacific region, where they can sometimes be found cast up on sandy beaches after storms. They usually congregate in schools, and habitually adopt a head-down vertical posture in the water.

▶ Common and widespread in the tropical Pacific, the spiny-back trumpetfish Aulostomus chinensis lies motionless in the water, looking very like a drifting yellowish stick, but it is galvanized into sudden action if a school of small fish should stray too close, when it dashes out to attack them.

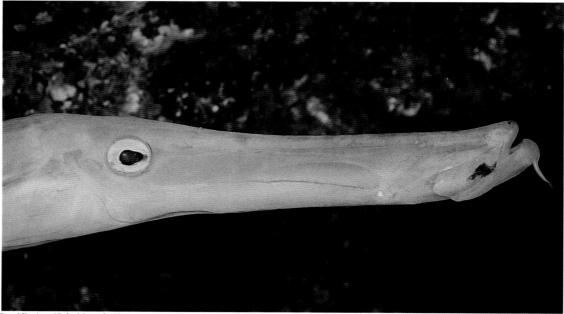

David Fleetham/Oxford Scientific Films

◄▲ Although all male seahorses present a bulging profile when brooding young in their pouch, the pot-bellied seahorse Hippocampus abdominalis (far left) of southern Australia has a permanently rounded abdomen. The Caribbean trumpetfish Aulostomus maculatus (top) often drifts alongside large herbivorous fish to sneak up on its prey. Its small, isolated dorsal spines are normally depressed as they are in this figure. Pipefishes, such as the banded pipefish Doryrhamphus dactyliophorus (near left), are similar to their close relatives the seahorses in that their bodies are encased in bony segments, they lack fin spines and pelvic fins, and it is the males that incubate the eggs.

system had long been uncertain until recent anatomical studies revealed its relationship to other pipefish families.

All syngnathoid families have specialized feeding modes related to their elongate snouts. Generally, individuals of the smaller species feed on invertebrates, while the larger species are voracious predators of larger prey such as fishes. Like most fishes, syngnathoids feed by sucking in water that contains prey but, because of their elongate snouts and other modifications in bone structure, they are capable of generating a pipette suction that is at once forceful and highly directional. Seamoths, the only family without strongly elongate snouts, are even more specialized, having the ability to form a tube-like mouth to slurp worms and other small aquatic creatures out of their burrows.

STICKLEBACKS

Sticklebacks and their allies (suborder Gasterosteoidei) are primarily a temperate group, ranging widely through the fresh waters of the northern hemisphere and in both the Atlantic and Pacific oceans. Credited with inspiring the aquarium hobby in Europe in the 1700s, the sticklebacks themselves (family Gasterosteidae) are familiar to all children of the northern hemisphere who have explored backyard ponds, streams, and marshes. The family includes five genera and seven species that are common in the coastal fresh waters of North America, Europe, and Asia. Their most striking characteristic is the strong isolated dorsal-fin spines, giving them the common name "sticklebacks".

Peter Scoones/Planet Earth Pictures

▲A seamoth, Eurypegasus draconis in the Red Sea. These bottom-dwellers have a body compressed from top to bottom, in contrast to the lateral compression of their relatives the shrimpfishes and seahorses, and derive their common name from their enlarged, somewhat mothlike pelvic fins.

schools of small fishes. Although superficially similar, trumpetfishes are compressed sideways and possess a series of small, isolated dorsal spines. Cornetfishes, on the other hand, are relatively flattened from top to bottom and have a long tail filament, often as long as the fish's body. Lined with sensory pores, this filament may be a long-range sensory system used to detect prey.

The five species of seamoths (family Pegasidae) are encased in thick bony plates, as are pipefishes and several other members of the group. Yet unlike all other members, seamoths are very flattened from top to bottom, have wing-like pectoral fins, and usually have a long rostrum jutting out in front of their highly specialized jaws. They are found in shallow, coastal tropical water, where they are closely associated with the sea bed. In fact, seamoths have been observed to "walk" across the bottom on their modified pelvic fins, using a specialized, articulated spine. The placement of this family in the classification

▶ *Flushed red in the breeding state, the male three-spined stickleback* Gasterosteus aculeatus *builds a nest of waterweed to which he seeks to attract as many females as possible to mate. Here he waits while a female examines his nest before consenting to spawn.*

Kim Taylor/Bruce Coleman Limited

David Thompson/Oxford Scientific Films

▲ *With eggs or young in his nest, the feisty male three-spined stickleback challenges all intruders onto his small territory, even an inoffensive watersnail.*

Another characteristic shared by the members of this family is both physiological and behavioral. The males construct nests of vegetation using secretions from their kidneys. They attract females to the nests and the females then lay their eggs for the males to guard. The highly ritualized behavior associated with this activity has been studied thoroughly since the turn of the century.

The tubesnouts (family Aulorhynchidae) include two genera, each having only one species, which are restricted to the marine coastal regions of the north Pacific Ocean. One genus, *Aulorhynchus*, occurs off the coast of North America, and the other, *Aulichthys*, is found in waters off Japan. The species are similar to each other, appearing like elongate sticklebacks with a long, thin body and a row of about 15 small spines in front of the soft dorsal fin. Details of their internal anatomy bear out the external similarities to sticklebacks. Sand eels (family Hypoptychidae) also occur only in marine waters off Japan. Little is known about their reproduction. Like the sticklebacks, these three species produce a sticky adhesive in their kidneys with which they attach their eggs to vegetation. *Aulorhynchus* attaches its eggs to kelp plants;

Hypoptychus, to sargassum seaweed. In contrast, *Aulichthys* deposits its eggs inside sea squirts. Despite the elongate snouts of many species, sticklebacks and their closely related families feed in ways typical of most fishes and not by the pipette suction mode of pipefishes. Their food usually consists of invertebrates and fish larvae.

THE PARADOX FISH

The smallest member of this assemblage of unusual fishes is aptly named for the uncertainty of its placement within the hierarchy of fishes. Discovered in 1926 in Lake Indawgyi, Burma, the paradox fish *Indostomus paradoxus* has been variously classified, but for this publication has been placed in the suborder Syngnathoidei. No definitive systematic position has been accepted by ichthyologists, but because it is externally similar to the pipefishes, the paradox fish is presently aligned with the pipefishes and their relatives. Except for its osteology (the structure and function of its bones), virtually nothing is known of the biology of this curious little fish. It has been observed feeding upon small invertebrates in aquariums. How it reproduces remains unknown.

JAMES WILDER ORR & THEODORE W. PIETSCH

SWAMPEELS & THEIR ALLIES

It is quite a sight for the uninitiated to encounter swampeels traveling over land after a tropical rainstorm. Swampeels are known to survive and remain active out of water for periods of up to six months, making them by far the most amphibious of living fishes. They thrive in even the most inhospitable deoxygenated waters by breathing atmospheric air. These primarily tropical and subtropical freshwater fishes are distributed widely in west Africa, Liberia, Asia, the Indo-Australian archipelago, Mexico, and Central and South America. Their order is called Synbranchiformes (fused gills) because the gill membranes from either side are united, producing a single gill opening or pore under the head or throat. Most swampeels begin life as females and remain so for up to four years when they reverse sex to male. Some males are born as male and remain smaller than those who were once female.

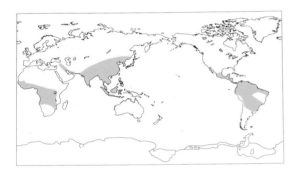

FINLESS AND AMPHIBIOUS FISHES

Swampeels resemble true eels superficially in body shape and have lost virtually all their fins. When a swampeel first hatches, however, it has huge pectoral fins, which propel a jet stream of oxygenated water from the surface along the length of the body. The larva's thin vascularized skin extracts the oxygen from this aquatic jet stream. As the larva grows, air-breathing organs begin to develop and the fish switches from water breathing to air breathing. About two weeks after hatching, it sheds the pectoral fins overnight. The dorsal and anal fins are nothing but rayless folds and the tail fin is either reduced or absent.

Finlessness may well be an adaptation for burrowing. Swampeels often live in habitats that undergo periodic droughts. When water levels drop, swampeels retreat into rather elaborate, deep burrows, in which some water remains and ensures the necessary high humidity. They dig such burrows by making corkscrew-like motions with their finless cylindrical bodies, which are mostly without scales. The fish remain in their burrows as long as water levels are low and emerge at the beginning of the rainy season.

During the wet season swampeels often engage in nocturnal migrations over land by undulating their bodies in a snake-like fashion. This amphibious life style is made possible by the development of air-breathing organs. The highly vascularized linings of the mouth and pharynx function as very efficient lungs. Consequently, swampeels can live in deoxygenated waters or on land for prolonged periods of time. Increased population density and dwindling resources are known to trigger swampeels to emerge from the waters and travel on land. But they do so only during wet conditions, thereby lending support to the hypothesis that the first land vertebrates evolved under similar warm tropical humid climates during the Devonian period (about 400 million years ago).

THE FOUR GENERA

The best-studied genus, *Monopterus*, is very widely distributed, from Liberia, India, China, and Japan to the Indo-Australian archipelago. It contains six species. The rice eel *Monopterus albus* is the most abundant in rivers, ponds, ditches, mudholes, and ricefields in the Malay archipelago and reaches a length of up to 90 centimeters (36 inches). The naked eel-like body is usually uniform and olive-brown, with the underside almost white. It has virtually lost its gills and, together with the Indian cuchia eel *Monopterus cuchia,* it is the most amphibious. *Monopterus* species are voracious predators and can consume their own body weight daily. Even though their eyes are quite small and covered with skin, they are very efficient crepuscular hunters and feed upon a great variety of larger aquatic animals. They locate prey by means of an acute sense of smell housed in a highly developed olfactory sac reminiscent of that of the true anguillid eels. Large *Monopterus* males have harems, and build and guard large floating bubble-nests in which

▲ *Tropical and subtropical in distribution, most swampeels live in stagnant pools and quiet river backwaters where they feed by night and usually conceal themselves in soft sediment by day; three species lack eyes and skin pigment and live in subterranean waters.*

▲ *Spiny eels have evolved a distinctive shape to suit their habit of burrowing into soft sand or mud. The white spotted spiny eel Mastacembelus armatus, popular in the aquarium trade, is an important food species in its native Asia.*

the eggs and larvae develop until the larvae begin to breathe air.

The second genus, *Synbranchus*, has four pairs of gill-bearing gill arches and breathes air with the entire lining of the mouth cavity and pharynx, which can distend like a balloon. *Synbranchus* contains two species inhabiting Mexico and Central and South America. They too make short sojourns on land and are voracious predators. The marbled eel *Synbranchus marmoratus* captures small prey by sucking them into the mouth, but large prey are dealt with very differently. *Synbranchus* seizes part of the body of the prey and twists it off by making corkscrew-like movements. In this way it can dislodge the legs of frogs or break up large prey.

The third genus, *Ophisternon* (one-gilled eels), resembles *Synbranchus* in many features. The six species are widely separated in their distribution, with two species in the New World and four in the Old World. The blind cave eel *Ophisternon candidum* lives in sink holes in northwestern Australia.

Finally, the genus *Macrotrema* may represent the most primitive condition among swampeels. It possesses fully developed gills and does not exhibit any special adaptations for air breathing. *Macrotrema*

caligans lives in both fresh and brackish waters in Thailand and the Malay Peninsula. It is the only swampeel that does not undergo sex reversal. Unlike other swampeels, *Macrotrema* is confined to a purely aquatic life. The unknown evolutionary origins of the swampeels may become clearer with a better knowledge of *Macrotrema*, since it still exhibits some ancestral features.

SPINY EELS

Some recent evidence indicates that spiny eels (suborder Mastacembeloidei) may be related to swampeels. They include about 75 species of Asian and African freshwater fishes in two families, Mastacembelidae and Chaudhuriidae. They have a posteriorly placed pectoral girdle that is not attached to the skull and an eel-like body adapted for burrowing into the sand or soft mud, sometimes for extended periods.

Unlike swampeels, spiny eels possess fins (including the isolated dorsal spines responsible for their common name), breathe aquatically, and do not change sex. Some African species have greatly reduced eyes.

KAREL F. LIEM

SCORPIONFISHES & THEIR ALLIES

The many species in this large order (Scorpaeniformes) share one unique feature—a bony strut across the cheek that connects bones under the eye with the front of the gill cover; the function is unclear, but in many species this serves as a site for head spines that may protect the eyes. Most scorpaeniforms have spiny heads. Body shape varies from pencil-shaped to bass-like to balloon-shaped. Most have rounded pectoral and tail fins. They are usually less than 30 centimeters (12 inches) in length. They occur worldwide, but are best represented in the northern hemisphere. Nearly all families are marine, with some sculpins and their allies being notable exceptions. Most live on the bottom in shallow water, but a few occur in the deep sea or in midwaters. They are predators, feeding on crustaceans and fishes. Some are egg-layers, while others have internal fertilization and bear live young. Evidence has been presented that these fishes are most closely related to some of the families within the Perciformes, but as the phylogeny of that order is not yet clarified, this book retains the Scorpaeniformes as a separate order.

SCORPAENIDS

The suborder Scorpaenoidea contains 7 to 12 families depending on the classification used. The large family Scorpaenidae has been divided into many subfamilies, and some of these have recently been treated as families. The subfamily Scorpaeninae contains about 100 species distributed primarily in tropical and temperate waters; many are reef species. They are important predators, feeding mostly on crustaceans and other fishes. While most species occur in shallow

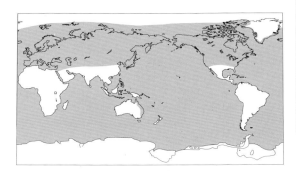

KEY FACTS

ORDER SCORPAENIFORMES
- *c.* 26 families • *c.* 275 genera
- *c.* 1,200 species

SMALLEST & LARGEST

Some velvetfishes (family Aploactinidae)
Total length: less than 2.5 cm (1 in)

LONGEST Lingcod *Ophiodon elongatus*
Total length: 152 cm (5 ft)
HEAVIEST Skilfish *Erilepis zonifer*
Weight: 91 kg (200 lb)

CONSERVATION WATCH
!!! The Boccacio rockfish *Sebastes paucispinus,* pygmy sculpin *Cottus pygmaeus,* and French sculpin *Cottus petiti* are listed as critically endangered.
!! The redfish *Sebastes fasciatus* is listed as endangered.
! The deepwater jack *Pontinus nigropunctatus* and 4 freshwater sculpins in the genus *Cottus* are listed as vulnerable.
■ Some of the Lake Baikal cottoids may be threatened by decreasing water quality.

◄ *Named for two black spots on its dorsal fin, the twin-spot firefish Dendrochirus biocellatus (family Scorpaenidae) is widespread in the western Pacific, from northern Australia to southern Japan. It feeds on crustaceans and small fishes, hunting by night and hiding by day.*

Kevin Deacon/Dive 2000

waters, the deepest living scorpaenids, the idiotfishes, (subfamily Sebastalobinae), are found on the bottom in depths to about 2,200 meters (7,000 feet) The only species occurring in oceanic midwaters belongs to the small subfamily Setarchinae; it shows typical midwater modifications, such as weak bones and soft flesh, and a thick fatty layer under the skin. The most important subfamily commercially is the Sebastinae, with over 100 species in the genus *Sebastes*—these are the rock cods of the North Pacific (wrongly called snappers locally) and the rosefishes of the North Atlantic that are important as commercial species for human consumption. A well-known subfamily (Pteroinae) contains about 18 species, including turkeyfishes and lionfishes (see page 180). Species of Choridactylinae and Inimicinae are coastal mud-bottom fishes that have several of the lowermost anterior pectoral rays as separate finger-like appendages which they use for creeping on the bottom. The stonefish subfamily (Synanceinae) contains about a dozen species, and the best-known four are the true stonefishes, genus *Synanceia*.

▼ *The scorpionfish* Scorpaena cardinalis *of New Zealand is common also in southern Australia, between the intertidal zone and a depth of about 100 meters (33 feet). It is valued by fishermen, but numerous venomous spines in the fins can inflict serious wounds if carelessly handled.*

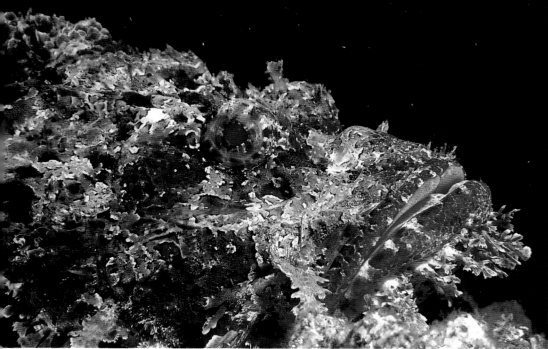

Carl Roessler

CORAL CROUCHERS

The coral crouchers (family Caracanthidae) are a small group with velvet-like skin. There are four or five species, and they have a compressed, oval body shape, small pectoral fins and tiny pelvic fins of one spine and three soft rays. These small fishes live among the branches of coral, especially of the genera *Acropora* and *Poecillopora*, in tropical regions of the Indo-Pacific. They apparently live only among coral. Their pectoral fins are used to wedge themselves between fingers of coral. They are almost never seen, except when coral is removed from the water and the fish fall out.

VELVETFISHES

Velvetfishes (family Aploactidae) are rare, small, benthic (bottom-dwelling) scorpaenid relatives with velvet-like skin similar to the coral crouchers. Most have a compressed body and a small fleshy pad under the head; in at least one species the pad is modified into a disk that is probably used to hold onto the substrate. All fin rays are unbranched, and the pelvic fins are small with a reduced number of rays (two or three). The body is usually covered with modified, isolated scales that form spinous points (giving a velvety feeling to the skin). They are found near the shore to a depth of about 100 meters (330 feet). About 40 species are known, many of which have been described only in the last 15 years. Some have venomous spines. As they are rarely seen, little is known of their biology.

PIGFISHES

The pigfishes or horsefishes (family Congiopodidae) comprise another small family (nine to ten species) restricted to cool south temperate and subantarctic seas of the southern hemisphere. They live on the bottom in depths to about 150 meters (500 feet). They have a projecting snout with a small mouth. They are scaleless, and the dorsal fin begins far forward. Pigfishes are occasionally taken in trawls but are otherwise rarely seen.

Two small, related families occur only in the temperate water of Australia: prowfishes (Pataecidae, 3 species) lack pelvic fins, have a long continuous dorsal fin that starts far forward on the head and is attached to the tail fin, and are scaleless. The red velvetfish (family Gnathanacanthidae) has pelvic fins, papillae in the skin, and a strongly notched dorsal fin.

SEA ROBINS

Gurnards or sea robins (subfamily Triglinae) and armored sea robins (subfamily Peristediinae) make up the family Triglidae; at least 120 species are included. They are bottom-dwelling fishes occurring primarily on continental shelves of tropical and temperate seas in depths to about 200 meters (660 feet). They are found mostly on sandy or muddy bottoms. Most are medium-sized, up to

WASPFISHES

The family Tetrarogidae is sometimes treated as a subfamily of the Scorpaenidae. They are known as waspfishes and are thought to be venomous. There are about 40 species in 15 genera, mostly confined to the Indo-West Pacific area. They occur on the bottom from shore to a depth of about 300 meters (980 feet). Some resemble members of the family Aploactidae (velvetfishes). Their dorsal fin begins on the head, and their scales are deeply embedded. Some species are very rare.

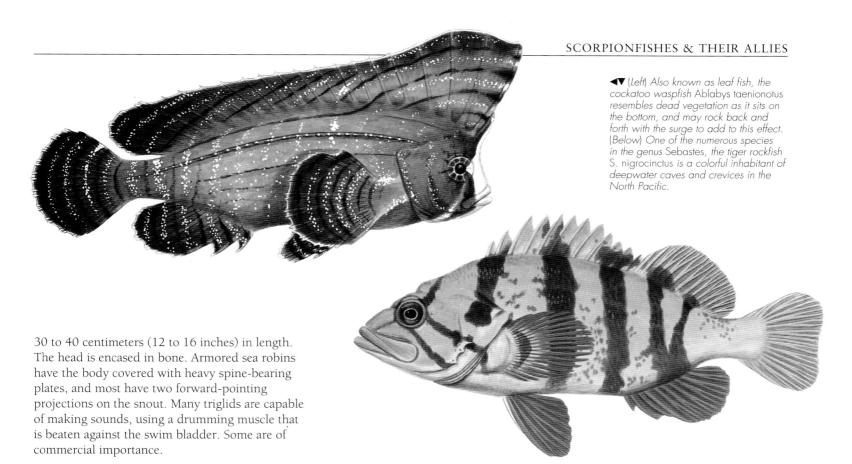

30 to 40 centimeters (12 to 16 inches) in length. The head is encased in bone. Armored sea robins have the body covered with heavy spine-bearing plates, and most have two forward-pointing projections on the snout. Many triglids are capable of making sounds, using a drumming muscle that is beaten against the swim bladder. Some are of commercial importance.

HELMET GURNARDS

Another armor-headed group (sometimes treated as its own order) are known as helmet gurnards (suborder Dactylopteriodea, family Dactylopteridae), now thought to be related to triglids (sharing with them a similar swim-bladder drumming muscle and a bi-lobed swim bladder). They were previously called flying gurnards because of their greatly extended pectoral fins. They are not capable of flight, however, and the newer name is helmet gurnards. They have bright colors on the underside of the pectoral fins, and it is possible that the flashing of such bright colors from a drab bottom-sitting fish might startle potential predators. Helmut gurnards have a long spine on the cheek (preopercle). Like triglids, they have the lowermost pectoral rays free from the remainder of the pectoral fin, but they form a unit and are not separate from each other. They can apparently walk on the bottom using their pelvic fins like legs. The body is covered with strong scales that are scute-like and form prominent rows. One or more anterior dorsal fin spines is free from the remainder of the fin. They live on the bottom at shallow to moderate depths. One species occurs in the Atlantic Ocean and seven in the Indo-Pacific. Some species have a long larval life and are often found far out at sea and in the stomach contents of oceanic fishes.

FLATHEADS

The suborder Platycephaloidea (families Bembridae, Platycephalidae, and Hoplichthyidae) contains 65 to 70 species that live on the bottom from inshore to moderate depths offshore (to 1,500 meters or 5,000 feet). Representatives of the suborder are found in tropical and temperate waters; there are no species in the New World. Their common name is flatheads. Deep-water flatheads (family Bembridae) are thought to be primitive members of the suborder; the head is only slightly depressed, and ridges on the head bear spines or serrations. They number about five species. The true flatheads (belonging to the family Platycephalidae) are more elongate, with a larger, duck-like mouth and flatter head. They number about 60 species in 12 genera. One group of species is found on reefs and rocky shores and another on mud and sand bottoms from shore to

Rudie H. Kuiter

David Doubilet

▲ *Waiting for hatching, a male cabezon guards his eggs. Largest of North American west coast sculpins, the cabezon Scorpaenichthys marmoratus (family Cottidae) is frequently caught by rock fisherman and supports a small-scale commercial fishing industry—the flesh makes very good eating. Cabezons feed mainly on mollusks and crustaceans.*

moderate depths of about 300 meters (990 feet). Many are good eating and they are an important food fish, especially in Australia. They are of moderate size, some reaching 70 centimeters (28 inches) or more. The spiny or ghost flatheads (family Hoplichthyidae) have a head so flattened that it looks as if it was run over by a steamroller; the body is scaleless and the lateral line is armed with spiny scutes. They are deeper living (to 1,500 meters or 5,000 feet), and most occur in either Japan or Australia, although new species are soon to be described.

COLD-WATER SPECIES
A small group of cold-water North Pacific species (suborders Anoplopomatoidei and Hexagrammoidei) contains two to four families, depending on the classification used. The sablefish and skilfish (family Anoplopomatidae) are large species, the skilfish reaching 183 centimeters (6 feet); they occur at moderate depths as adults. The sablefish supports commercial fisheries; adult fishes are known from 300 meters (990 feet) to as deep as 1,830 meters (5,950 feet) and it is a dominant species in deeper water. Greenlings and the lingcod (family Hexagrammidae, eight species) are inshore kelp and rock reef inhabitants. The

lingcod reaches 152 centimeters (5 feet) in length and is one of the most important recreational species off the Pacific coast of the United States. Combfishes (family Zaniolepididae) have sandpaper-like skin made up of tiny rough scales. They live on mud bottoms at 35 to 250 meters (100 to 800 feet).

SCULPINS
The second very large division in the order consists of sculpins and their relatives (suborder Cottoidei). Most species are in the sculpin family Cottidae, which is sometimes split into more families. Primarily North Pacific in distribution, the sculpin family has over 200 species. Most are found in shallow water, some at moderate depths, and a few to about 1,000 meters (3,300 feet); some occur in fresh water. They are a dominant tide-pool species, and most are under 10 centimeters (4 inches) in length. The cabezon (*Scorpaenichthys*) is a larger species, usually more than 46 centimeters (18 inches) and reaching 72 centimeters (29 inches) and 11 kilograms (24 pounds); it is important commercially and as a sport fish. It is good eating but its eggs are poisonous and will make humans violently ill. Homing has been demonstrated in some tidepool

species—when removed they can find their way back to their home tidepool. A few species occur in the southern hemisphere

The families Comephoridae and Abyssocottidae, consisting of about 26 species, and 6 cottids occur only in and near Lake Baikal, Russia. Some of these species are threatened because of habitat destruction.

FATHEADS

The family Psychrolutidae (including the Cottunculidae) are not easily distinguished from sculpins. Some are called fatheads because of their very large, fat, globular heads; the skin is loose in most species. Adults are bottom-dwelling and occur in depths of 100 to 1,600 meters (330 to 5,200 feet). There are about 30 species.

POACHERS

Poachers (family Agonidae) are mostly long, pencil-shaped fishes about 15 to 20 centimeters (6 to 8 inches) in size. They are covered with fused bony plates, and have a small mouth. They are benthic and marine and occur in the North Pacific, Arctic, and North Atlantic; a few species occur inshore, but most live at moderate depths to about 1,280 meters (4,200 feet). There are about 50 species, one occurring in the South Atlantic off the tip of South America. Most are brownish above, paler below. They are rarely seen by the public, but are most likely to be observed as part of the stomach contents of other fishes. The diet of most poachers is crustaceans.

SNAILFISHES AND LUMPFISHES

Snailfishes (family Liparidae) and lumpfishes (family Cyclopteridae) are mostly cold-water fishes found in the Arctic and neighboring Pacific and Atlantic and in the Antarctic Ocean. They have a sucking disk on the throat (involving the pelvic

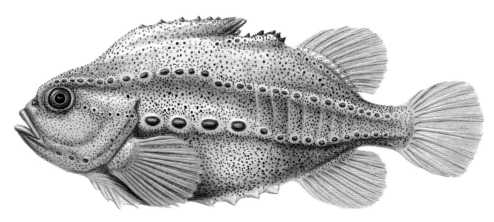

fins); the disk is reduced or lost in some deepsea species, especially those residing in midwaters. Shallow-living species use the disk to cling to the substrate in areas with heavy wave action. Snailfishes have a tadpole-like shape, with a large head and tapering body, and they lack scales. They occur in tide pools to very deep water (over 7,500 meters or 24,000 feet). The number of species is over 150, in about 15 genera. This group has been the most successful one in the order in colonizing and moving about in the deep sea, reaching hadel depths and including at least one deep pelagic species. The eggs are demersal and adhesive and fecundity (number of eggs) is low. The eggs of some species are deposited in the gill chambers of crabs, and mouth brooding may be involved in some species. Unlike most members of the order, liparids apparently have benthic larvae. Their cold-water ancestry and the absence of the pelagic larval stage probably explains the wide distribution of liparids in deep water. Lumpfishes are confined to the North Atlantic, Arctic, and North Pacific. Many are covered with conical plates or tubercles. They are usually found in rocky areas, clinging to the rocks with their disk.

WILLIAM N. ESCHMEYER

▲ *Lumpsucker* Cyclopterus lumpus *eggs, sold commercially as inexpensive caviar, are guarded by the male fish rather than the female. As the eggs are often laid on the edge of the shore at the low tide level, the male maintains his position on the rocks with pelvic fins that have been modified into a sucking disk.*

◀ *Most abundant of its genus, the red Irish lord* Hemilepidotus hemilepidotus *(family Cottidae) occurs from Kamchatka through Alaska to California, along rocky coasts from the intertidal zone to around 50 meters (160 feet).*

THE MOST DANGEROUS ONES: STONEFISHES & TURKEYFISHES

▼ *Several groups of fishes have members that are prone to hiding in shallow waters where unwary humans may tread on them and receive a wound from venomous spines, at best excruciatingly painful and at worst fatal, such as stingrays and stonefishes. Turkeyfishes, sometimes known as lionfishes or firefishes, (genus Dendrochirus, below), on the other hand are brightly colored, warning of their danger. Some are aggressive, known on occasion to approach and threaten humans in the water.*

Most people have heard of the deadly stonefishes (genus *Synanceia*), since they are the most venomous fishes, capable of causing death in humans. There were two species of stonefishes known until 1973 when two more, rarer species were described. Although there are less venomous "look-alikes" in the Atlantic and eastern Pacific, stonefishes proper are confined to the Indo-Pacific. Two species are widespread and two have very restricted ranges.

The venom glands of these dangerous species are located at the base of the fin spines, and the spines have grooves on each side, with a duct running to the spine tip. Very little is known about how the stonefishes use this protective feature in nature. They have no active method of delivering the venom—injection depends on pressing down the spine and forcing a drop of venom into the wound. If a predator (such as a skate or ray) is fast-acting, it might reject the stonefish, thereby saving its life, or the venom may kill the predator, or the predator may learn from the experience.

Stonefishes are rarely listed as the stomach contents of other species. But we do know that the venom is very potent for mammals, although this is obviously not why the venom evolved in the first place. Stonefishes have a habit of sitting on the bottom in shallow water. They are ambush predators and depend on camouflage; frequently they are covered with algae and other growths. The problem is that humans step on them. The venom is denatured by heat, so placing the foot or affected area in hot water will render the venom powerless. Stonefish antivenenes have been developed in Australia and some research has been done on the venoms, as powerful biological chemicals can be a source of possible pharmaceutical products.

Turkeyfishes and lionfishes differ from stonefishes in having a less powerful venom and no duct. Unlike stonefishes, which conceal themselves in mud or sand, turkeyfishes sometimes advertise their presence, although they are mostly active at night and live in caves and crevasses during the day. They will turn their dorsal fins towards an approaching object. Turkeyfishes are rarely reported as stomach contents except in other turkeyfishes; injection of their venom shows a much lower response in turkeyfishes than in other species (which the venom readily kills when they are experimentally injected).

Marty Snyderman

PERCHES & THEIR ALLIES

KEY FACTS

ORDER PERCIFORMES
• 15 suborders • *c.* 160 families
• *c.* 1,500 genera • *c.* 9,500 species

SMALLEST & LARGEST

Trimmatom nanus (a goby)
Total length: 1 cm (½ in)
* See gobies chapter, page 218.

Black marlin *Makaira indica*
Total length: 4.5 m (14 ft) +
Weight: up to 900 kg (2,000 lb)

CONSERVATION WATCH
!!! 59 species are listed as
critically endangered, including:
giant sea bass *Stereolepis gigas*;
apron *Zingel asper*; totoaba
Totoaba macdonaldi; and 38
species in the family Cichlidae.
!! 30 species are listed as
endangered.
! 138 species are listed as
vulnerable.

The composition and internal classification of the order Perciformes remain controversial. As currently recognized, it represents by far the largest and most diverse order of vertebrates. The name "perciform", meaning perch-like, is somewhat misleading, as many members of this diverse and probably unnatural assemblage bear no resemblance to the true perches (family Percidae), on which the name is based. Perciforms are found in almost every kind of aquatic habitat, from high-mountain freshwater streams to the deep ocean. Most are marine, but a few groups dominate the fish faunas of some tropical and subtropical fresh waters. The disparate size of perciforms is equally impressive, ranging from tiny gobies that mature at less than 1 centimeter (½ inch) to the great pelagic tunas and billfishes that grow to well over 4 meters (13 feet).

AN IMMENSE RADIATION

The evolutionary radiation of perciform fishes, which resulted in more than 150 different families, has produced an extraordinary array of forms; they range from deep-bodied, slab-sided fishes, like the spadefishes (family Ephippidae) and the moonfish (family Menidae), to elongate, eel-like fishes such as the eelpouts (family Zoarcidae), convict blennies (family Pholidichthyidae), and cutlassfishes (family Trichiuridae), with a wide variety of shapes in between. As might be expected, there is no single morphological feature or combination of features that characterizes perciform fishes, but certain generalizations can be made. Most perciforms have both spines and soft rays in the dorsal and anal fins. The pelvic fins occupy a position directly below or anterior to the pectoral fins and have one spine and five or fewer rays. The tail fin has 17 or fewer principal rays. Most perciforms have ctenoid (spiny) scales, although cycloid (smooth) scales are not uncommon and several groups lack scales altogether. The ctenoid scales of most perciforms differ from those of the more primitive beryciforms (squirrelfishes and their allies) in that the cteni (or teeth) are separate from, not fused to, the scale plate.

In the classification followed here, the order Perciformes comprises 15 suborders. Three of these—Labroidei, Blenniodei, and Gobioidei—are treated separately in subsequent chapters and are not considered further here.

PERCOIDS

The suborder Percoidei is the largest and most diverse of the perciform suborders, containing about 70 families and well over 2,500 species. Percoid classification has undergone considerable

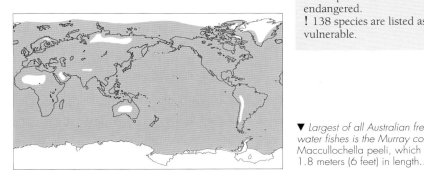

▼ *Largest of all Australian fresh-water fishes is the Murray cod Maccullochella peeli, which reaches 1.8 meters (6 feet) in length..*

G.E. Schmida

modification in the last two decades and continues to change. Like the order Perciformes, the suborder Percoidei is difficult to characterize, and there is no scientific evidence to defend it as a natural group. It has traditionally served as a convenient taxonomic repository for the more generalized perciform families but still encompasses extensive morphological and ecological diversity.

Most percoids inhabit the nearshore marine environment and dominate the coral reef fish fauna of tropical and subtropical seas. A few families are epipelagic (shallow oceanic) or mesopelagic (living in the twilight zone). Association with brackish water occurs in many nearshore marine families, some of which have one or more exclusively freshwater members, but only three families are primarily restricted to fresh waters—the north temperate perches (family Percidae) and sunfishes (family Centrarchidae), and the south temperate basses (family Percichthyidae, with one brackish water species).

Although many percoids have a generalized bass-like or perch-like form, extremes in body shape range from deep and flat to long and thin. Add to this the diversity in fin conformation, scales, jaw configuration, ornamentation of head bones, and internal skeletal features, and the Percoidei present a remarkably heterogeneous collection of fishes. There is yet another dimension of morphological diversity: many percoids have larval stages that are highly specialized for a planktonic existence, and thus the larvae may differ strikingly in form and structure from the adults. The morphology of these larvae can often provide clues to evolutionary relationships.

Percoid families range in size from a maximum of close to 400 species in the Serranidae to a single

species in several families (for example, the families Enoplosidae, Menidae, Pomatomidae). Approximately 50 percent of the species are contained in the five largest marine families: Serranidae (treated in a separate chapter on pages 197–201), Sciaenidae, Haemulidae, Apogonidae, and Lutjanidae.

Only selected percoid families, including several of the largest, are covered here. Adult serranids (sea basses and soapfishes) are treated in a separate chapter, but their larvae are worth mentioning here as an example of how the diversity of larval morphology may exceed that of the adults. Serranid larvae range from unspecialized forms, such as those of the subfamily Serraninae, to those of the Anthiinae, many of which have elaborate ornamentation on the head bones and spiny scales. The specialized morphology of larval anthiines has provided more information about the classification of genera and species than the less varied external and internal anatomy of the adults. Both larvae and adults of the subfamily Epinephelinae (groupers, soapfishes, and allies) are diverse in form, but a uniquely shared feature of the larvae demonstrates that this is a natural group; they all have an elongate dorsal spine supported internally in a distinctive way. The long spine ranges from sharp-pointed and saw-like in groupers to a fancifully ornamented filament many times the length of the body in the liopropomin basslets.

Drums and croakers

There are over 250 species and about 70 genera of drums and croakers (family Sciaenidae) distributed worldwide, mainly in shallow continental waters of the tropics and subtropics, with a few freshwater species. Distinguishing external features include two or fewer anal spines (most percoids have three), a distinct notch separating the spinous and soft portions of the dorsal fin, and lateral-line scales that extend to the end of the tail fin. Considerable diversity in feeding habits among sciaenids is reflected in a wide range of mouth size and form: from large and oblique in the fast-swimming, fish-eating sea trouts (genus *Cynoscion*) to small and inferior in many species, such as the black drum *Pogonias chromis*, which feeds on hard-shelled invertebrates that it crushes with enlarged molar-like teeth in the pharyngeal jaws.

The common names "drum" and "croaker" refer to the ability of sciaenids to produce sound by resonating the highly specialized swim bladder with special muscles. In association with this, there is considerable diversity in the form of the swim bladder and otoliths (ear bones) among sciaenids. Sciaenids are one of the most important families of food fishes around the world. Juveniles of many species utilize estuarine waters as nursery grounds.

▼ The jack knife fish Equetus lanceolatus (family Sciaenidae) differs from other drums in its striking shape, with a greatly elongated dorsal fin. Found only in the Caribbean, and growing to a length of about 50 centimeters (2 feet), this fish is nocturnal, hiding in coral caves by day.

Marty Snyderman

Kevin Deacon/Dive 2000

Cardinalfishes

The cardinalfishes (family Apogonidae) include close to 200 species in about 23 genera of mostly small, brightly colored, warm water fishes, primarily associated with coral reefs and lagoons. Some enter brackish and fresh waters. In Australia, New Guinea, and on several Pacific islands, a few species live exclusively in fresh water. Most cardinalfishes are recognizable by their distinctive body form—the mouth is large, the short spinous and soft dorsal fins are fully separated, and there is a relatively long caudal peduncle. There are only two anal spines. Most species are smaller than 10 centimeters (4 inches), although a few reach more than twice that size. Oral incubation of eggs (mouth brooding) by males has been documented for many species and may occur in all of them. Apogonids are mostly nocturnal and spend the day hovering in dark caves and crevices. A few species have a bacterial luminescent organ associated with the gut. Abdominal light organs also characterize some species of sweepers (Pempherididae), another family of small, nocturnal, reef-associated percoids. Although eaten in some areas, apogonids are not important food fishes.

Snappers and fusiliers

There are about 21 genera and 125 species of snappers and fusiliers (family Lutjanidae). Snappers are moderate to large (up to 1 meter or 39 inches); they are perch-like predators that are common in most shallow tropical and subtropical waters and are usually associated with the bottom. Most have a characteristic look in which the body is moderately compressed, the upper profile of the head slopes steeply downward, the upper jaw is

Peter Scoones/Planet Earth Pictures

▲ *Most of the cardinalfish (family Apogonidae) are nocturnal, and a few are notable for their transparent bodies, as these cardinalfishes of the genus Rhabdamia.*

◄ *Snappers in the Red Sea. The snappers (family Lutjanidae) are a large group of some 125 species, many of which are so similar that identification in life is difficult.*

183

D. Parer & E. Parer-Cook/AUSCAPE International

▲ *Diagonal-banded sweetlips*
Plectorhinchus goldmanni. Known also as
grunts, these fishes have loose, rubbery
lips well adapted for "vacuuming" the
sandy bottom for worms and molluscs.
They feed by day, so vigorously that silt
is swept back through the gill arches to
emerge as a distinct cloud in the water.

covered by the lacrimal bone when the mouth is closed, and there are conspicuous canine teeth in the jaws. Some of the deeper-living species (up to 450 meters or 1,500 feet) in the subfamily Etelinae lack the steep triangular head and have longer bodies. Snappers are important food fishes wherever they occur, although some are known to cause the tropical fish poisoning, ciguatera.

The fusiliers (subfamily Caesioninae) are smaller (up to 60 centimeters or 24 inches), free-swimming fishes with cylindrical, streamlined bodies and highly protrusible upper jaws adapted for picking plankton. Although adult fusiliers bear little external resemblance to snappers, the larvae of both are almost indistinguishable in form and head spination; other anatomical evidence confirms that fusiliers are plankton-eating derivatives of the more sedentary snappers.

Plankton-eaters

Specialization for planktivory is a common theme in percoid evolution. Numerous unrelated families of primarily bottom-associated percoids have given rise to small, cylindrical-bodied fishes with highly protrusible mouths that take advantage of the abundant food resource in the plankton. Because they appear so similar, several of these unrelated planktivores were once classified in a single family, the rovers (family Emmelichthyidae).

Subsequent comparisons of the protrusible jaws in these "rovers" revealed many structural differences, and further anatomical study clarified their true relationships. Only three of 10 genera remain in the Emmelichthyidae. The western Atlantic genus *Schultzea* was shown to be a plankton-eating serranid, and the Indo-Pacific *Dipterygonotus*, a fusilier. Two other western Atlantic planktivores, called bonnetmouths (*Inermia* and *Emmelichthyops*), are derived from the

grunts (family Haemulidae). With about 17 genera and 175 species, the grunts are a large circumtropical family of mainly benthic (bottom-dwelling) predators named for their ability to produce sound by grinding their pharyngeal teeth. Two other genera were shown to be centra-canthids, a small family of eastern Atlantic planktivores thought to be related to the porgies (family Sparidae); the porgies are another fairly large group (about 30 genera and 100 species) of tropical to temperate shallow water bottom-dwelling carnivores, notable particularly for the molarlike teeth possessed by most of them. Grunts and porgies were once thought to be closely allied to snappers, which they somewhat resemble, but there is no evidence supporting such a relationship.

Butterflyfishes and angelfishes

The butterflyfishes (family Chaetodontidae) are small (9 to 30 centimeters, or 3½ to 12 inches), conspicuous reef fishes distributed throughout tropical, subtropical, and warm temperate marine waters, the majority occurring in the Indo-Pacific. Their bright and varied color patterns have often inspired artists and designers. Many species have a false eye spot near the tail with the real eye concealed by a dark bar or other marking, presumably a pattern that could confuse would-be predators. The approximately 115 species in 10 genera have a similar disk-like shape with densely scaled dorsal and anal fins, but there is considerable diversity in size and shape of the jaws, correlated with different feeding modes. Some, such as the ornate butterflyfish *Chaetodon ornatissimus*, have short jaws used in nipping off the live coral polyps on which they subsist. Others, such as the long-nosed butterflyfish *Forcipiger flavissimus*, have elongate jaws used like forceps to pick small invertebrates from among sea urchin spines and coral crevices. The name "chaetodont" refers to the bristlelike teeth possessed by all members of the family.

Butterflyfishes have a specialized prolonged larval stage that may persist in the plankton for two or more months. All these larvae are characterized by having several head bones modified and expanded as thin bony plates; the considerable variation in form, extent, and ornamentation of these plates should provide important information about relationships among butterflyfishes.

The closest relatives of the butterflyfishes are the angelfishes (family Pomacanthidae, 9 genera and about 74 species), another circumtropical family of brightly colored, slab-sided reef fishes, characterized by a strong spine at the angle of the preopercle. The color pattern of juvenile angelfishes may differ strikingly from that of adults. The larvae are distinctive in having spiny scales and notably lack the bony head plates and variation in form of larval butterflyfishes.

Dottybacks

The approximately 130 species of dottybacks (family Pseudochromidae) occur around reefs throughout the tropical and subtropical Indo-Pacific. They are characterized by a long-based dorsal fin with only three or fewer spines and an interrupted or incomplete lateral line. Many are brightly colored and, because some exhibit pronounced sexual differences in color, males and females have sometimes been described as separate species. Although most dottybacks are more or less perch-like in form, members of the subfamily Congrogadinae (eel blennies) are very elongate and have been classified incorrectly with the blennioids. Dottybacks are among the smallest of percoids (most are less than 10 centimeters or 4 inches), with some species attaining sexual maturity at less than 2 centimeters (1 inch). Even smaller are certain species of *Lipogramma*, members of a closely related western Atlantic family, Grammatidae (basslets), which mature at less than 1.5 centimeters (³⁄₅ inch).

Another closely related marine family, the Indo-Pacific Plesiopidae (roundheads), are well represented in Australian waters, with such

familiar and brightly colored species as the hulafishes (*Trachinops*), devilfishes (*Paraplesiops*), and longfins (*Plesiops*). Another group of small Indo-West Pacific fishes, the spiny basslets, so called because of the numerous spines in the dorsal and anal fins, are now classified as a subfamily (Acanthoclininae) of the family

▲ *Meyer's butterflyfish* Chaetodon meyeri, *feeds exclusively on coral polyps.*

▼ *Angelfishes (Pomacanthidae) are notable for the striking differences in color and pattern between adults and juveniles of the same species. These are adult French angelfishes* Pomacanthus paru; *juveniles are black with five bold vertical bands of golden yellow.*

Marty Snyderman

Plesiopidae. The elongate bearded snakeblennies (family Notograptidae) of Australia and New Guinea were once thought to be blennioids, but recent studies suggest that they may be most closely related to spiny basslets, with which they share high numbers of dorsal and anal fin spines. Many of the brightly colored dottyback, basslet, and roundhead species are popular aquarium fishes.

Kev in Deacon/Dive 2000

Ponyfishes

The slipmouths or ponyfishes (family Leiognathidae) are named for their highly protrusible mouth which, depending on the genus, can be extended straight forward, forward and down, or forward and up. There are 3 genera and about 25 species of these small, silvery, laterally compressed fishes found in marine and brackish waters of the Indo-West Pacific. Ponyfishes have bacterial light organs in the throat region from which light is projected through the belly. They also have a spine-locking mechanism in the dorsal and anal fins that is analogous to, but mechanically different from, that of surgeonfishes. Whereas surgeonfishes lock the basal grooves in one spine into a ratchet on the first fin support, the ratchet mechanism in ponyfishes works by interlocking the basal grooves in several adjacent spines with one another. The spines are disengaged with a muscle that pulls the fin forward. Mojarras or silver biddies (family Gerreidae) are also silvery compressed fishes with a highly protrusible mouth and consequently were once classified with ponyfishes. More detailed comparisons of their osteology, including the structure of the jaws, has shown the similarities to be convergent (not resulting from a common ancestor).

▼ A southern Australian representative of a world-wide group, a blue-spotted goatfish Upeneichthys vlamingii (family Mullidae) uses its long sensory barbels to feel for food.

Bill & Peter Boyle/AUSCAPE International

Goatfishes

The approximately 6 genera and 60 species of goatfishes (family Mullidae) occur around the world in tropical to temperate shallow marine waters and are highly prized as food. They are found around reefs and on sandy or muddy bottoms. Goatfishes are relatively elongate with two separate dorsal fins and are often brightly colored. Their most distinctive feature is a pair of long chin barbels with which they probe the bottom in search of small invertebrates or other food. The slender, silvery postlarva lives in surface waters and may reach 5 to 6 centimeters (2 to 2½ inches) before settling to the adult habitat.

Snooks and giant perches

Until recently, the family Centropomidae included the neotropical snooks (12 *Centropomus* species) and the African and Indo-Pacific giant perches (9 *Lates* species and 1 each of *Psammoperca* and *Hypopterus*). The giant perches do not appear to be closely related to the snooks, and have now been assigned to a separate family, Latidae. Members of both families are found in coastal marine, estuarine, and fresh waters in tropical and subtropical regions. They are characterized by a relatively high, triangular spinous dorsal fin and a lateral line that extends well onto the tail (some species have three sensory lines on the tail), though neither feature is unique among percoids. Centropomids range in size from about 35 to 120 centimeters (about 14 inches to 4 feet) and latids range from about 20 centimeters to 2 meters (about 8 inches to 6½ feet). Many of the larger species, such as the barramundi *Lates calcarifer*, are popular sport fishes. For this very reason, the Nile perch *Lates niloticus* has been translocated within Africa; in some cases this has had disastrous consequences (see cichlids chapter).

The Asiatic glassfishes (family Ambassidae) were also once classified with the centropomids, but there is no convincing evidence to support such a relationship. The small, compressed glassfishes superficially resemble cardinalfishes and, like them, are popular with aquarists. Some ichthyologists have proposed that the closest relatives of centropomids are the north temperate basses (family Moronidae), which also have three lateral lines on the tail, but no features unique to these two families have yet been found.

Freshwater perches

With over 160 species, the North American and Eurasian perches (family Percidae) are the largest freshwater percoid family, and only four marine families are larger. Percids are not easily characterized but, like the apogonids and sciaenids, they are one of only a few percoid families with fewer than three anal spines. The North American darters (genus *Etheostomatini*) constitute the bulk of the family, with about 150 species of small bottom-dwelling, goby-like fishes

The glassfishes (family Ambassidae) are a group of small freshwater species popular among aquarium enthusiasts. One of these, the glass perch Chanda ranga inhabits brackish lowland rivers of India and Burma. Seldom exceeding about 5 centimeters (2 inches) in length, they are sociable and well-behaved with most other species in an aquarium, though they are sensitive to change and will feed only on live animal food such as Cyclops, Daphnia, or mosquito larvae.

that live mainly in the rapids of streams. They are small, from 4 to 20 centimeters (1½ to 8 inches), with most less than 10 centimeters (4 inches). The males of most species are brightly colored during the breeding season. Because they may have highly restricted distributions, many darter species are threatened or endangered. The snail darter *Percina tanasi*, for example, was made famous in the 1970s by debate over the completion of the Telico Dam and the resulting inundation of the Little Tennessee River, USA, then believed to support the only population of the snail darter.

Among the larger percids are popular angling fishes that have been introduced to other parts of the world; the redfin perch *Perca fluviatilis* is one such introduction to temperate Australia. Accidental introductions include the recent appearance of the ruffe *Gymnocephalus cerruus* in the Great Lakes of North America, presumably from the ballast water of ships from Eurasia.

Sunfishes

The approximately 8 genera and 30 species of sunfishes (family Centrarchidae) occur in North American temperate fresh waters. Their most distinctive characteristic is behavioral. Males of most species construct circular pits of gravel or vegetation in which they care for the eggs and young. Sunfishes are similar in form and range in size from the 8-centimeter (3-inch) blackbanded sunfish *Enneacanthus chaetodon* to the 1-meter (39-inch) largemouth bass *Micropterus salmoides*. The larger sunfishes are popular angling species, many of which have had their ranges extended by translocation throughout North America and introduction to other temperate parts of the world, such as Europe and Japan. The six species of the diminutive (3 to 5 centimeters or 1¼ to 2 inches) North American pygmy sunfishes (genus *Elassoma*) were once included in this family, but studies show that they are not perciforms and may be allied to atherinomorphs and their relatives.

The perch Perca fluviatilis (Percidae) is native to European freshwaters, where it reaches a length of 50 centimeters (20 inches). This species has also been introduced to Australia.

▲ *Commensal medusafish of the genus Nomeus (family Centrolophidae) live in close association with jellyfishes, especially the giant medusa Physalia. They feed on other fishes killed by the medusa's stinging cells, or nematocysts, but it is not clear how they avoid being stung themselves.*

Temperate basses

Some scientists have included other marine and northern hemisphere freshwater fishes in the temperate bass family (Percichthyidae), but they are currently restricted to about 11 genera and 25 mainly freshwater species from temperate Australia and South America. The classificatory history of the Australian percichthyid genera is another example of the confused state of percoid classification. The pygmy perches (*Edelia*, *Nannoperca*, and *Nannatherina*) have been incorrectly classified with the aholeholes and flagtails (Kuhliidae), a small family of tropical Indo-West Pacific marine to freshwater species. The river blackfishes (*Gadopsis*) have been treated as a separate family allied variously with blennies, weeverfishes, and ophidiiforms; they have even been placed in their own order, Gadopsiformes.

Recent anatomical studies indicate that these elongate fishes are specialized percichthyids, most closely related to the Western Australian nightfish (genus *Bostockia*). Percichthyids range in size from the 8-centimeter (3-inch) pygmy perches to the 180-centimeter (6-foot) Murray cod *Maccullochella peeli*, the largest of Australia's freshwater fishes. Many of the larger Australian species are popular angling fishes. Pygmy perches are often highly colored and make ideal aquarium fishes.

Archerfishes

The six species of archerfishes (family Toxotidae) are aptly named for their ability to spit a jet of water at insects and other prey, knocking them into the water from overhanging vegetation. This is accomplished with rapid compression of the oral cavity, which forces water through a narrow tube formed by pressing the tongue against the specially grooved palate. The aim is very accurate at distances of at least a meter (about 3 feet). Archerfishes are distributed throughout the Indo-Australian region where they typically occur in brackish coastal waters, frequently in association with mangroves; some species penetrate well into the upper reaches of freshwater streams. Although most species are small (less than 15 centimeters or 6 inches), the common archerfish *Toxotes chatareus* reaches 40 centimeters (16 inches) and is angled and eaten in some areas.

Nurseryfishes

The nurseryfishes (family Kurtidae) are named for their unusual brooding behavior. Males carry the eggs in a bi-lobed cluster looped around a prominent bony hook that projects forward on top of the head as part of the supraoccipital bone. Nurseryfishes have a large upturned mouth and compressed body with a long-based anal fin and single short-based dorsal fin in which the few very short spines are immovably fixed to their internal supports. The two species live in murky fresh and brackish waters of Indo-Malaysia (*Kurtus indicus*) and New Guinea and northwestern Australia (*K. gulliveri*).

BUTTERFISHES AND THEIR ALLIES

Butterfishes and their relatives (suborder Stromateoidei) have a distinctive head and snout. According to one expert, "It is a fat-nosed, wide-eyed, stuffed-up look, smug and at the same time apprehensive." The six families, butterfishes (Stromateidae), driftfishes (Nomeidae), medusafishes (Centrolophidae), ariommatids (Ariommatidae), squaretails (Tetragonuridae), and amarsipids (Amarsipidae), encompass about 16 genera and 65 species that occur worldwide in tropical and temperate coastal and oceanic waters. All but one of the six families (Amarsipidae) have toothed pharyngeal sacs in the gullet behind the last gill arch, presumed to function in the shredding and processing of food. Body shape ranges from relatively slender to deep and compressed, fin spines are often feeble, scales tend to be weak and easily shed, and the skin is covered with tiny pores that open into a mucous canal system beneath the skin. Young stromateoids are pelagic and are often found under floating objects or seaweed. Many associate with jellyfishes, which apparently not only provide protection from predators but also serve as a source of food.

JACKS AND THEIR ALLIES

There are five families in the suborder Carangoidei, united by their possession of small, adherent cycloid scales and an unusual anterior extension of the nasal sensory canal on the snout. The largest family, Carangidae (jacks, scads, and pompanos), contains about 140 species in 32 genera, distributed worldwide in tropical to temperate seas. Many are important food fishes. Although they range in body shape from very deep and compressed to fusiform (spindle-shaped), carangids are usually recognizable by having the first two anal spines isolated in front of the fin. Lateral-line scales are modified into large, keeled scutes in many, and some have detached finlets behind the dorsal and anal fins, similar to those in tunas and mackerels.

The most unusual of the carangoid fishes are the remoras (family Echeneidae), in which the spinous dorsal fin has migrated forward and has been transformed into a flat sucking disk. Ridges on the disk operate like slats in a venetian blind to generate a vacuum that remoras use to attach themselves to sharks, other bony fishes, sea turtles, and marine mammals.

The closest relatives of the remoras are two popular open-water food and game fishes, the cobia (family Rachycentridae) and dolphins (two species, family Coryphaenidae), both found around the world in warm seas. Dolphins are well known for their beautiful colors. There is little

▼ *Jacks (family Carangidae) circling a shimmering school of anchovies in the Coral Sea. Large individuals of some species of this family sometimes contain ciguatera toxin, which can cause severe food poisoning.*

Carl Roessler

▲ *The importance of studying the larval phases of fishes is highlighted when comparing the dolphin Coyphaena hippurus (left) to the cobia (center) and remora (far right). Although the similarity between the latter two fishes is evident and the dolphin appears to be unlike either, the larva of the dolphin bears a resemblance to that of the cobia, suggesting a relationship. The slender remora Echeneis naucrates bears a striking resemblance to a juvenile cobia Rachycentron canadum, with the exception that the remora's first dorsal fin has been modified into a sucking disk. When the disk is held against the body of a shark or other large marine animal and the ridges in its center raised, the remora creates suction to attach itself for a free ride and ready access to scraps from its host's meals. The word "remora" is Latin for "hindrance", for it was once believed these fishes slowed ships down by attaching in large numbers.*

superficial resemblance between adult cobia and dolphins, but their larvae are remarkably similar in body form and head spination and their bodies and eyes are covered by minute crown-shaped spicules, a unique condition suggesting a close relationship between the two families.

TUNAS AND THEIR ALLIES

As traditionally classified, the suborder Scombroidei comprises six families of oceanic fishes, including some of the most important species in commercial and recreational marine fisheries.

Snake mackerels and cutlassfishes

The snake mackerels (Gempylidae), with 22 species in 16 genera, are swift, oblong to elongate predators with large, fang-like teeth and reduced pelvic fins. They occur worldwide in relatively deep waters. Two species, the escolar and oilfish, are known for their extremely oily flesh, which has purgative properties. The cutlassfishes (Trichiuridae) are closely related to the snake mackerels and probably should be included within that family. They have very elongate, ribbon-like bodies, usually with a silvery metallic sheen, and large, fang-like jaw teeth. Both the pelvic and tail fins are reduced or absent, and cutlassfishes swim much like eels. There are 9 genera and at least 18 species occurring in deep shelf and slope waters of all tropical and temperate seas.

Tunas and mackerels

The tunas and mackerels (family Scombridae) are probably some of the best known fishes in the sea, both from a popular and scientific standpoint. Because of their value in commercial fisheries and their remarkable adaptations as oceanic predators, the anatomy, physiology, and life history of scombrids have been studied extensively for many decades. With 15 genera and 49 species, they are

found worldwide in tropical to temperate waters. Some species are restricted to coastal waters, while others roam the open ocean.

Scombrids have a characteristic fusiform body, beak-like non-protrusible upper jaws, a separate spinous dorsal fin, and a series of separate finlets following the second dorsal and anal fins. They are literally "swimming machines", with every aspect of their external morphology designed for maximum hydrodynamic efficiency, from the streamlined body shape to the grooves and depressions into which various fins can be tucked away to reduce turbulence. The tail fin is stiff and sickle-shaped for maximum thrust at high speeds. Scombrids have a high proportion of red muscle, the type that is metabolically suited to continuous activity, and they depend on continuous swimming to ventilate the gills. Unlike most other fishes, tunas further increase performance efficiency by maintaining a body temperature higher than that of the surrounding water, using a complex countercurrent circulation of blood. The speed and stamina of tunas are demonstrated in their unparalleled long-distance migrations that have been documented by tag and recapture studies. In one such recapture, a northern bluefin tuna traveled at least 7,700 kilometers (4,774 miles) across the Atlantic Ocean in just 119 days. In addition to their capacity for sustained speed, scombrids can attain remarkable speeds in brief bursts of activity. The wahoo reaches at least 75 kilometers per hour (47 miles per hour), a speed that may be exceeded only by another scombroid group, the billfishes.

Billfishes

The large pelagic billfishes comprise two families, the commercially important swordfish (Xiphiidae), with one genus and species, and the sailfish, spearfishes, and marlins (Istiophoridae), with 3 genera and 11 species, most of which are very

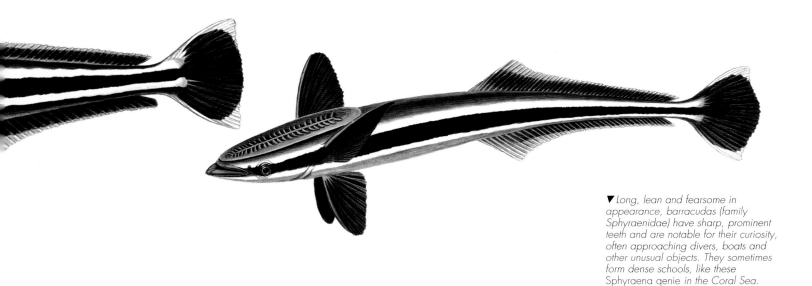

▼ Long, lean and fearsome in appearance, barracudas (family Sphyraenidae) have sharp, prominent teeth and are notable for their curiosity, often approaching divers, boats and other unusual objects. They sometimes form dense schools, like these Sphyraena qenie in the Coral Sea.

popular sport fishes. Billfishes are found in all oceans and are highly migratory. They are close relatives of the tunas and mackerels and share with them many of the same specializations for rapid, continuous swimming. No other fish is known to swim faster than the sailfish, which has been clocked in excess of 110 kilometers per hour (68 miles per hour) for short periods.

Studies of developing billfish larvae have shown that the long sword-like or spear-like beaks of adults are essentially extensions of the shorter beak-like upper jaws that characterize the scombrids. The wahoo has a longer beak than other scombrids and also shares with all billfishes a unique condition of the gill filaments, in which there are cartilaginous connections interspersed with tooth plates. Some ichthyologists believe this proves that the wahoo is the closest living relative of the billfishes, while others believe these conditions evolved independently in the two groups. There are many such disagreements about hypothetical relationships among perciform fishes, and the study of these relationships continues to provide a fascinating and exciting arena for scientific research.

SURGEONFISHES AND THEIR ALLIES

The suborder Acanthuroidei includes tropical marine fishes comprising about 110 species in 4 families. Most are herbivores, and all but the oceanic louvar are shore fishes, primarily associated with coral reefs. Acanthuroids have a distinctive specialized post-larval stage, the "acronurus", during which the fish remains pelagic for extended periods before settling to the adult habitat. Like many perciforms, acanthuroid larvae have quite a different appearance from the adults, and those of all four families share specialized features that demonstrate their close relationship.

The rabbitfishes (Siganidae), so named for their

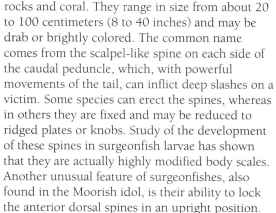

Doug Perrine/Planet Earth Pictures

▲ *Among the elite of fishes from the marine big-game angler's viewpoint, the sailfish Istiophorus platypterus (family Istiophoridae) is a high-speed cruiser of the open ocean, maiming or stunning prey with its rapier-like beak, then turning to scoop it up with toothless jaws. The large dorsal fin fits snugly into a groove down the back as it sails through the water at high speed.*

▼ *The only member of the family Zanclidae and perhaps the most widely recognised of all reef fishes, the Moorish idol Zanclus cornutus resembles (superficially) the butterflyfishes (family Chaetodontidae) but is in fact closely related to the surgeonfishes.*

rabbit-like snout and mouth, include one genus and 28 species distributed from the eastern Mediterranean throughout the tropical Indo-Pacific. The most unusual features of these small algal browsers are the presence of two spines in each pelvic fin and the venom glands associated with the very strong dorsal and anal spines.

The family Zanclidae contains one species, the Moorish idol *Zanclus cornutus*, occurring throughout the tropical Indo-Pacific. It is a strikingly beautiful coral-reef fish, with broad vertical black bars on a yellowish-white background, an extremely deep body, and a long filamentous third dorsal spine. The protruding snout with small mouth and bristle-like teeth is used to forage among cracks and crevices in the coral.

There are 6 genera and about 70 species of surgeonfishes (Acanthuridae), occurring in all tropical seas. Surgeonfishes have deep, compressed, almost disk-like bodies and small mouths with incisor-like teeth suited to nibbling or scraping small animals and plants from the

rocks and coral. They range in size from about 20 to 100 centimeters (8 to 40 inches) and may be drab or brightly colored. The common name comes from the scalpel-like spine on each side of the caudal peduncle, which, with powerful movements of the tail, can inflict deep slashes on a victim. Some species can erect the spines, whereas in others they are fixed and may be reduced to ridged plates or knobs. Study of the development of these spines in surgeonfish larvae has shown that they are actually highly modified body scales. Another unusual feature of surgeonfishes, also found in the Moorish idol, is their ability to lock the anterior dorsal spines in an upright position.

The large, epipelagic louvar (family Luvaridae) spends its entire life in the open sea feeding on jellyfish and other plankton and may grow to 2 meters (6½ feet) and 150 kilograms (330 pounds). For many years, scientists believed that the louvar was related to the pelagic tunas, mackerels, and billfishes because of similarities in the structure of its tail skeleton and fin. Recent studies of larval morphology and the developing skeleton prove that the louvar is actually a highly divergent acanthuroid that has independently adapted to the pelagic realm. The louvar undergoes an extraordinary morphological transformation from larva to adult, and in the earliest stages it exhibits features found elsewhere only in the larvae of the Moorish idol and surgeonfishes. Furthermore, although adult louvars lack dorsal spines, these spines are present in the larvae and have a rudimentary locking mechanism just like that of the surgeonfishes.

MULLETS

The suborder Mugiloidei comprises a single family with about 80 species widely distributed in tropical to temperate marine and brackish waters, with a few freshwater representatives.

Mullets (family Mugilidae) lack a lateral line and have small triangular mouths, two widely separated dorsal fins, and pelvic fins placed well behind the pectorals. They feed primarily on detritus—most species have a specialized pharyngeal apparatus for processing it and very muscular stomachs and long intestines to facilitate digestion. Because of the abdominal pelvic fins and separate dorsal fins, mullets were once thought to be relatives of the barracudas (Sphyraenidae), now placed in the Scombroidei, or threadfins (Polynemidae). Recent studies suggest a closer relationship to atherinomorphs and several other non-perciform fishes. Mullets are important food fishes throughout the world and are frequently cultured in ponds.

THREADFINS

There is a single family of threadfins (Polynemidae), placed in the suborder

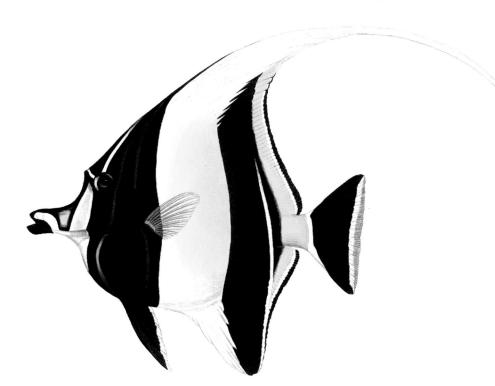

Polynemoidei, comprising 7 genera and over 30 species. They occur worldwide in warm, shallow continental waters, including rivers. These fishes have a very distinctive protuberant, anchovy-like snout and inferior mouth. The upper rays of the divided pectoral fin are connected by membrane, while the lower three to seven rays are filamentous and free. They have two widely separated dorsal fins, and the pelvic fins are placed posterior to the pectorals. Although usually treated as a separate suborder and previously thought to be related to mullets or barracudas, threadfins are probably highly specialized percoids and may be related to drums and croakers, to which their excellent flesh has been likened. The larvae of threadfins and sciaenids are remarkably similar and have distinctive large star-shaped melanophores (spots of pigment); the adults share a specialized sensory system on the snout.

HAWKFISHES AND THEIR ALLIES

The suborder Cirrhitoidei is made up of about 70 species in 5 families of small to moderately large marine fishes that are mainly bottom-associated. They are characterized by three features—unusual pectoral fins in which the lower 4 to 8 rays are unbranched and thickened, cycloid scales, and only 15 principal caudal rays. The often brightly colored hawkfishes (family Cirrhitidae) are found worldwide in rocky and coral-reef habitats of tropical and subtropical waters. The remaining families are mostly confined to temperate and subtropical areas of the southern hemisphere, with greatest diversity in southern Australia. One group of species within the family Cheilodactylidae, the magpie morwongs (genus *Goniistius*), is exceptional in having an antitropical distribution that includes Hawaii, Taiwan, China, and Japan in the northern hemisphere, and southern Australia and the South Pacific in the southern hemisphere. All cirrhitoids have an extended larval period, and in all but the hawkfishes this includes a distinctive "paperfish" stage. The silvery, deepbodied, strongly compressed paperfish confused some early naturalists who, for example, mistook a larval trumpeter (Latrididae) for a new genus of jack (Carangidae) and a larval morwong for a new family of fishes. Some morwongs and trumpeters are important food and recreational fishes, and some of the colorful hawkfishes are popular with aquarists.

WEEVERFISHES AND THEIR ALLIES

The 11 families united in the poorly defined suborder Trachinoidei are quite diverse in form, structure, and habitat, and it is unlikely that they constitute a natural group. At one extreme is the torrentfish (family Cheimarrhichthyidae), which inhabits torrential freshwater streams in New Zealand. At the other extreme are the deepsea swallowers (family Chiasmodontidae), which are found in the bathypelagic depths of the ocean and are able to swallow prey equal to or larger than their own size. Most other trachinoids are bottom-dwelling marine and estuarine fishes distributed throughout the world. Many are ambush predators that lie partly concealed in soft mud or sand. The eyes of some are located on top of the head and directed upwards, so that they are not concealed when the fish is buried; this has led to the common name "stargazer" for the family Uranoscopidae. Similar body form and habits have evolved independently within several other fish groups including toadfishes, blennies, and gobies. Members of the Creediidae, Trichonotidae and Ammodytidae families are known as sandburrowers, sanddivers, and sandlances because of their habit of diving headfirst into the sand. Weeverfishes (Trachinidae) and stargazers have venom glands variously associated with the head and dorsal-fin spines.

GOURAMIES AND THEIR ALLIES

The pikehead and four families of gouramies make up the traditional anabantoids, a group of about 70 species of small African and southeast Asian freshwater fishes. They are characterized by their ability to breathe air by means of an accessory respiratory organ located immediately above each gill chamber. This unique organ is formed of highly vascularized, convoluted tissue, supported by an enlarged dorsal element of the gill arches, and has led to the common name "labyrinth" fishes. The ability to breathe air is perhaps best developed in certain species of the family Anabantidae, some of which are able to travel over land between bodies of water by "crawling" with their pectoral fins and mobile, notched subopercular bones. The common name, "climbing gouramies", alludes to the supposed ability of species of the southeast Asian anabantid, *Anabas*, to climb trees; however, recent experiments suggest that *Anabas* may only be able to climb over obstacles of about half its body length.

Anabantoids range in size from the 2.5-centimeter (1-inch) lesser licorice gourami (*Parosphromenus parvulus*, Belontiidae) to the 1-meter (39-inch) giant gourami (*Osphronemus gouramy*, Osphronemidae). They are popular aquarium fishes because of their small size and the bright coloration of many species, as well as their interesting reproductive behaviors. Males of most species brood the eggs, either in their mouths or in floating bubble nests.

The "percoid" leaffishes (Nandidae, Badidae, Pristolepidae) and the channoid snakeheads (Channidae) have been variously included in the Anabantoidei by some scientists. Evidence for a relationship between these families and the traditional anabantoids includes the unusual presence of a tooth patch on the parasphenoid bone of the skull.

Kevin Deacon/Dive 2000

▲ Common in coral reefs of the Indo-Pacific region, unicornfishes (family Acanthuridae) have two large, knife-bladed scales on each side of the base of the tail, and some have a horn-like growth on the head. This is Vlaming's unicornfish *Naso vlamingii* on the Great Barrier Reef, here accompanied by a cleaner wrasse on its dorsal fin.

▼ Cirrhitidae is a small family of coral-reef fishes restricted to the Indo-Pacific, distinguished by the fact that the lower rays of the pectoral fins are thickened and extended. This is the longnose hawkfish *Oxycirrhites typus*, common in New Caledonian waters, and found throughout the tropical Indo-Pacific.

Kevin Deacon/AUSCAPE International

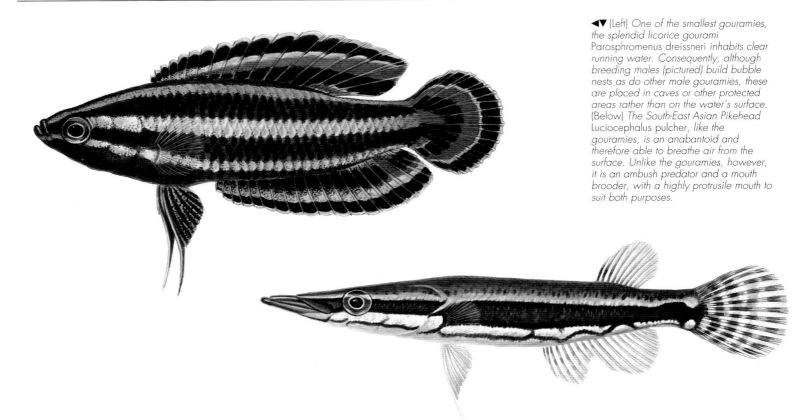

▼▼ (Left) *One of the smallest gouramies, the splendid licorice gourami* Parosphromenus dreissneri *inhabits clear running water. Consequently, although breeding males (pictured) build bubble nests as do other male gouramies, these are placed in caves or other protected areas rather than on the water's surface. (Below) The South-East Asian Pikehead* Luciocephalus pulcher, *like the gouramies, is an anabantoid and therefore able to breathe air from the surface. Unlike the gouramies, however, it is an ambush predator and a mouth brooder, with a highly protrusile mouth to suit both purposes.*

EELPOUTS AND THEIR ALLIES

The subfamily Zoarcoidei includes nine families and about 280 species of marine fishes characterized by an elongate body and anteriorly positioned (jugular) pelvic fins. Primarily because of these two features they have sometimes been classified with the blennioids, from which they differ in many skeletal and other features, including having one rather than two pairs of nostrils. Most zoarcoids are restricted to coastal areas (including tide pools) of the cooler northern parts of the Atlantic and Pacific oceans. The eelpouts (Zoarcidae) are a noteworthy exception, being mainly deepwater fishes that are distributed worldwide. Although zoarcoids are mainly small fishes (less than 30 centimeters or 12 inches), some wolffishes (Anarhichadidae) may reach over 2 meters (6½ feet).

ANTARCTIC FISHES

The 6 families and approximately 100 species of the suborder Notothenioidei are distinguished from other perciforms by having three rather than four basal radials in each pectoral fin. Notothenioids are restricted to high latitudes in the southern hemisphere and effectively dominate the Antarctic and sub-Antarctic fish faunas. Not surprisingly, then, notothenioids exhibit considerable morphological and ecological diversity, ranging from 10-centimeter (4-inch) benthic species that feed on invertebrates to 2-meter (6½-foot) pelagic species that eat fishes. Some species can live at temperatures close to the freezing point of sea water (−1.9°C or 28.6°F) due to the presence of glycopeptide "antifreeze" in their blood.

The crocodile icefishes (family Channichthyidae) are also noteworthy in that they lack red blood cells. With the exception of the congolli (*Pseudaphritis urvillii*, family Bovichtidae), which enters fresh water in Tasmania and southeastern Australia, notothenioids are essentially marine.

G. DAVID JOHNSON & ANTHONY C. GILL

▼ *One of a score or more rather similar species found around the world (family Uranoscopidae), the New Zealand stargazer* Kathetostoma laeve *has a compressed face suggesting that of a bulldog and eyes on the top rather than the sides of the head.*

Gary Bell/Planet Earth Pictures

GROUPERS, SEABASSES & THEIR ALLIES

KEY FACTS

ORDER PERCIFORMES
SUBORDER PERCOIDEI
FAMILY SERRANIDAE
• 67 genera • *c.* 400 species

SMALLEST & LARGEST

Plectranthias longimanus
Total length: 3.5 cm (1½ in)
Weight: 0.7 g (³⁄₁₀₀ oz)

Giant grouper *Epinephelus lanceolatus*
Total length: 2.5 m (8 ft)
Weight: 300 kg (660 lb)
* Unconfirmed reports of much larger size.

CONSERVATION WATCH
!!! The 3 critically endangered species are: speckled hind *Epinephelus drummondhayi;* jewfish *Epinephelus itajara;* and Warsaw grouper *Epinephelus nigritus.*
!! The Nassau grouper *Epinephelus striatus* is listed as endangered.
! There are 13 species listed as vulnerable.

T he grouper and seabass family (Serranidae) is a large and diverse group of fishes. They are difficult to characterize because they lack the kind of specializations that often provide distinctive traits for other families of fish. To define the family, several characteristics need to be considered together.

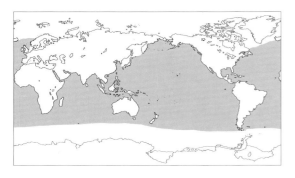

ROBUST PREDATORS

Most groupers and their allies have a robust body, although it may be notably compressed. There are three (rarely two) spines on the opercle (largest bone of the gill cover), and the edge of the preopercle nearly always has notches or small spines. If these fishes are not handled carefully, wounds may result from these spines. The mouth is large, and the upper jawbone is fully exposed on the cheek when the mouth is closed. Typically there are several rows of sharp teeth in the jaws. Most species also have prominent canine teeth at the front of the jaws. Scales are small with tiny spines on the edges (ctenoid), or secondarily rounded (cycloid). Nearly all have a continuous dorsal fin with 6 to 13 spines in the front part (except New World soapfishes of the genus *Rypticus* which have two to four spines), and an anal fin preceded by three stout spines. All are carnivorous, although many small species feed only on zooplankton.

Wayne Sterrie/AUSCAPE International

◄ The harlequinfish Othos dentex of inshore waters of southern and western Australia is variable in color but characterized by a row of bold yellow spots along the flanks, giving rise to its alternative name, "Chinese lantern". This individual is being groomed by a western cleaner clingfish Cochleoceps bicolor.

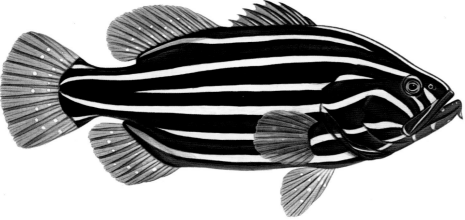

▲ *The young of the giant, or Queensland, grouper Epinephalus lanceolatus (top) are attractively marked with black bands on a yellow background but this pattern breaks up as they grow, eventually becoming a uniform drab olive. Despite its common name, the sixline soapfish Grammistes sexlineatus (above) may have anywhere from 3 to 12 stripes, the number increasing with size and age, and in very large individuals may even break up into dashes. The skin of this fish is very slimy and secretes the toxin grammistin.*

FOOD VALUE

Groupers are among the most important fishes living on the sea bed in tropical to warm temperate seas. Many are of commercial value and highly prized as food. For example, the most important commercial fish in the Bahamas is the Nassau grouper *Epinephelus striatus*. The fish of greatest commercial value to the nations bordering the Persian Gulf is the orangespotted grouper, *E. coioides* (often misidentified as *E. tauvina*). However, some tropical groupers are among the worst offenders in causing ciguatera poisoning when eaten. Examples are the species of the Indo-Pacific genus *Plectropomus* (called coral trout in Australia) and those of the American genus *Mycteroperca*.

Groupers are readily caught by hook and line, and many of the species are easily approached underwater, so they are easy targets for spearfishers. Because they are at or near the upper end of food chains, there are few at any one locality. Therefore, any one reef may have only one or two very large groupers in residence. These species are among the first fishes to have their populations drastically reduced as a result of overfishing. Because large groupers are believed to very old, perhaps more than 50 years, it takes a long time for them to be replaced.

▶ *The potato grouper Epinephelus tukula is an Indo-Pacific member of a cosmopolitan warm-water group that includes some of the largest of all reef fishes. Most are highly prized as food for humans, but some are prone to cause ciguatera poisoning. This pair is engaged in courtship behavior.*

Alby Ziebell/AUSCAPE International

There is another important biological reason why grouper populations are particularly vulnerable to depletion by humans. They start out life as females and switch sex to males when much older. If they are all caught before they become males, the effect on the population is obvious.

FEEDING HABITS

Groupers feed primarily on fishes and crustaceans. The species that eat mainly fishes are the ones most likely to cause ciguatera poisoning; they can usually be spotted as piscivores (fish eaters) by the very large canine teeth in their jaws.

Groupers tend to feed by ambush. They lurk in the reef and quickly lunge to engulf prey that venture near. Their capacious mouth and ability to expand the mouth cavity to a great size (thus creating strong suction) are obvious adaptations for such a mode of feeding. Their anterior canine teeth are used to seize large prey; the smaller teeth depress inwards, hence allowing for easy inward movement of prey but not outward. As experienced fishers know, the best time to go fishing for groupers is at dusk, for that is when these fishes do most of their feeding. It is a time of confusion on the reef, with the light level low, diurnal fishes seeking a place to hide in the reef to sleep at night, and nocturnal fishes emerging from their day-time shelter to feed. In the sea-bed communities of warm seas, groupers and seabasses are among the most important predators.

Groupers are often disruptively colored so they can blend with the coral reef or rocky bottom, and many are also able to change their color rapidly. If one moves from a reef over open sand, it will become noticeably paler. Color change can take place when a fish exhibits aggression to another of its own species, when being serviced by a cleaning shrimp or fish, or during courtship and spawning.

SPAWNING

Some groupers and seabasses, such as the common dwarf spotted grouper *Epinephelus merra* of the Indo-Pacific region, spawn in pairs at dusk. Others such as the Nassau grouper are known to form large aggregations at specific localities at spawning time. Most of the fishes at these sites had to migrate considerable distances for spawning. Unfortunately, fishers have learned some of these localities and the spawning time, and are able then to take a heavy toll of these fishes.

FIVE SUBFAMILIES

The family Serranidae can be divided into five subfamilies, the most primitive being the Niphoninae which consists of a single species, the ara *Niphon spinosus*; it ranges from southern Japan to the Philippines. The best-known and largest subfamily in number of species (and also in having the species of largest size) is the Epinephelinae.

Kev Deacon/AUSCAPE International

These fishes are called groupers (generally known as rock cod in Australia). The fishes in the subfamily Grammistinae occur in the tropical Indo-Pacific region, except for the species of the genus *Rypticus*, which are found in the tropical Atlantic and east Pacific. The subfamily Serraninae occurs in the Mediterranean Sea, western Africa, the east coast of the Americas from south Carolina to Uruguay, and the eastern Pacific from the Gulf of California to Peru. The fifth subfamily, Anthiinae, contains many species and has a circumglobal distribution in tropical to warm temperate seas.

EPINEPHELINAE

The groupers of the subfamily Epinephelinae are among the most important food fishes of inshore waters. The group contains 15 genera, the largest being *Epinephalus* with 97 species. They are protogynous hermaphrodites, meaning they are females at maturity, changing sex to males later in life.

GRAMMISTINAE

The subfamily Grammistinae is divided into four tribes. Two of them, the Grammistini and Diploprionini, collectively known as soapfishes, form a small group that has at times been classified as a family. These fishes look like groupers but are able to secrete a poisonous substance called

▲ *Often known as rock cod (in Australia) or hinds (in America), members of the genus Cephalopholis are small groupers that are widespread in tropical waters, especially on coral reefs. This is the six-spot rock cod C. sexmaculatus of Indo-Pacific waters.*

Gerard Soury/Jacana/AUSCAPE International

▲ *The butter hamlets, such as Hypoplectrus guttavarius, are characterized by their extraordinary breeding behavior: each individual carries a uniquely modified sexual organ that produces both sperm and eggs, and mating pairs cross-fertilize each other. The ten species are almost identical in structure but they differ in color pattern and utter distinctive sounds when spawning.*

the small ones are very cryptic, hence infrequently observed. The largest genus is *Liopropoma*. This genus has 23 species, some of which are very colorful. Because of their retiring nature, however, they are not popular aquarium fishes.

SERRANINAE

Most of the species of the large subfamily Serraninae are also small. The largest genus is *Serranus*, with about 21 species. These occur in the Mediterranean Sea, western Africa, the east coast of the Americas from South Carolina to Uruguay, and in the eastern Pacific from the Gulf of California to Peru.

The most important fishes of the subfamily Serraninae from a commercial standpoint (both as food fishes and as game fishes for anglers and spearfishers) are the species of the eastern Pacific genus *Paralabrax*, such as the kelp bass *P. clathratus* and the cabrilla *P. humeralis*. These fishes are unusual for serranids in their occurrence in temperate waters; the kelp bass ranges north on the Pacific coast of the United States to the Columbia River, and the cabrilla is known from the cool coastal waters of Peru and Chile.

Butter hamlets

The small fishes called butter hamlets (genus *Hypoplectrus*), a tropical western Atlantic genus of about 10 species, also belong in the subfamily Serraninae. Although they are very different in color—some are vividly hued—no characteristics of form or structure have been found to separate them. As a result, some ichthyologists have regarded them only as color forms of a single species. These fishes (and those of the genus *Serranus*) have a very unusual mode of reproduction. They are synchronously hermaphroditic, meaning that each mature individual is both male and female at the same time. Instead of having an ovary or a testis, they have what might be termed an ovotestis; ripe eggs and functional sperm are found in different parts of the same organ. Butter hamlets are generally seen in pairs. When spawning, they rise above the bottom with bodies intertwined and cross-fertilize. Further evidence has come from recent and ongoing research which shows that these fishes make distinctive low-frequency sounds before and

grammistin from their skin when frightened. The toxic mucus is very bad-tasting (touch some of it to your tongue and you will know how bad). Predators soon learn to avoid these fishes. The best known in the Indo-Pacific region is the sixline soapfish *Grammistes sexlineatus*, a showy black fish with yellow stripes; in the Atlantic and eastern Pacific, the greater soapfish *Rypticus saponaceus* (from the Latin *sopa*, meaning soap) is the largest and most common.

The tribe Pseudogrammini, with the common name of podges, is a small group that has also been given family status in the past. Only 14 species are known, all less than 10 centimeters (4 inches) total length. These small fishes, most of which are drab-colored, are so well hidden in the reef that they are rarely seen by divers.

The Liopropomini is the fourth grammistin tribe. The larger species occur at greater depths than are normally penetrated by scuba divers, and

▶ *All the purplequeens Pseudanthis tuka start life as females (illustrated here) but can change sex. When a male dies, the dominant female of his harem becomes male and takes over his role.*

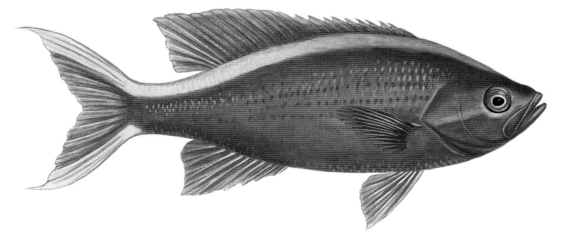

during spawning. Although these sounds cannot be heard by divers, they can be recorded by hydrophone and analyzed. It is believed that different spawning sounds will demonstrate that these fishes are distinct species, not color phases of one species.

ANTHIINAE

The fifth subfamily, the Anthiinae, has also been regarded at times as a family. It is a large group of 23 currently recognized genera; however, the generic classification is not yet stabilized. The largest genus was *Anthias*, and this generic name is often used as the common name in the same way that *Chromis* of the damselfish family has become a common name. Recent taxonomic research has resulted in the restriction of *Anthias* to six Atlantic species. The largest genus is now the Indo-Pacific *Pseudanthias*, which has 45 described species. Some unnamed species are languishing as specimens on museum shelves waiting for their scientific descriptions, and undoubtedly still more, especially those from deeper water, remain to be discovered.

The fishes of this subfamily are among the most beautifully colored in the world, with pink, magenta, lavender, and yellow predominating. Many of the species exhibit strikingly different color with sex. It is believed that all are females initially, capable of undergoing sex reversal to males later in life. Males maintain a large harem of females, and some species exhibit color change in courtship. Field experimentation has shown that the removal of the male results in a dominant female beginning to act like a male. Within a few days she will change her sex and her color pattern to that of a male and take over the harem.

The small Anthiinae, such as the species of *Pseudanthias*, form large aggregations which rise from one to a few meters (3 to 20 feet) above the bottom to feed on zooplankton; with the approach of danger, they quickly descend to the shelter of coral or rocky reefs. Some of these fishes are extremely abundant on reefs, forming in their numbers, if not in total biomass, a major component of the reef fish fauna. They are preyed upon by roving predaceous fishes such as jacks and by resident reef predators such as groupers, scorpionfishes, and morays.

CONSERVATION

Conservation measures are needed to preserve the stocks of the larger groupers. Fortunately, this need is now recognized for the largest members of the family, such as the giant grouper *Epinpheluls lanceolatus* (Queensland grouper in Australia; brindle bass in South Africa), the potato grouper *Epinephelus tukula* (potato cod; potato bass), and the Atlantic jewfish *Epinephelus itajara*. These huge fishes are now protected in some areas.

JOHN E. RANDALL

Kev Deacon/Dive 2000

▲ *The scalefin anthias* Pseudanthias squamipinnis, *one of the inhabitants of Indo-Pacific coral reefs.*

CICHLIDS

KEY FACTS

ORDER PERCIFORMES
SUBORDER LABROIDEI
FAMILY CICHLIDAE
• *c.* 200+ genera
• *c.* 2,000+ species

SMALLEST & LARGEST

Tilapia snyderae from Lake Bermin, Cameroon
Total length: 2.5 cm (1 in)

Boulengerochromis microlepis
Total length: 90 cm (36 in)

CONSERVATION WATCH

!!! There are 38 species listed as critically endangered, including: dikume *Konia dikume;* konye *Konia eisentrauti;* kotso *Paretroplus petiti;* pindu *Stomatepia pindu.*
!! 13 species are listed as endangered.
! 35 species are listed as vulnerable.

Although cichlids (family Cichlidae) are essentially freshwater fishes, their nearest relatives are the marine surfperches (family Embiotocidae), damselfishes (family Pomacentridae), wrasses (family Labridae), and parrotfishes (family Scaridae). These five families are grouped as the suborder Labroidei. Current estimates for the number of cichlid species vary from about 900 to more than 2,000, making it the second—or third—largest family of bony fishes. The exact number is uncertain, because there are many new species still awaiting formal description.

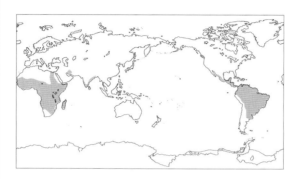

APPEARANCE AND ANATOMY

Although there is much variation in body form, most cichlids have a typical perch-like body; that is, moderately deep and somewhat compressed, with fins that are obvious but not excessively large. Some departures from that body shape are the long, slender, and streamlined bodies of many fish-eating predators and the deep, very compressed bodies of certain South American species such as the freshwater angelfish *Pterophyllum*, which also has unusually high fins. Species living in fast-flowing waters have distinctive, goby-like bodies, long and nearly cylindrical, with small scales that are often deeply embedded in the skin, and enlarged, thick-skinned pelvic fins. These probably help to hold the fish against the current.

Other external distinguishing features of a cichlid are the presence of a single nostril on each side of the head, and the division of the lateral-line scale series into a long upper and a shorter lower section. This lower section is usually single but can be double or even treble. Other characteristic features are mainly anatomical, the main ones being those concerned with the bones and muscles forming the "pharyngeal jaws", effectively a second set of jaws in the throat (see page 210). In an evolutionary sense, the development of the pharyngeal jaws has allowed the oral jaws to specialize in catching food. This has led to a variety of possible feeding habits and may well have played an important role in establishing the evolutionary success of cichlids. These fishes are now often dominant in tropical freshwater ecosystems, particularly in certain African lakes.

A great variety of tooth forms is found in the pharyngeal jaws of different species, ranging from long, slender, and hooked teeth, through stouter, knife-like forms, to large, molar-like teeth with flat

► *Cichlids* (Boulengerochromis microlepis) *surrounded by swarms of fry, Lake Tanganyika. Parental care is universal in cichlids, and many are mouth-brooders—the tiny young are periodically released to feed but gathered up into the parent's mouth at the first sign of danger.*

Peter Scoones/Planet Earth Pictures

Laboi-Lanceau / AUSCAPE International

crowns. The first type gives a mechanism for hooking finely particulate food, such as microscopic plants, which are first trapped in mucus secreted in the mouth and then raked backward into the gullet. The knife-like teeth serve to slice up prey such as fishes, and the third type act as grinders to crush hard-shelled mollusks. A more robust and less strongly hooked version of the fine teeth provides a general purpose dentition able to break up insects, small crustaceans, and fishes; when combined with a few slightly enlarged and flat-topped teeth this type of dentition can also deal with small or thin-shelled mollusks.

DISTRIBUTION AND HABITAT

The fossil record for cichlids is poor, with the oldest specimens found in the tropical regions of North and South America possibly dating from Eocene times (57 to 37 million years ago) and specimens from East Africa deposited about 24 million years ago in the late Oligocene and early Miocene epochs. Because these fossils are of species which were relatively specialized it is likely that the family evolved at an even earlier date, possibly before the break-up of the super-continent Gondwana during the late Jurassic and Cretaceous periods (146 to 65 million years ago).

Cichlids live mainly in the lowland tropical and subtropical regions of South America, Central America, the West Indies, Africa, Madagascar, Iran, Syria, Israel, coastal India, and Sri Lanka. One species is found in North America (in southern Texas), and at least two species are found in the temperate parts of South Africa.

While it is understandable that no cichlid genus occurs in more than one continent, a remarkable feature of the family, especially in Africa, is the large number of species that occur only in a single location. For example, in the major African lakes, up to 99 percent of the cichlids in a particular lake occur there and nowhere else. The number of species in these lakes is also noteworthy. In East Africa, at least 300 cichlid species have been recorded in Lake Victoria, about 170 in Lake Tanganyika, and further south there are thought to be possibly as many as 1,000 species in Lake Malawi.

Since South America has only a few lakes and these are small, the estimated 350 cichlid species there live mainly in rivers. In Africa probably no more than 150 species are found only in rivers. Consequently, in African and South American rivers, cichlids are out-classed, in diversity and ecologically, by species from other families. In

▲ *One of the most popular of aquarium fishes for the home hobbyist, the angelfish Pterophyllum scalare of Amazonia is silver with black stripes in the wild; the color variety here has been developed by selective breeding in captivity. This species is a solicitous parent, laying its clutch of several hundred eggs on the leaf of some suitable aquatic plant (having first meticulously cleaned it), then fanning them constantly until they hatch after three days. Once hatched, the fry are guarded carefully, and gathered up into a dense pack every evening for protection.*

India and Sri Lanka too, cichlids are less important elements of the freshwater fish faunas, but they are dominant in Madagascar.

Although cichlids are ranked as a freshwater family, several are known to tolerate salinity of varying degrees. The best studied of these are certain species of the African genus *Oreochromis*, some of which can live and breed in brackish or even sea water. One species, translocated to a Pacific Ocean atoll, has now established a population in the sea. A group of *Oreochromis* species in East Africa is only found in the highly alkaline salt lakes of Kenya and Tanzania. The water temperature in some parts of these lakes exceeds 40°C (104°F); although the fish do not live in these areas, this group can apparently tolerate surprisingly high temperatures.

Because this family has so many species that have not been fully studied, it is difficult to generalize on such matters as temperature and salinity tolerance, oxygen requirements, and acceptable water qualities. Some species of the African genera *Tilapia* and *Oreochromis* can withstand oxygen concentrations as low as 0.1 parts per million. A few species of other genera, mostly small fishes living in swampy areas, make use of the thin, well-oxygenated surface layer of the water. At the opposite extreme, some deepwater cichlids in Lake Tanganyika can survive short excursions into the permanently deoxygenated deeper waters of that lake. Judging from the experiences of aquarium keepers, however, it seems that the tolerance of many species is limited, especially with regard to water quality, oxygen availability, and temperature. Nevertheless, taking the family as a whole, there are few freshwater habitats or niches not occupied by one or more cichlid species.

Labat-Lanceau/AUSCAPE International

▲ *The sole member of its genus,* Cyphotilapia frontosa *of Lake Tanganyika displays a humped head characteristic of a few species of cichlids, particularly older individuals.*

▲◄ *Two of the African Rift Lake cichlids: the Lake Malawi zebra cichlid* Pseudotropheus zebra *(above) and the striped Julie* Julidochromis regani *(left), endemic to Lake Tanganyika.*

FEEDING

Members of the cichlid family can eat a wide variety of foods. The pharyngeal jaws and their role in food processing have already been mentioned. Food capture is centered on the oral jaws. Some jaws, usually those of fish-eaters, can be thrust forward (protusion) and widened; others are used for nibbling by browsers and grazers, or for sucking or scooping by detritus-feeders. Some jaws act like forceps, removing prey from holes and crevices, while some are stout and powerful and can crush hard-shelled mollusks. Apart from slight modifications to these various jaw types, many species have what could be called a "standard" jaw type—one capable of moderate protrusion and widening, which, depending on the teeth, is able to deal with a variety of food.

It is impossible to describe briefly the full variety of tooth forms among cichlids, or the different ways these teeth are distributed in the jaws. However, overall in combination the oral and pharyngeal jaws provide a feeding mechanism that has allowed different cichlids to use virtually every available plant, animal, and even bacterial food source in many different habitats, a claim that can be made for few other families.

REPRODUCTION

Male cichlids usually perform an elaborate courtship display before spawning. This involves a variety of body postures and fin movements, and an intensification of the male's color. Mouth brooders particularly show marked differences in the color of the sexes; usually the male is brightly colored and the female drab. In these species male coloration is apparently an important device, enabling the female to recognize a male of her own species. Field studies in African lakes show that despite the large number of species in a lake, no two species have identical breeding coloration.

In central Africa, the males of over 50 species in several genera build elaborate spawning sites on the sandy bottom of Lake Malawi. The constructions are aptly named "bowers" because, since they are built to attract the females, they are similar in function to those made by bowerbirds. Fish bowers may be deep pits more than 3 meters (almost 10 feet) in diameter, or simple or sandcastle-like mounds up to a meter (39 inches) in diameter at the base, yet all are produced by individuals often less than 15 centimeters (6 inches) long. It seems

▲ *Also known as the velvet cichlid for its velvety appearance, the oscar* Astronotus ocellatus *is a large South American cichlid made popular by the aquarium trade. The prominent eye spot on the base of its tail, combined with the rounded shape formed by its thick, leathery fins, give this fish a false second "head" to confuse predators. Consequently, young piranhas, known to nip the tail of other fish for an easy meal, appear to leave oscars alone.*

▼ Neolamprologus brevis *is a member of a remarkable "species-flock" of cichlids that inhabits Lake Tanganyika.*

Labat-Lanceau/AUSCAPE International

Richard W. Beales/Planet Earth Pictures

▲ *One of the west African species of* Hemichromis. *Both parents cooperate in guarding the young.*

▼ *Several* Oreochromis *species are remarkable for their crowded "bowers". They are excavated out of the sandy bottom of shallow waters, usually by mouth.*

when the eggs hatch, the young are collected and brooded in the mouth of one or both parents. They are released at intervals but gathered back into the mouth at any threat of danger.

The most advanced form of parental care is continuous mouth brooding: non-adhesive eggs are laid and usually fertilized after the female has taken them into her mouth. Fertilization takes place when the female nuzzles the male near his genital opening and draws the ejaculated sperm into her mouth. In some species the eggs are fertilized before being taken into the mouth. The developing embryos and young are kept in the mouth and may be brooded for up to three weeks or longer. During the later part of that time the young are released to forage, but are taken back into the mouth for safety and at night. Usually it is the female that broods, but in a few species it is the male, and in some, brooding is shared by both parents.

CONSERVATION

Cichlids have long been of interest to evolutionary biologists, who are particularly fascinated by the so-called flocks of endemic, closely related species that have evolved in the African lakes during a relatively short period of geological time. In one, Lake Victoria, a flock of more than 300 species evolved in less than 200,000 years and perhaps even in as little as 14,000 years according to some estimates. This phenomenon, often referred to as explosive evolution and speciation, occurs rarely among other vertebrate animals, aquatic or terrestrial. In no other case is it of such proportions, and rarely does it have such important ecological ramifications.

Regrettably, much of the species-flock in Lake Victoria is probably nearing extinction. This disaster has come about partly through overfishing but mainly as the result of introducing the Nile perch *Lates niloticus*, a predator from other African lakes; this introduction occurred in the early 1960s. Now, in most of Lake Victoria many of the cichlid species (and those of other families too) have virtually disappeared. Inevitably the removal of this diverse and dominant element from the ecosystem has had profound affects on the lake. Lake Victoria now appears to be well on the way to becoming an aquatic near-desert in which neither the Nile perch nor many cichlids can survive.

However, cichlids are not only of importance to ecologists and biologists. Particularly in Africa, cichlids are important contributors to commercial and subsistence fisheries. Species of two African genera, *Tilapia* and *Oreochromis*, are widely used in aquaculture, especially in Israel and Asia, and have been introduced into many other countries. Many African and South American species are popular aquarium fishes, and thus are also of economic importance.

P. HUMPHRY GREENWOOD & MELANIE L. J. STIASSNY

that the form of a bower is not only characteristic for particular species, but can influence the female's choice of a mate. The males of bower-building and certain mouth-brooding species congregate during breeding seasons in large colonies, described as leks or mating arenas. A single lek of one Lake Malawi species can have as many as 50,000 bowers in it.

All Asiatic and most Central and South American cichlids are bottom spawners. Most African species are mouth brooders, but a substantial number of both lake and river dwellers are bottom spawners that use a variety of places to hide their eggs and young. In Lake Tanganyika, for instance, the female of one species, *Neolamprologus brevis* (pictured on page 203), lays her eggs inside an empty snail shell where she too lives. The male stays outside and guards her and their young.

All cichlid species show some form of parental care. The least advanced kind is that of the bottom spawners. These species lay adhesive eggs on a prepared spawning site where the eggs are fertilized and then guarded by both parents. Once hatched, the larvae remain attached to solid objects, still under parental guard. At a slightly later stage of development, the young of some species are carried in the mouth of one or both parents to a different, often more protected site. Here, too, the parents remain on guard, and continue doing so after the young are capable of swimming and foraging on their own.

An elaboration of this system is also known:

DAMSELFISHES

KEY FACTS

ORDER PERCIFORMES
SUBORDER LABROIDEI
FAMILY POMACENTRIDAE
• 28 genera • 335 species

SMALLEST & LARGEST

Rolland's demoiselle *Chrysiptera rollandi*
Head–body length: 5.5 cm (2⅕ in)

Garibaldi *Hypsypops rubicundus*
Head–body length: 36 cm (14 in)

CONSERVATION WATCH
! There are 3 species listed as vulnerable: *Chromis sanctaehelenae*; *Stegastes sanctaehelenae*; and *Stegastes sanctipauli*.

D amselfishes belong to the family Pomacentridae, which is one of the largest families of reef fishes, both in number of species and number of individuals. Indeed, they are perhaps the most conspicuous members of the daytime coral reef fish community throughout the vast Indo-Pacific region. In some areas, nearly every square centimeter of the bottom is occupied by damselfish territories, while huge schools of midwater swimming damselfishes swarm above. Because of the large range of ecologically diverse species, their unusual territorial habits, and interesting reproductive biology, damselfishes have long attracted the attention of research scientists. Indeed, more than 200 technical and popular articles have been published since 1960. They also enjoy a popular following among aquarium keepers, thanks largely to the colorful anemonefishes.

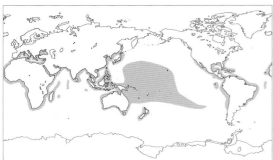

DAZZLING DAMSELS
The coloration of damselfishes is highly variable and includes vivid shades of orange, yellow, red, and blue. More somber hues of brown, black, and gray are also well represented. Many damsels have brightly colored juvenile stages, frequently yellow with one or more neon-blue stripes on the upper part of the head and back. Their conspicuous coloration fades gradually with increasing size, often resulting in a nondescript pattern that bears

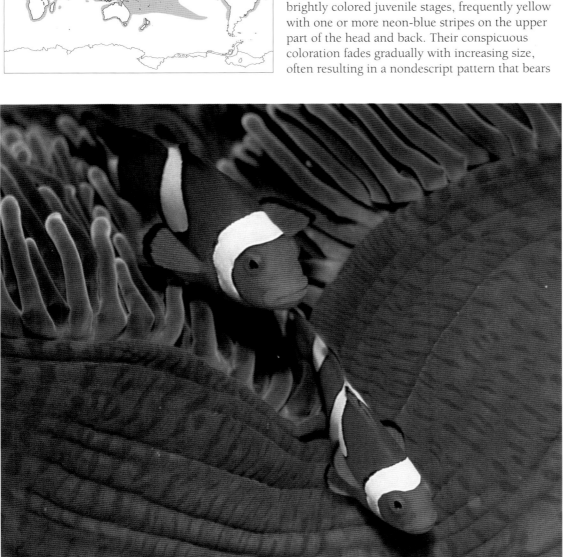

◄ *Perhaps the most famous and best-known of the family Pomacentridae are the 27 species of the Indo-Pacific genus Amphiprion, the anemonefishes, that live in a constant symbiotic association with sea anemones. The fish is immune to the anemone's stinging tentacles, and keeps the anemone clean and free of debris while enjoying its protection from predators. This species is the clown anemonefish A. percula.*

Norbert Wu

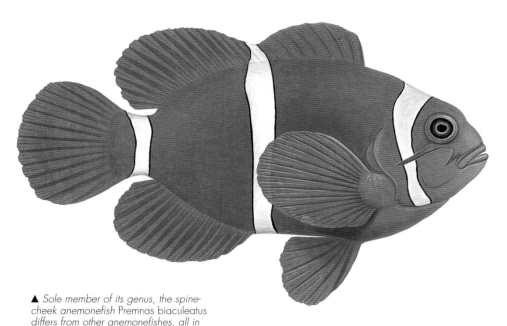

▲ *Sole member of its genus, the spine-cheek anemonefish* Premnas biaculeatus *differs from other anemonefishes, all in the genus* Amphiprion, *by the presence of two spines beneath each eye.*

▼ *Largest of the damselfishes, the garibaldi* Hypsypops rubicundus *of Californian and west Mexican waters occasionally reaches 36 centimeters (14 inches) in length. Here the male tends his red algal patch where eventually spawning will take place.*

10 to 18 flexible segmented rays, and an anal fin containing two spines and 10 to 15 segmented rays. The head, body, and fin bases are covered with medium-sized scales that usually have microscopic serrations along the exposed margins. The small jaw teeth are organized in one or two rows and range from a stout conical shape to a columnar or spoon shape.

DISTRIBUTION

The family occurs worldwide in tropical and warm temperate seas. Three species, which are mainly estuarine dwellers, are sometimes found in fresh water. By far the overwhelming majority (268 species) occur on coral or rocky reefs and associated habitats in the Indo-west and central Pacific region, the vast area stretching from the Red Sea, and shores of East Africa eastward to the islands of Polynesia. Relatively few are found in the eastern Pacific (22 species) and Atlantic (31 species). Australia has more species than any other part of the world, with 132 or 41 percent of the total species.

Most damselfishes have relatively narrow habitat zones that are best defined in terms of bottom type and depth. Some are restricted to shallow areas exposed to vigorous wave action and minimal shelter, while many others frequent deeper water where there is an abundance of either live or dead coral. Some species live in sheltered sandy lagoons while others are confined to steep outer reef slopes. The majority of damselfishes occur in depths ranging from 2 to 15 meters (6 ½ to 50 feet), but a few penetrate below 100 meters (330 feet). Many species are entirely dependent on the presence of live coral, either for shelter or, in the case of the single species of *Cheiloprion* and some members of the genus *Plectroglyphidodon*, for nourishment from eating the polyps. The food of damselfishes consists of algae, small planktonic plants and animals, and a variety of tiny bottom-dwelling invertebrates. Generally the midwater swimming species, such as those belonging to the genus *Chromis*, consume plankton, and bottom-dwelling species have either omnivorous or herbivorous diets.

little or no resemblance to its original colors. The pattern for a single species may also vary, to a lesser degree, according to geographic locality, or more rarely, in response to its surroundings.

Damselfishes are characterized by a combination of features which include an oval to somewhat elongate, laterally compressed body, a single dorsal fin composed of 8 to 17 spines and

BEHAVIOR AND REPRODUCTION

The most fascinating aspect of damselfishes is their behavior. Aside from midwater schooling species of *Azurina, Chromis, Lepidozygus, Neopomacentrus, Pomachromis,* and *Teixerichthys,* most damsels are bottom dwellers occupying well-defined territories. In many cases the focal point of the territory is some form of shelter which serves as a retreat when danger threatens or as the place in which eggs are deposited during the breeding season. Although not closely tied to the bottom, even the midwater schooling species mentioned have a rock or coral retreat. Territories are guarded zealously by individual damselfishes,

Norbert Wu

Carl Roessler

◄ In breeding, clownfishes prepare
special sites on the bottom in the
protective shelter of their sea anemone
hosts. The male may be visited by
several females to spawn and guards
the eggs until they hatch.

Rudie H. Kuiter

▲ The white ear Parma microlepis is a
temperate species of southeastern
Australia, where it is common around
rocky reefs. The drab adult has only a
white spot on the operculum.

particularly some members of the genus *Stegastes*.
Experiments have shown that their aggressive
behavior towards other fishes encourages the
development of a thick algae mat in their territory
which they use as food. The algae is far more
luxuriant within their territory than outside,
where it is cropped to a low stubble by wandering
grazers such as parrotfishes and surgeonfishes. In
essence, the territorial damsels cultivate their own
algal gardens! Two Indo-Pacific species of
damselfishes (*Stegastes lividus* and *S. nigricans*) are
particularly pugnacious. If a snorkeler or scuba
diver dares to enter their domain, they unleash an
attack that is accompanied by biting and clearly
audible grunting noises. Similarly, nesting
anemonefishes (genus *Amphiprion*) will boldly
attack anyone coming near their nest during the
incubation period. Damselfishes exhibit a rather
stereotyped mode of ritualized reproductive

behavior. Several days prior to spawning the male
begins preparing the nest site. This usually
consists of clearing a rocky surface of filamentous
algae and invertebrates which may be attached to
it. The male, sometimes accompanied by the
female, vigorously works over the surface area
with his mouth during this period.

The garibaldi *Hypsypops rubicundus* of
Californian seas shows an intriguing variation of
this theme. Males culture a patch of red alga by
selectively weeding out other algae. The red alga
then serves as the nest site. In conjunction with
nest preparation activities, the male begins
displaying toward prospective mates. These
displays consist of rapid swimming bursts,
chasing, nipping, and stationary hovering in
which the dorsal, anal, and pelvic fins are fully
extended. Another common display is known as
"signal-jumping". It consists of rapid up and down

Rudie H. Kuiter

▲ The juvenile white ear, in contrast, has
the striking electric blue markings typical
of many juvenile damselfish species. The
young white ears can be found in water
as shallow as inter-tidal pools.

swimming movements as if the fish were on a roller coaster. This behavior is rather sporadic when courtship commences, but as the time of spawning nears these displays are frequent and intense. Apparently most of this male-oriented activity is for the sole purpose of attracting one or more gravid females to the nest site.

Eventually the pair skims over the nest surface. The female leads with a slow deliberate motion, exuding a trail of eggs that stick to the bottom of the nest. The male follows close behind while fertilizing the spawn. In some species, for example *Chromis* and *Dascyllus*, a single male will fertilize the eggs of two or more females at the same nest. Depending on the species, each female lays between about 50 to 1,000 or more eggs. The eggs then incubate for approximately two to seven days until hatching.

Care of the nest is largely the responsibility of the male. He frequently fans the nest with his pectoral fins and removes dead eggs and debris with his mouth. The nearly transparent hatchlings are about 2 to 4 millimeters ($1/12$ to $1/6$ inch) in length. Parental duties terminate upon hatching except in *Acanthochromis polyacanthus* of the Indo-Australian Archipelago, which continues to guard its brood in a fashion similar to some cichlid fishes (family Cichlidae). The larvae of all other damselfishes live in the open sea for their first weeks, drifting at the mercy of tides, waves, and currents. Once they are swept into shallow reef areas the young settle to the bottom and juvenile colors are quickly assumed. There is scant information on damselfish longevity, but some species may live up to 10 to 12 years in nature and 18 years or more in captivity.

GERALD R. ALLEN

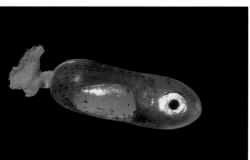

Paulo Oliveira/Planet Earth Pictures

▲ *Amphiprion frenatus*: a photo-micrograph of a clownfish embryo within the egg, 7 days old.

COLORFUL CLOWNS OF THE SEA

The 27 species of damselfishes in the genus *Amphiprion* and the single species of *Premnas* are the most colorful and interesting representatives of the family Pomacentridae. Collectively they are known as anemonefishes or clownfishes. Virtually all large public aquariums have at least one anemonefish display, and these animals have been at or near the top in aquarium fish sales for the past three decades.

The fascination with these fishes stems from their habit of living in association with large tropical sea anemones. Anemonefishes and their invertebrate hosts have delighted the Western world since 1881 when the first captive specimens were kept in a tub of seawater. However, it was not until the middle of the twentieth century that the intimate relationship of these tropical animals became well known. With the advent of scuba diving and the establishment of commercial air routes to equatorial destinations in the Indian and Pacific Oceans, pristine coral reefs became accessible to an increasing audience. Skin-diving tourists, sport divers, naturalists, and marine scientists have all helped contribute to underwater discoveries, among them the fascinating natural history of anemonefishes.

What is the mechanism that enables the fish to live in harmony with its anemone? The stinging tentacles of the anemone are a deterrent to all other fishes, but anemonefishes continuously frolic among them and sleep deep within the anemone's clutches at night. Two factors contribute to the fishes "immunity." Apparently their distinctive undulating swimming behavior combined with special chemicals in the external mucous coat prevent the anemone from firing the microscopic stinging cells on the surface of its tentacles (called nematocysts). Anemonefishes gain security from

Becca Saunders/AUSCAPE International

predators due to their association and the anemones themselves are protected from predators such as butterflyfishes that feed on them. The fishes also keep their hosts free of debris and experiments indicate that the fishes "preen" the tentacles, which keeps them in a robust condition. Nevertheless, nearly all the host species are sometimes found without fishes and appear able to survive in their absence. On the contrary, the fishes are absolutely dependent on the association and are never found in nature without their host. This has been dramatically demonstrated in field experiments by releasing the fish some distance away from the anemone. They quickly succumb to voracious predators, such as groupers or snappers.

▲ A red and black anemonefish *Amphiprion melanopus* nestles amid the tentacles of its sheltering sea anemone in Truk Lagoon, Micronesia. The anemone's tentacles secrete a chemical film that inhibits their triggering mechanism, thus preventing the tentacles from stinging each other: anemonefish have evolved a similar chemical in their own body secretions, that enables them to share their protection without harm.

WRASSES & PARROTFISHES

KEY FACTS

ORDER PERCIFORMES
SUBORDER LABROIDEI
FAMILIES LABRIDAE, SCARIDAE,
& ODACIDAE
• 88 genera • *c.* 520 species

SMALLEST & LARGEST

Minilabrus striatus
Total length: 4 cm (1½ in)
Weight: 1 g (¹⁄₂₅ oz)

Humphead wrasse *Chelinus
undulatus*
Total length: 2.5 m (8 ft)
Weight: 191 kg (421 lb)

CONSERVATION WATCH

! The species listed as vulnerable
are: humphead wrasse *Cheilinus
undulatus*; hogfish *Lachnolaimus
maximus*; *Thalassoma ascensoinis*;
Xyrichtys virens; rainbow
parrotfish *Scarus guacamaia*.

The wrasses (family Labridae) and parrotfishes (family Scaridae) are abundant and diverse groups of reef-associated fishes characterized by the possession of a modified pharyngeal apparatus that acts as a second set of jaws in the throat. With their striking color patterns (which may vary with the sexes), these groups are among the most conspicuous fishes of coral reefs. Although both families are found in large numbers in the tropics, the labrids in particular are also well represented in cooler waters. Southern temperate waters also harbor the closely related butterfishes or rock whitings (family Odacidae). Recent work suggests that the parrotfishes and odacids may represent highly modified forms within the Labridae.

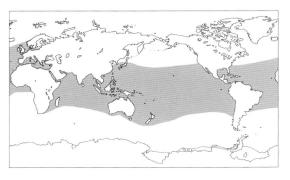

widespread *Chlorurus microrhinos* are capable of excavating significant amounts of reef, reducing it to sediment by the action of their pharyngeal jaws. As a consequence, such scarids may contribute substantially to the natural erosion of the reef structure. One species, the spectacular bump-headed parrotfish *Bolbometopon muricatum*, departs from the normal pattern of herbivory and consumes significant amounts of live coral.

Parrotfishes occur in large numbers in coral reef environments, especially on reef flats and the exposed reef crests. Besides contributing to the

▼ *Clad in bold zebra stripes, a cleaner* Labroides dimidiatus *works over its parrotfish "client" in the Red Sea.*

PARROTFISHES

The parrotfishes, so named for the fusion of their teeth into a distinctive beak, represent a moderately diverse group of fishes with 10 genera and 84 species. They differ from the families Labridae and Odacidae in that they are almost exclusively tropical. The great majority of parrotfish species occur on coral reefs. The exceptions are a small number of species which occur in seagrass beds and over rocky reefs. The species that occur with seagrass are most abundant in the Caribbean, whereas species such as *Nicholsina denticulata* and *Sparisoma cretensis* occur over rocky reefs in the Gulf of California and the Mediterranean respectively. In addition, some widely distributed tropical species, such as the blue-banded parrotfish *Scarus ghobban*, may be found in subtropical waters and some distance from reef environments.

With few exceptions parrotfishes are herbivores and their jaw structure allows them to graze intensively on the small plants that cover the exposed calcareous matrix of coral reefs. Almost all the grazed plant material is made up of small, densely packed, thread-like turf algae and calcareous (high in calcium carbonate) algae. In most instances parrotfishes do not select individual plants but graze the surface, removing plant material, detritus, and associated sediment. Some members of Scaridae, especially the

Peter Scoones/Planet Earth Pictures

▶ *Some parrotfishes have the extraordinary habit of producing a loose cocoon of mucus about themselves before settling in for the night. Open at both ends to allow a current flow, this unique sleeping bag takes about 30 minutes to construct. The coating is foul-tasting and may deter nocturnal predators, which hunt by scent.*

Christian Petron/Planet Earth Pictures

process of bioerosion, parrotfishes play an important role in the dynamics of reef ecosystems, converting plant material into fish flesh by their continual grazing of the turf algae. Their feeding can modify the reef environment by removing some species of coral and larger algae when they are newly established or small.

Most parrotfish species appear to have relatively rapid growth rates, reaching maturity in 2 to 4 years and with lifespans in the vicinity of 5 to 20

AN UNUSUAL CHEWING MECHANISM

The labrids share with the scarids and odacids a highly modified pharyngeal apparatus. In these groups, this apparatus acts as a second set of jaws in the throat which chops, crushes, grinds or otherwise processes the food. The structure and arrangement of the teeth in these jaws is related to the diet. Species that feed on hard-shelled prey, such as echinoderms and mollusks, have large rounded teeth while species feeding on softer prey have more finely pointed teeth.

In cichlids (family Cichlidae), the presence of a set of powerful jaws specifically for food processing is believed to have released the oral jaws from this task, thus enabling them to specialize in various methods of obtaining food. The same seems to have happened in the labrids, scarids, and odacids, where a wide range of feeding modes have evolved.

Most odacids follow the general labrid plan. However, in two New Zealand species the pharyngeal jaws have sharp-toothed ridges, that chop the kelp on which these species feed into small pieces. The gut contents look like seaweed coleslaw.

The most highly specialized pharyngeal jaws are those of parrotfishes, in which the teeth in the pharyngeal bones are arranged in rows. As the upper and lower jaws grind against each other during feeding, the ingested material (turf algae and associated pieces of coral) is turned into a fine paste. In this way the algal cells are broken down and the cellular contents released for digestion. As a by-product, a large amount of fine sediment is produced. Parrotfishes are a major source of sediments on reefs. A single large specimen of the Indo-Pacific species *Bolbometopon muricatum* consumes approximately 1 cubic meter (35 cubic feet) of coral skeletons per year, releasing the material again as a fine silt. These bump-head parrotfishes graze in large schools like underwater buffalos, and divers often hear their distinctive "crunch" long before the fish become visible. Feeding on either live or dead coral, their actions are believed to have a significant impact on the fine-scale topography of coral reefs.

Hugh Jones/Planet Earth Pictures

▶ *A parrotfish* Chlorurus gibbus *grazes algae-covered coral in the Red Sea. Some species forage in water so shallow their tails break the surface.*

years. The sizes achieved by parrotfishes are fairly uniform, in the vicinity of 30 to 40 centimeters (12 to 16 inches) in length. One or two species, such as the bump-headed parrotfish, may reach 1 meter (3¼ feet).

WRASSES

The closely related wrasses provide numerous contrasts to the parrotfishes. They constitute the myriad of usually small, brightly colored carnivorous fishes on coral reefs. They are also the most abundant of the three groups considered, with approximately 425 described species. Unlike the parrotfishes they extend into temperate and even sub-Arctic waters, with large numbers of exclusively coldwater species that make up an important element of temperate reef faunas. Some are found as far north as Norway.

Wrasses are perhaps the most efficient carnivores among reef fish, actively seeking a wide range of small invertebrates. Most feed upon hard-shelled prey. These are captured with a strikingly diverse range of feeding structures which enable wrasses to extract camouflaged prey from a variety of shelter sites. Like the parrotfishes, they feed only during the day. A substantial number of wrasses are plankton-feeders, gathering in active groups at reef fronts, reef gaps, and other areas where plankton is concentrated. The best known feeding pattern among the wrasses is probably that of the

with the largest labrid, *Cheilinus undulatus*, reaching 2.5 meters (about 8 feet) in length. Small wrasses are quick to exploit any disturbance to the reef bottom that exposes normally camouflaged invertebrates. In this fashion many small labrids show feeding parallels with birds such as cattle egrets, exploiting the disturbances created by larger species when feeding. On reefs small wrasses often follow the feeding trails of larger scarids and labrids, picking up the crustaceans and mollusks exposed by the activities of their larger relatives.

BUTTERFISHES AND ROCK WHITINGS

The third group within this trio of labroid fishes, the Odacidae, are very distinctive. They differ from their more diverse relatives in that the family is comprised of only 12 species restricted to the temperate reefs of Australia and New Zealand. Despite their scarcity they represent a fascinating offshoot of the mainstream pattern of wrasse and parrotfish evolution. The feeding patterns within the family are diverse, ranging from pickers of small invertebrates to herbivores grazing on large brown algae. It is the three species of the latter group that establish the unique features of odacids. They represent one of the few groups of fishes that are able to feed on the prolific but chemically unpleasant kelps and some algae that dominate the colder reef waters of the southern hemisphere.

J. HOWARD CHOAT & DAVID R. BELLWOOD

Georgette Douwma/Planet Earth Pictures

▲ *The Napoleon wrasse* Cheilinus undulatus *has a distinctive hump and thick-lipped jaws. Members of this genus are among the largest in the family, sometimes growing to 2 meters (about 6 feet) or more and weighing more than 150 kilograms (330 pounds).*

◀▼ *The rainbow cale* Odax acroptilus *(left), an Australian member of the small family Odacidae. Closely related species feed on pieces of kelp and seaweed torn off with their parrot-like beaks. The striking harlequin tuskfish* Choerodon fasciatus *(below), an Indo-Pacific wrasse, uses its strong jaws and stout, tusk-like teeth to grasp shellfish.*

cleaner wrasses (genus *Labroides*), which remove parasites, mucus, and scales from the bodies of large fishes.

Wrasses are generally smaller than parrotfishes. The majority of wrasse species in tropical waters have an average length of less than 20 centimeters (8 inches), and many species are smaller. There is also a greater variation in size than in parrotfishes,

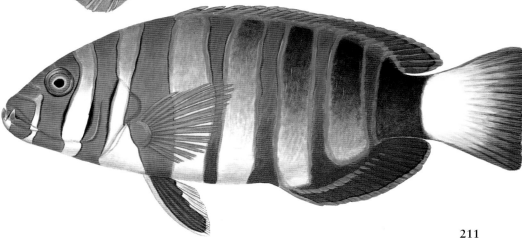

Fredrik Ehrenstrom/Oxford Scientific Films

▲ *The clown coris or twin-spot wrasse Coris aygula has a broad Indo-west Pacific distribution in association with coral reefs. The initial phase females are characterized by a broad white bar around the body and maroon spots on the head, as shown here.*

▼▶*The highfin parrotfish Scarus altipinnis—same fish, three uniforms. Most wrasses and parrotfishes travel through three color phases. Moreover, gender changes as well: most juveniles are female; terminals are males.*

SEX REVERSAL IN PARROTFISHES & WRASSES

Thirty years ago researchers were surprised to discover sex reversal in a number of reef fishes. Today, detailed studies have revealed that sex reversal appears to be the norm for a large proportion of reef fishes. Of these, the parrotfishes and wrasses demonstrate these sex changes in the most spectacular way. They are both herma-phrodites (one individual capable of being male or female) and sexually dichromatic (the sexes have different color patterns). In many species, the color patterns of the sexes are so different that the males and females have been regarded as different species for more than two hundred years.

However, this is not just a simple case of sexual dichromatism. Parrotfishes and wrasses also have one of the most complex and bizarre patterns of reproduction.The most complex system is as follows. In each species there may be up to three distinct color phases. These are termed the juvenile phase, the initial phase, and the terminal phase. Immature fishes have a juvenile phase color pattern. This may be similar to the initial phase but is often quite distinct. As juveniles mature they take on the initial phase (IP) coloration. This phase is usually relatively drab, with colors

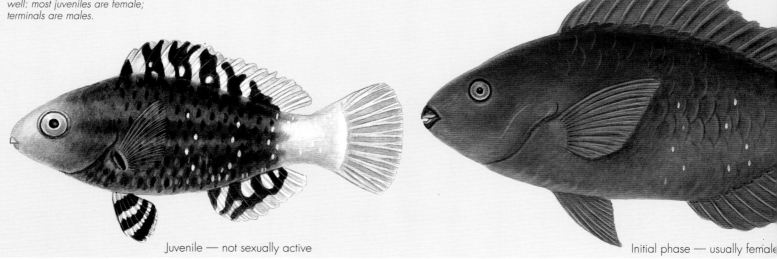

Juvenile — not sexually active

Initial phase — usually female

dominated by grays, reds, and browns. Most fishes in this phase are female. In some species, however, some individuals in this phase are male. These males are termed primary males and often look identical to the females. The only way to distinguish them is by internal examination or to watch them during the process of spawning, when they take on a distinctive breeding pattern. After a length of time, initial phase individuals may change color and take on the terminal phase (TP) coloration. For the females this color change is associated with a change of sex. These new males are termed secondary males. Primary males simply change their colors.

Successful terminal phase males usually dominate reproductive activity. They may have a harem of females or hold temporary reproductive territories. In these cases, spawning usually involves just one TP male and one IP female. Occasionally primary males will try to "sneak-spawn". Having approached a TP male's territory, presumably masquerading as a female, the primary male will rapidly follow a pair of spawning fish into the water column and at the point of spawning will release a large quantity of milt (sperm and seminal fluid). Its large testes presumably enable it to swamp the efforts of the TP male which has smaller testes and therefore a more limited capacity to fertilize eggs. TP males rely on aggression rather than volume of milt to dominate egg fertilization. The other main spawning mode involves both TP and IP individuals in mass spawning events.

Some species have no primary males, while others display no color changes. The presence, absence, and relative number of primary males appear to be related to a complex set of behavioral and demographic factors.

Sex change in the cleaner wrasse *Labroides dimidiatus* has been shown to be dependent on behavioral interactions, with the aggression of a dominant male keeping the other individuals in the harem female. Loss of the male results in the dominant female taking over the aggressive male role within hours, with a complete sex change to a functional male within days.

◄ A pair of three-striped wrasses Stethojulis trilineata (male above, female below). In some species of wrasses a single male may have a harem of four or five females so different that the two sexes have been described as separate species.

Rudie H. Kuiter

◄ The clown coris or twin-spot wrasse takes its common name from the beautiful coloration of the juvenile, shown here. As it grows, the two black and orange spots are lost, while the initial phase coloration (illustrated opposite) develops.

Max Gibbs/Oxford Scientific Films

Terminal phase — always a mature male

BLENNIES

KEY FACTS

ORDER PERCIFORMES
SUBORDER BLENNIOIDEI
• 6 families • c. 140 genera
• c. 800 species

SMALLEST & LARGEST

Medusablennius chani
Total length: 1.7 cm (⅝ in)

Xiphasia setifer
Total length: 55 cm (22 in)

CONSERVATION WATCH
! Tayrona blenny *Coralliozetus
tayrona, Protemblemaria punctata,*
and Bot River klipfish *Clinus
spatulatus* are vulnerable.

The Blenniidae (combtooth blennies) is the largest family in the suborder Blennioidei, which is made up of 6 families and about 800 species of small, shallow-water and bottom-dwelling fishes. Except for a few combtooth blennies that occur in fresh or brackish water, all blennioids are primarily marine inhabitants. Combtooth blennies, like most other blennioids, deposit their eggs on hidden surfaces, especially in holes, where they are fertilized and guarded by the males.

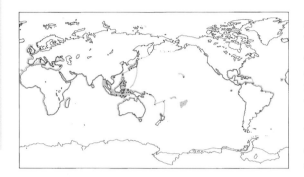

CRYPTIC COLORS

Blennioids are distinguished from other fishes mainly by modifications of their skeletons. Externally, they can be recognized by a combination of characters. The pelvic fins are small, placed in front of the pectoral fins, and each consists of a hidden spine and two to four unbranched soft rays. The anal fin consists of zero to two weak spines and only unbranched soft rays, and the dorsal fin is long, extending from over or just behind the head almost to the tail, or even joined to the tail in some species. Many species have hairlike filaments, called cirri, distributed variously on the nostrils, eyes, and nape. Different sexes of the same species are often strikingly different in appearance.

Most blennioids are cryptically colored and secretive. As they are also small, they may not be seen easily by the casual observer. They are easy to maintain in aquaria and some are of spectacular appearance, which has made them quite popular with aquarists.

COMBTOOTH BLENNIES

The Blenniidae can be distinguished from the other blennioid families by a combination of characteristics: lack of scales; having fewer dorsal-fin spines than soft rays; and having incisor-like teeth that are close-set in a single row in each jaw (hence the common name, combtooth blennies). Depending on the sex or species, they may have an enlarged, conical tooth curving back at the rear of each side of the lower jaw and, in some species, on each side of the upper jaw.

Combtooth blennies vary in color pattern from almost uniformly black (*Atrosalarias fuscus*) to almost uniformly red (*Lipophrys nigriceps*) or uniformly yellow (*Meiacanthus ovalauensis*). Many species are drab, brown-banded or mottled; others are striped; and many have red or iridescent blue spots. Some species exhibit two or three color patterns so strikingly different that each pattern has been described as a different species.

There are several tribes of combtooth blennies, each of which has a well-defined center of distribution. One tribe, the Parablenniini, although distributed right around the world, has its greatest concentration of species in and around the Mediterranean Sea. The largest tribe, Salariini, with almost half the species, is also distributed around the world, but most of its species are concentrated in the tropical Indo-Pacific. The salariins are distinguished by the nature of their teeth, which are small and held loosely in a band of tissue in each jaw. Most salariins feed by scraping algae and small organisms off the surface of rocks and coral.

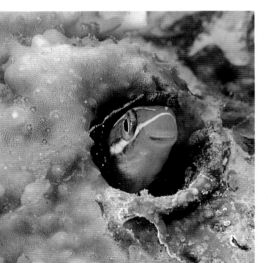

Brent Hedges

▲ A blue-lined saber-toothed blenny *Plagiotremus rhinorhynchos* peers from a tube-worm tunnel. Saber-toothed blennies feed by darting out at other fishes and removing skin, mucus, or scales with their large lower canine teeth.

▶ Among the most attractive blennies, the black-headed blenny *Lipophrys nigriceps* lives in small dark caverns and crevices in the Mediterranean.

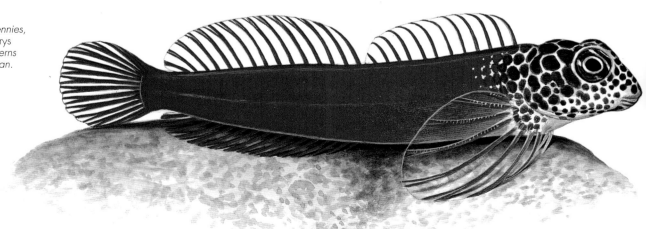

Georgette Douwma/Planet Earth Pictures

Except for one species in the eastern Pacific, the tribe Nemophini, or saber-toothed blennies, is restricted to the tropical Indo-Pacific. Although the tribe is defined primarily by certain skeletal characters, it takes its common name from a very large pair of recurved canine teeth in the rear of the lower jaw. These canines are used defensively against potential predators, and the fish can, if handled, inflict a painful wound to the finger. Several species of saber-toothed blennies feed by nipping the fins and skin of other fishes. To fool their prey, these blennies have evolved into mimics of harmless or socially desirable species.

OTHER BLENNIOID FAMILIES

The members of the unscaled family Chaenopsidae, or tube blennies, with 11 genera, about 65 species, and the largest species 16 centimeters or 6 ¼ inches in length, are found only in the warm seas of the western hemisphere. Many male tube blennies are more brightly or contrastingly colored and have much higher dorsal fins than females. These differences are associated with mating behavior. Males rhythmically jerk in and out of their tubes and erect their fins to attract females into the tubes for mating. Among the tube blennies is one of the smallest vertebrate species,

▲ A salariin combtooth blenny (genus Ecsenius) resting among honeycomb coral. The tribe Salariini comprises almost half of all combtooth blenny species, most of which are tropical reef dwellers.

▶ *The yellow-crested weedfish Cristiceps aurantiacus is a clinid, or kelp blenny, from southern Australia. Like other clinids living in this region, it bears live young. All clinids are adapted for life among seaweed, where their shape and cryptic coloration render them virtually invisible.*

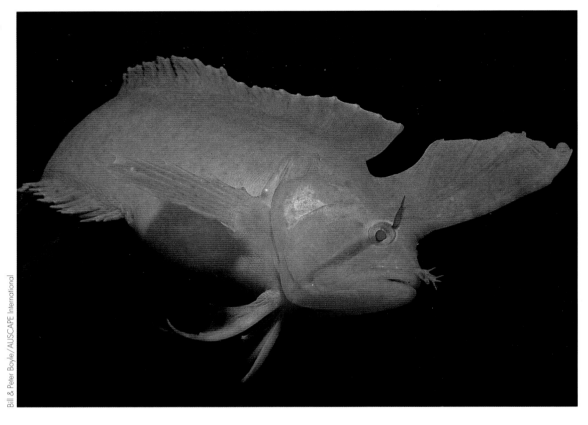

Bill & Peter Boyle/AUSCAPE International

▼ *Triplefins are poorly known blennioids named after the division of their dorsal fins into three parts. This is the banded triplefin Tripterygion segmentatum.*

Kevin Deacon/Dive 2000

the Caribbean *Acanthemblemaria paula*, females of which may be mature at 1.3 centimeters (½ inch).

The scaled, sand-dwelling Dactyloscopidae, or sand stargazers (9 genera, 41 species, largest 17 centimeters, or 6¾ inches), are drably colored and are restricted to the warm seas of the western hemisphere. Many species have peculiar tubular eyes that project above the sand in which they bury themselves for protection. Little is known of

their behavior, but males of some species hold a clump of eggs in the axilla ("armpit") of the pectoral fin on each side, a unique method of parental guardianship among fishes.

The scaled reef-dwelling Labrisomidae, or labrisomids (15 genera, 102 species, largest about 35 centimeters, or 14 inches), is primarily distributed in the warm waters of the western hemisphere, but a few species are present on the coasts of west Africa, Korea, Japan, and Taiwan. The eastern Pacific species of one genus, *Starksia*, give birth to live young, but the western Atlantic species of the same genus apparently do not.

The scaled Clinidae, or kelp blennies (27 genera, about 80 species, largest 61 centimeters, or 24 inches), have a distribution essentially restricted to temperate coasts north and south of the equator, and are closely associated with seaweeds. Many are bizarre in shape, with greatly constricted caudal peduncles (tail stalks) and dorsal fins that begin high on the head. The scales of kelp blennies are much smaller than those of other scaled blennioids, and are imbedded in the skin. Most species are found on the coasts of southern Africa and southern Australia. All the Indo-West Pacific species bear live young, whereas all of the seven or eight species that occur variously in the eastern Pacific, Mediterranean, and Atlantic are egg-layers. The giant kelpfish of California *Heterostichus rostratus* is the largest blennioid, attaining a length of about 61 centimeters (24 inches). Its coloration matches its background, varying from brown to red to green. It is often mottled and has irregular bands and stripes. Mature males are always brown, and only females may be red. Except during mating, mature males and females occupy different habitats. This is believed to account for the differences in coloration.

The scaled Tripterygiidae, or triplefins, are the least known of the blennioids. There are more than 20 genera and over 150 species, many undescribed. Their dorsal fin is divided into three parts. All but about three species of triplefins, and no other blennioids, have rough-edged (ctenoid) scales. The species are quite small, the largest attaining 20 centimeters (8 inches). They are distributed around the world, but are noticeably absent from most of the Atlantic coasts of South America and Africa. Most species live on coral reefs, but one species is known from the Antarctic Peninsula, the southernmost occurrence of any blennioid.

VICTOR G. SPRINGER

MIMICRY, BLENNY-STYLE

One of the best known mimetic associations in the world of fishes involves the saber-toothed mimic blenny *Aspidontus taeniatus* and its model, the cleaner wrasse *Labroides dimidiatus* (family Labridae), which picks parasites or damaged pieces of flesh from wounds of large, predatory fishes. The blenny and wrasse are almost impossible to tell apart in their natural habitat. Even geographic variations in color pattern exhibited by one species are paralleled by the other. In the Indian Ocean, both species have a black spot at the base of the pectoral fin; the spot is absent in Pacific individuals. In the southwest Pacific, both species may have a bright orange blotch on the side posteriorly; the blotch is absent in individuals from other areas. The blenny's similarity to the wrasse fools potential prey into thinking it is going to be cleaned, but the unsuspecting target is surprised by a painful nip instead. The blenny is a veritable wolf in sheep's clothing!

The saber-toothed blenny genus *Meiacanthus* is unique among fishes in that its enlarged canines are deeply grooved anteriorly. At the base of the groove is a gland that releases venom into the groove and from there into a wound caused by the bite of these fishes. *Meiacanthus* species feed on small invertebrates, and do not menace other fishes. However, aquarium experiments have shown that if a predatory fish makes the mistake of trying to eat a *Meiacanthus*, the predator is apparently bitten inside its mouth and immediately spits out the *Meiacanthus* unharmed. Sometimes the foolish predator quickly opens its jaws wide and keeps them open, while an unconcerned *Meiacanthus* leisurely swims out. If, however, the tips of the fangs of the *Meiacanthus* have been removed surgically the predator eats the *Meiacanthus* with impunity.

Meiacanthus species serve as models for mimics from various fish families, but the beautiful blue and yellow black-lined saber-toothed blenny *Meiacanthus nigrovittatus,* endemic to the Red Sea, serves as a protective model for another blenny, the salariin *Esenius gravieri,* also endemic to the Red Sea. These two species are just as indistinguishable in nature as the mimic blenny and the cleaner wrasse, but the mimetic relationship is different. The

V. G. Springer

◀▲ *The Red Sea mimic blenny Ecsenius gravieri (left) has gained impunity from predators by evolving an appearance similar to that of the black-lined saber-toothed blenny Meiacanthus nigrovittatus (above). Predators attempting to eat the saber-toothed blenny are usually bitten and thus learn to avoid this blenny and all fishes resembling it.*

V. G. Springer

Ecsenius is a harmless, delectable species that gains protection from predators by looking like the *Meiacanthus.* Aquarium studies have shown that predators that have had a bad experience trying to eat *Meiacanthus nigrolineatus* do not eat *Ecsenius gravieri.* This type of mimicry is called Batesian mimicry, after the English naturalist Henry W. Bates, who first described it from observations of butterfly mimics and models.

GOBIES

Trimmatom nanus
Total length: 1 cm (½ in)

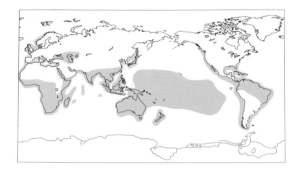

Marbled sleeper *Oxyeleotris marmorata*
Total length: 66 cm (26 in)

CONSERVATION WATCH
!!! There are 5 critically
endangered species: Elizabeth
Springs goby *Chlamydogobius
micropterus*; Edgbaston goby
Chlamydogobius squamigenus;
Glossogobius ankaranensis;
dwarf pygmy goby *Pandaka
pygmaea*; and Poso Bungu goby
Weberogobius amadi.
! There are 18 species listed as
vulnerable.

Gobies belong to the one of the largest groups of fishes, but this group is probably one of the least known suborder of fishes. They are basically small fishes with two fins on the back, a tail fin, a fin along the midline of the lower part of the body, two ventral fins (which in most gobies are fused into a cupshaped disk), and paired fins on the side. Most have scales, although some species may have only a few scales near the tail and a few have no scales at all. They lack the sensory lateral line on the body found in most other fishes. Gobies are so diverse that it is hard to define them without using a lot of technical terms relating to their bones. They first appeared in the fossil record in freshwater deposits during the Eocene, around 30 to 50 million years ago.

THREE FAMILIES
Gobies are often divided into two to six families, and the classification of the group is currently being studied to resolve this disparity. The latest classification recognizes three families: the

Rhyacichthyidae, containing two species found in fast-flowing streams on islands in the western Pacific; the Odontobutidae, a small group of freshwater fishes from China, Korea, and Japan; and the Gobiidae from most parts of the world, containing over 99 percent of the species, which includes the sleepers or gudgeons (formerly the family Eleotridae). Current estimates suggest there are probably 2,000 species of gobies, which is almost one out of every ten species of fish. The group is so poorly known that 10 to 20 new species are described each year.

HABITAT AND FOOD
Gobies occur in most parts of the world, except the Arctic and Antarctic oceans and in the deep sea. Slightly less than half the species occur on

▶ *Gobies include the smallest of all fishes—in fact the smallest of all vertebrates—especially species in the genera Pandaka and Trimmatom, shown here. Nearly half of the estimated 2,000 species in the group inhabit coral reefs.*

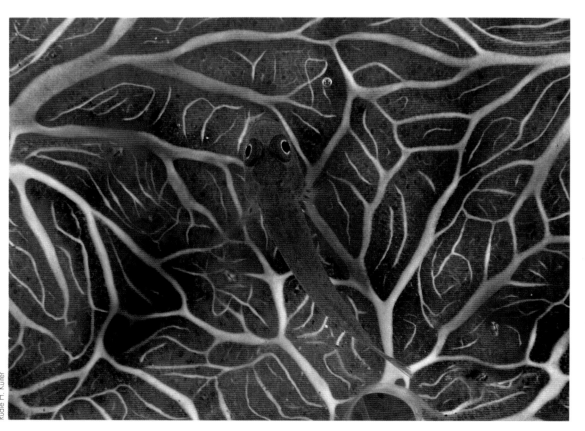

Rudie H. Kuiter

coral reefs, about a third in estuaries or muddy bays, about one-tenth in fresh water, and the rest on rocky reefs, sandy beaches, or on the continental shelf. In fresh water some species have been found at altitudes of over 2,000 meters (6,000 feet), and in the ocean at depths of around 800 meters (2,600 feet). Most occur in the tropics, but a small number range into cooler regions. For example, some are known from Alaska, Norway, and southern South America, to latitudes of about 55° from the equator. Most gobies spend the majority of their time sitting on the substrate, but some species are active swimmers and may occur in schools of up to 100 fish. They feed mostly on small invertebrates, although some with large mouths may eat other fishes, and a few feed on algae. Many are selective and feed by attacking an individual prey item. Others take a large mouthful of mud or sand and sift out invertebrates or minute algae. The free-swimming species often feed on tiny plankton.

A VARIETY OF SIZES

Gobies come in a variety of sizes. The smallest known vertebrate is a small goby, reaching a maximum size of 1 centimeter (½ inch), found on coral reefs of the western Pacific Ocean. When printed, its scientific name, *Trimmatom nanus*, is longer than the fish itself. At the other extreme are a few gudgeons and freshwater gobies which reach a size of half a meter (20 inches), or 50 times larger than the smallest goby. Most adult gobies are in the range of 40 to 100 millimeters (1½ to 4 inches).

LIFE CYCLE

Most, if not all, gobies have a very similar life cycle. The females can lay from five to a few hundred eggs, attaching them to vegetation or to a bit of shell, rock, or coral. The male then fertilizes the eggs. The female departs and the male is left to guard the eggs and keep them clean. The eggs hatch in one to a few days into a small transparent larval stage of 2 to 10 millimeters (1/12 to ½ inch) long. The larvae are dispersed into the water column and swim for 3 to 20 days, depending on the species. The larvae then settle into a suitable habitat and rapidly develop coloration to match their surroundings. In warm waters the fish grow rapidly and reach maturity in a few months. Most of these probably live only a year. In cooler areas, the gobies grow more slowly and reach maturity after one to two years. Some of these species are thought to live for two to ten years.

▲ *Its relationship to true gobies uncertain, the decorated fire-goby* Netmateleotris decora *(far left) is often placed in a separate family, the Microdesmidae. This small but spectacular fish is always found near its burrow, into which it darts when threatened. Although its colors suggest that it might be a tropical reef dweller the bluebanded goby* Lythrypnus dalli *(top) is found around Catalina and other islands off southern California. It is associated with open rocky areas but quickly dives into a crevice or hides among urchin spines when frightened. Doria's bumblebee goby* Brachygobius doriae *(right) is a diminutive brackish water goby of Southeast Asia. Like other true gobies, its pelvic fins form a sucking or perching disc with which it can cling to rocks or support itself on soft sediment.*

Norbert Wu

▲ A goby on bubble-coral. Most species of goby are tiny, and have the ventral fins to form a sucker-like disk; they spend most of their time on the bottom or even partly buried in mud or sand. They are extremely common on coral reefs, where they may be the most abundant of fishes.

cases the goby takes advantage of its invertebrate host by sharing its home. The goby may return the favor in another way. It has been known for a long time that gobies live in burrows in mud or sand. When researchers began digging up the burrows, they found they were home to a variety of invertebrates, including crabs, shrimps, and worms. For example, the blind goby of California spends its whole life in a burrow constructed by a small shrimp.

One of the best-known associations is that of gobies with snapping shrimps on coral reefs. The shrimps occur in pairs and construct extensive burrows which zig-zag back and forth in sand, reaching a depth of about 1 meter (3 feet) from the surface of the sand. The shrimp piles the sand on its large claw, taking it out of the burrow, and dumping it a short distance away from the entrance. The goby sits just outside the entrance doing sentry duty. When the shrimp is ready to dump its load, it places one of its long antennae on the tail of the fish. If it is safe to come out, the fish wiggles its tail in a certain way. But if there is a predator around, the goby will not give the signal, and if a predator comes too close the fish turns and dives into the hole. The shrimp feeds on detritus and tiny bacteria and algae in the sand. When the shrimp brings out its load, the sand often contains small crabs or other invertebrates, which become food for the goby. This cooperative association allows both animals to survive in a habitat where one would not survive on its own.

Another common association is that of gobies with invertebrates that are fixed to the substrate, such as corals, sponges, giant clams, soft corals, or seawhips. In these associations, the gobies spend their whole lives in a single coral or a small group of invertebrates, such as seawhips. Those fishes that live in corals gain a good hiding place, but it is not clear what benefit the coral derives from the association. The sponges and seawhips enable the fish to move higher up in the water column and feed on plankton. In these types of associations it is common to find one male in each invertebrate clump and two or more females. If the male dies or is eaten, there is a problem; to solve it, one of the female gobies changes sex.

On coral reefs some gobies may breed several times a year, but in colder regions they are thought to breed only once or twice in a year. In most marine species, the larvae and adults usually occur in the same vicinity. However, in many freshwater species the adults breed in fresh water and the larvae are then carried downstream to estuaries and into the sea. After a few weeks to months, the larvae migrate into fresh water. In some cases these migrations involve such large numbers of fishes that they are caught for food, particularly in the Caribbean and Philippines.

USEFUL ASSOCIATIONS

Within the large number of species in the goby group, some unusual life styles have developed. One of the most interesting is the association of gobies with various marine invertebrates. In most

SEX CHANGE

Changing sex from female to male also occurs in some other coral reef gobies, which are not associated with invertebrates. By having mostly females, the fish are able to produce more eggs, thus increasing the species' chance of survival. Males then have to breed with several females and also guard the eggs. Because of this disadvantage for the males, no one thought gobies would change sex, and the sex change has been discovered only in the last few years. Although sex change is known in some species of gobies, most species are thought not to change sex.

◀ *Two mudskippers, Boleophthalmus caeruleomaculatus (family Gobiidae), in confrontation on a tropical Australian mudflat. Mudskippers live in burrows in mangrove swamps, emerging at low tide to move about on the mud and catch insects. Strongly territorial, they challenge and sometimes fight trespassers.*

Jean-Paul Ferrero/AUSCAPE International

MUDSKIPPERS

Another strange feature among gobies is found in the mudskipper (genus *Periophthalmus*). During low tide these fishes spend more time out of water than in it. They live in muddy mangrove swamps, and are able to skip across the mud, perching on the roots of the mangroves or on the mud itself. At high tide they retreat into burrows. These fishes have special adaptations to allow them to live on land. Their skin contains numerous blood vessels, which are able to extract oxygen from the air. Their eyes are perched high on the head, enabling them to see their food of flies and small insects, and to avoid birds, which prey on them. They also have a very muscular tail and side fins to help them "skip" over the mud.

AVOIDING PREDATORS

Being small, gobies are often preyed upon by larger fishes, sea snakes, and shore birds, and they

▼ *Many goby species live in very intimate symbiotic relationships with other animals such as corals, mollusks, and crustaceans, like this goby Amblyeleotris guttata sharing a burrow with an Alpheus shrimp.*

Kev Deacon/Dive 2000

Rudie H. Kuiter

▲ *Signigobius biocellatus, a courting pair. The female lays up to several hundred eggs (depending on species) which are fertilized and then guarded by the male. Many goby species consist mainly of females, adults only occasionally changing gender to males as they get older.*

have developed a number of adaptations to reduce the chance of being eaten. Many gobies spend most of their time in burrows and come out only to feed. Others bury themselves in sand. Around coral reefs, many avoid predators by living in the branches of corals and others by living in the dark caves, often swimming upside down or resting upside down on the roof of the cave. A few gobies have even developed a powerful poison called tetrodotoxin, which also occurs in pufferfishes and a species of salamander. One of the most common forms of defense is to develop cryptic coloration. For example, species living on sand develop a speckled coloration that matches the sand. On coral reefs, many gobies are largely transparent, with a few colored spots that match the coral or other invertebrate on which the fishes live.

ENDANGERED SPECIES

Gobies are generally so common and abundant that one might not expect any need for conservation, but this is not the case for some freshwater species. Many freshwater species in Australia and New Guinea are confined to a single lake or river system. Development and agricultural practices can result in the decline of these species. The Poso Bungu goby *Webergobius amadi* of Sulawesi is thought to be endangered due to introduced fishes. A second goby, *Knipowitschia*

thessala in Greece, is also thought to be endangered. Another 22 species are listed as very rare or vulnerable. Gobies are common on small islands and recent studies are finding that many species are restricted to one or a few islands. A number of new species have been found recently in rivers of New Guinea, Sulawesi, the Philippines and Madagascar. In some of these places fishes not native to the area have been introduced, threatening the native species. Some goby species may become extinct even before they are discovered.

Gobies have also been a culprit in introductions. Unlike most introduced species, gobies appear to be carried in ballast water or in the intake pipes of ships. As a result several species have been transported in sufficient numbers for populations to be established outside their normal range. The most successful dispersal by this method are two marine gobies native to Japan, Korea, and China. These found their way to California between 1960 and 1963 and to several parts of Australia between 1970 and 1980. Both are known from Sydney and Melbourne and one is known from Perth.

More recently, a European goby was found in the Great Lakes, along with the zebra clam. The clam is causing considerable havoc there, but as yet little is known of the effect of the goby introduction.

DOUGLASS F. HOESE

FLATFISHES

Flatfishes are as strange-looking as they are delectable. Adults have deep bodies that are compressed sideways and have both eyes oddly situated on the same side of the body. This remarkable departure from the normal symmetrical body plan of most fishes is the final outcome of spectacular changes that occur during larval development. Members of this highly diversified group are found in all oceans of the world, with a few species inhabiting brackish or fresh waters. Flatfishes are found mainly on the continental slopes of oceans but can also be captured in deep waters. Numerous species, such as the sole, halibut, turbot, plaice, and brill, are well known to seafood fanciers and are vital components of commercial fisheries in several countries.

KEY FACTS

ORDER PLEURONECTIFORMES
- 3 suborders • 7 families
- 13 subfamilies • 117 genera
- *c.* 540 species

SMALLEST & LARGEST

Pigmy whiff *Tarphops oligolepis*
Total length: 4.5 cm (1¾ in)
Weight: 2 g (⁷⁄₁₀₀ oz)

Atlantic halibut *Hippoglossus hippoglossus*
Total length: 2.5 m (8 ft)
Weight: 316 kg (697 lb)

CONSERVATION WATCH
!! Atlantic halibut *Hippoglossus hippoglossus* is listed as endangered.
! Yellowtail flounder *Pleuronectes ferrugineus* is listed as vulnerable.

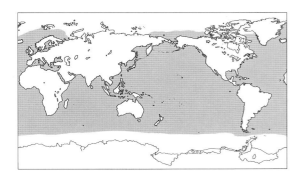

CHAMELEONS OF THE SEAS

All flatfishes (order Pleuronectiformes) are asymmetrical and probably share a unique and common ancestor with a still unidentified group of symmetrical fishes. The oldest fossils of flatfishes are otoliths from the Paleocene epoch (65 to 57 million years ago) and skeletons from the Eocene epoch (57 to 37 million years ago). Flatfishes are also the chameleons of the seas. Several species can modify the pigmentation pattern of their eye side to mimic the environment. They usually cover their body with a fine layer of sand and stay immobile for varying periods of time, with only their protruding eyes probing the surrounding area for prey or predators.

The pigmentation pattern of the eye side varies greatly between species but the blind side is usually very pale or unpigmented. In the soleid genus *Zebrias* and some related genera, differences in the patterns of vertical dark bands on the eye side of the body allows for proper species identification.

SPINY FLATFISHES

The primitive suborder Psettodoidei contains one family, the Psettodidae, with one genus, *Psettodes*, and three tropical and subtropical species. Except for the twisted eye region, this group does not

◄ *The flatfishes typically lie quietly on the bottom, hidden below a light sifting of sand with only their eyes protruding; even the angle of their oblique mouths reveals their adaptations to a bottom-living lifestyle. This is the topknot Zeugopterus punctatus of European waters.*

Jim Greenfield/Planet Earth Pictures

Ken Lucas/Planet Earth Pictures

▲ *A speckled sanddab* Citharichthys stigmaeus *of America's Pacific Coast. Some flatfishes are the chameleons of the fish world, with a remarkable ability to modify their skin colour to match the environment. Growing to about 17 centimeters (7 inches) in length, this species prefers sandy bottoms at depths to about 90 meters (300 feet).*

America the same common name is given to a less valued flounder, *Microstomus pacificus.*

Most flounders are divided into a right-eyed group (family Pleuronectidae) and a left-eyed group (family Bothidae). This distinction is a convenient, but not phylogenetically informative, way of representing the diversity within flatfishes. Another small family, the Citharidae, contains a right-eyed subfamily (Brachypleurinae) and a left-eyed subfamily (Citharinae) that are thought to be evolutionary intermediates between the spiny flatfishes and other groups of flounders.

The right-eyed flounders are diverse in form and structure. Several species, such as halibut, flounder, and plaice, support important fisheries in the Arctic, Atlantic, and Pacific oceans. The Atlantic halibut *Hippoglossus hippoglossus* holds the world record for flatfish size, with a specimen of 2.5 meters (about 8 feet) in length, weighing 316 kilograms (697 pounds). After intensive fishing in recent decades, specimens weighing over 90 kilograms (200 pounds) are now quite rare in Canadian waters. Very large specimens are estimated to be 30 to 35 years old and are always females. A 90-kilogram (200-pound) female lays over 2 million small spherical eggs. Because the buoyancy of the eggs is neutral, they usually float in midwater and sink gradually as the larvae develop.

In the left-eyed flounders, the genus *Paralichthys* also contains some important commercial species, such as the California halibut and the Japanese halibut. Some grow to 1.5 meters (5 feet) in length. Interestingly, in the same family we find the miniature genus *Tarphops*, with sexually mature females of only 4.5 centimeters (1¾ inches) in length. In several left-eyed flounders,

show obvious signs of asymmetry. It is the only group with spines in its dorsal and anal fins, hence the common name, "spiny flatfishes". This characteristic and other skeletal features may indicate that the order had a perciform ancestor.

FLOUNDERS

The 300 or more species of the suborder Pleuronectoidei are commonly referred to as flounders. Although the group contains several species that are called soles, this is not entirely accurate since soles are members of the suborder Soleoidei. For example, in Europe, the tasty Dover sole *Solea vulgaris* is a true soleid, while in North

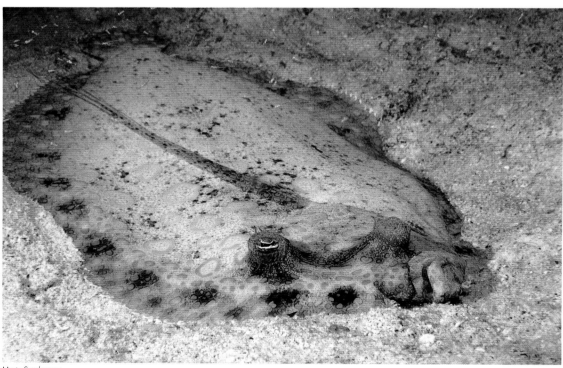

▶ *Inhabiting warmer Atlantic waters between Florida and Brazil, the peacock flounder* Bothus lunatus *is one of the relatively few strikingly marked or colored flatfishes. It grows to about half a meter (18 inches) in length.*

Marty Snyderman

RIGHT-EYED (DEXTRAL) FLATFISH

larvae — normal eye position   left eye migrates to top of head   adult — eyes both on right side

◀ *At hatching young flatfishes differ little from other fishes (far left), but at an early age the larva begins to lean to one side (middle) as one eye (the left eye in dextral flatfishes, the right in sinistral flatfishes) migrates across the crown to take up a position beside the other eye (left). The "down" or blind side turns whitish. At the same time the front of the skull twists to bring the jaws into an oblique sideways position.*

males have wider spaces between their eyes than females. They also have spines at the tip of the snout and near the eye socket, and elongated fin rays in the pectoral, pelvic, and/or dorsal fins.

SOLES

The 220 or more species of the suborder Soleoidei are most often referred to as soles. They include three families: the right-eyed Achiridae (American soles), the Soleidae (true soles), and the left-eyed Cynoglossidae (tongue soles). Soles form a group of highly specialized flatfishes with several asymmetrical features in addition to eye orientation: a small twisted mouth with small teeth; small eyes; and often poorly developed and highly asymmetrical pectoral and pelvic fins. In addition, the tongue sole has a long hook on the snout that overhangs the mouth region and lacks pectoral fins and the eyed-side pelvic fins.

MIGRATING EYES

All flatfishes lay eggs. After hatching, the larva has the general features and behavior of a symmetrical fish. At a particular stage of its development, one eye will begin a slow but dramatic journey that will take it across the top of the head toward the other side of the body, near the other eye. The adult fish, which now looks more like a deformed creature from a Picasso painting, comes to lie on the ocean floor on its "blind" side.

The distance of this unique eye migration varies greatly among flatfishes, from about 5 millimeters (1/5 inch) in some American soles to more than 12 centimeters (4 3/4 inches) in some left-eyed flounders belonging to the deepsea genus *Chascanopsetta*. The migration route also varies among flatfish groups. In some flounder larvae, the migrating eye follows a depression that forms between the eyes. This depression is in front of the developing dorsal fin. After the passage of the eye, the fin extends forward to take its normal adult position. In other larvae, the eye will pass through a slit or space between the already well-developed dorsal fin base and the skull.

Most flatfish species have their eyes on either the left side (sinistral) or on the right side (dextral). Only the spiny flatfishes (genus *Psettodes*) have an equal proportion of right-eyed and left-eyed specimens. The starry flounder *Platichthys stellatus*, however, is noteworthy. This species is found in the North Pacific, from Japan to southern California. Almost 100 percent of the fishes from Japanese waters are left-eyed. Off Alaska, the ratio becomes 70:30 in favour of left-eyed specimens while in the waters off California the ratio is 50:50. Since the normal state for the starry flounder is to have its eyes on the right side, all specimens off Japan are considered to be abnormal. This is confirmed by the orientation of the optic nerves linking the eyes to the brain. In the right-eyed starry flounder, and in most other right-eyed flounders, the optic nerve of the migrating left eye always crosses dorsally to the nerve of the right eye. The optic nerve in the left-eyed starry flounder shows a unique and abnormal twisted condition.

LARVAL FORMS

After hatching, the elongated flatfish larva will gradually transform itself into a more rounded, egg-shaped or oval fish. Most flatfish larvae have an opaque body with moderately or heavily pigmented musculature. The larvae of the left-eyed flounders (family Bothidae), however, are extremely thin and have translucent flesh sparsely

▼ *The European lemon sole Microstomus kitt is actually a member of the right-eyed flounder family Pleuronectidae. Growing to a maximum length of 65 centimeters (26 inches), this species is a highly popular food fish, prominent in northern European fisheries.*

Yves Gladu/AUSCAPE International

▶ *In many flatfishes, such as this turbot* Scophthalmus maximus, *the mouth is asymmetrical. The upper or eyed side is toothless and forms a siphon for more efficient respiration.*

Marty Snyderman

Bill Wales/Planet Earth Pictures

▲ *A plaice larva shown at a fairly advanced stage of the extraordinary movement of one eye relative to the other during early development. The fish will end up lying permanently on its side, with the mouth and eyes distorted to bring them to the new dorsal position.*

decorated by yellow, red, orange, pink, and black pigments. The condition, size, and position of color pigments, together with the body shape, play an important role in species identification.

Most flatfish larvae are armed with spines of various sizes on several parts of the body (including the head, the bones protecting the gill, and some bones of the pelvic and pectoral fins). The pattern of spines is particularly extensive on the head and lower margins of the body. On the topknot *Zeugopterus punctatus*, strong spines are found on the cranium and operculum (the rear bone protecting the gill), where they expand laterally in a wing-like fashion. The larva of *Syacium ovale* has an antler-like spine at the angle of the front part of the gill cover, and a single extended horn-like spine on the cranium. The Indo-Pacific bothine larvae of *Crossorhombus* and *Engyprosopon* species are strongly serrated on the front and lower margins of the belly. These spines may provide protection for vital body organs. They are lost with several other body features, such as the swim bladder, during metamorphosis.

Several flatfish larvae have elongated fin rays in the front portion of the dorsal fin. In *Arnoglossus* and *Laeops* (subfamily Bothinae), the second ray of the dorsal fin is very long and ornamented. In *Arnoglossus japonicus* the ray is one and a half times as long as the body and supports seven elongated flag-like membranes. In *Laeops*, *Chascanopsetta*, and *Cynoglossus* species, the liver and the gut are oddly expanded outward as if the body had been trampled. All these features are thought to be floating devices that increase the ability of the species to disperse during the long planktonic stage of their life cycle. These

adaptations allow for dispersal over a wide area, and bothine larvae are usually caught very far from land. Larvae of other flatfish species, such as the right-eyed flounders, lack the spectacular attributes found in the Bothinae and go through a short planktonic life before settling on the ocean floor. Not surprisingly, their larvae are always caught close to shore.

FOOD AND FEEDING

Primitive flatfishes, belonging to families such as the Psettodidae, Bothidae, and Pleuronectidae are, as a rule, fish eaters. They have large, nearly symmetrical mouths with well-developed dentition. These fishes will often forage in midwaters and migrate between feeding, wintering, or spawning grounds. As a consequence, they are not entirely limited to life on the ocean bottom.

More specialized flatfish groups (some pleuronectids and soleids) are mainly worm and mollusk eaters while other groups (such as tongue soles) feed mainly on crustaceans. Because these fishes are almost exclusively bottom dwelling and so have one side almost always lying on sand, their asymmetry seems to have evolved to a higher degree. In some species, the jaws have teeth only on the blind side while the eyed-side jaws have evolved a secondary function: they form a siphon that increases water intake into the mouth. This adaptation facilitates respiration, allowing water without sand to be taken in. A remarkable flatfish is the Hawaiian deepsea bothine *Chascanopsetta crumenalis* which has a long lower jaw projecting at the front, forming a deep pouch in the throat that probably aids in the capture of prey.

FRANÇOIS CHAPLEAU & KUNIO AMAOKA

TRIGGERFISHES & THEIR ALLIES

KEY FACTS

ORDER TETRAODONTIFORMES
• 10 families • 100 genera
• *c.* 350 species

SMALLEST & LARGEST

Rudarius excelsus
Total length: 2 cm (less than 1 in)

Ocean sunfish *Mola mola*
Total length: up to 3 m (10 ft)

CONSERVATION WATCH
! Queen triggerfish *Balistes vetula*;
Canthigaster rapaensis; and
Sphoeroides pachygaster are listed
as vulnerable species.

Triggerfishes and their allies (order Tetraodontiformes) are one of the most specialized groups of the teleost fishes. The order contains approximately 350 species that are exceptionally diverse in structure, shape, size, and way of life—from triggerfishes and boxfishes to pufferfishes, porcupinefishes, and ocean sunfishes. The distinguishing features of tetraodontiforms are a small mouth with either relatively few teeth that are often enlarged (as in members of the suborder Balistoidei, which have individual teeth protruding from the jaw bones) or massive beak-like tooth plates (seen in the suborder Tetraodontoidei, where the teeth are incorporated within the jaw bones); a small gill opening restricted to the side of the head; and a low number of vertebrae. They also usually have modified and enlarged scales that form some sort of protective covering over the body. Most species are bottom dwellers living in the coastal waters of temperate to tropical seas at depths of less than 200 meters (about 660 feet).

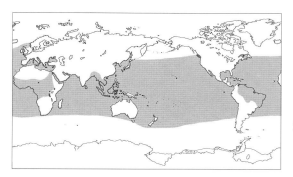

SUBORDER BALISTOIDEI

Members of this suborder have separate individual teeth protruding from the jaw bones. Spikefishes (family Triacanthodidae) are reddish-pink in color, and are found in the deep waters of the continental shelves and upper slopes of the Atlantic and Indo-west central Pacific, usually below 100 meters (about 330 feet). They are the most primitive members of the order and are a good example of the variety of often strange and

Kevin Deacon/AUSCAPE International

◀ *A titan triggerfish Balistoides
viridescens of the Indo-Pacific region.
Among the most garishly colored and
patterned of all coral reef fishes, the
triggerfishes are named for the long,
strong, first spine of their dorsal fin that
can be locked into an erect position by
the action of the much shorter spine
behind it.*

Peter Scoones/Planet Earth Pictures

▲ *A Red Sea triggerfish* Pseudobalistes
fuscus *positions itself to perform its special
party trick. Several triggerfish species can
defeat the spiny defenses of the sea
urchin by blowing a stream of water at it,
which overturns the sea urchin to expose
its vulnerable underside.*

weird forms that are found in most of its families.
For example, the species of *Halimochirurgus* and
Macrorhamphosodes have greatly extended tube-like
snouts, which in the latter genus become
increasingly twisted to one side or the other as the
individuals grow larger. Such asymmetry is rare
among vertebrates, although an even more striking
example occurs in the flatfishes. The species of
Macrorhamphosodes have an equally unusual
feeding mechanism; spoon-shaped teeth in the
lower jaw lean against a fleshy pad in the upper
jaw, allowing scales to be scraped off other fishes.

Triplespines

Triplespines (family Triacanthidae) are the closest
relatives of the spikefishes but have a slender and
powerful caudal (tail) peduncle and caudal fin for
strong swimming along the bottom of their
shallow water habitats in the tropical Indo-west

Pacific. They are silvery gray with green and yellow
markings. They use their heavy incisor-like teeth
to feed on hard-shelled invertebrates, such as
mollusks and crustaceans.

Triggerfishes and filefishes

Triggerfishes (family Balistidae) and their more
specialized relatives, the filefishes or leatherjackets
(family Monacanthidae), are common in shallow
tropical and temperate seas throughout the world.
Both families are unique in having a ball and
socket mechanism by which the second spine in
the dorsal fin can lock the large first spine in an
erected position (hence the name "trigger"). Also
restricted to these families is a shaft-like pelvic
bone at the end of which is a rudimentary pelvic
fin covered with a sheath of specialized articulated
scales. The skin between the pelvis and the anus
can be stretched and the pelvis can be rotated

downward to flare out a loose piece of skin (dewlap). This increases the apparent size of the body so as to discourage predators. Should that not be effective, these fishes can dart into a crevice or hole in a coral or rocky reef and erect both the dorsal spine and the pelvis, securely wedging themselves into their shelter. A predator would have difficulty dislodging such a potential meal, while their thick or prickly scales give them protection from bites while in their refuge.

Triggerfishes are often brightly colored and strikingly patterned. They use their massive incisor-like teeth to feed on mollusks, echinoderms, and crustaceans. They can use sources of food unavailable to other fishes, such as spiny sea urchins, which they turn over with their snouts to expose the unprotected underside.

Filefishes are more diversified than triggerfishes in body shape, habitat, and way of life. Some species of *Rudarius* are as small as 2 centimeters (less than 1 inch) at maturity, while *Aluterus scriptus* reaches a length of about 1 meter (3¼ feet). Many filefishes are inhabitants of coral reefs, but species of *Paramonacanthus*, *Thamnaconus*, and *Stephanolepis* occur on sandy and muddy coastal flats with seagrasses and rocky reefs. Filefishes have more delicate incisor-like teeth than do triggerfishes and usually specialize in feeding on small invertebrates. Medium-to-large-sized filefishes, such as species of *Stephanolepis*, *Thamnaconus*, and *Aluterus* are of commercial importance in Asia and other tropical areas. Most filefishes have colorations and body shapes or ornamentations that help them blend into their surroundings and so avoid the attention of predators. One filefish, *Paraluteres prionurus*, mimics the shape and coloration of a poisonous pufferfish, *Canthigaster valentini* (see below).

Boxfishes

Boxfishes or trunkfishes are a remarkable group of tetraodontiforms that have evolved a protective covering of enlarged, thickened, and sutured scale plates as complete as any of the now extinct armored fishes of the Paleozoic era (more than 245 million years ago). There are two families in this group, a more primitive one (family Aracanidae) that lives in the waters of the Indo-west Pacific at depths of over 200 meters (660 feet), and another (family Ostraciidae) that is found in shallow temperate and tropical waters, usually at depths of less than 50 meters (165 feet), throughout the world, often around coral reefs. The hard covering or shell that encloses most of their body makes adult boxfishes relatively

Kevin Deacon/AUSCAPE International

▲ *Among the smallest of all the triggerfish tribe, a minute filefish Rudarius minutus is superbly camouflaged amidst soft coral.*

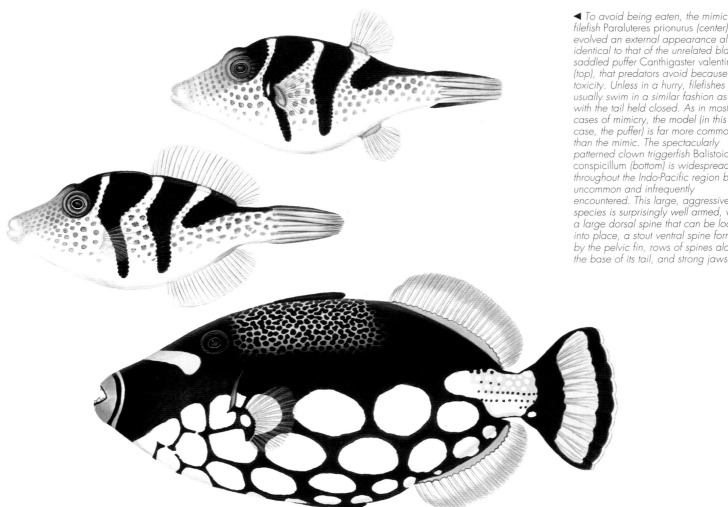

◄ *To avoid being eaten, the mimic filefish Paraluteres prionurus (center) has evolved an external appearance almost identical to that of the unrelated black-saddled puffer Canthigaster valentini (top), that predators avoid because of its toxicity. Unless in a hurry, filefishes usually swim in a similar fashion as well, with the tail held closed. As in most cases of mimicry, the model (in this case, the puffer) is far more common than the mimic. The spectacularly patterned clown triggerfish Balistoides conspicillum (bottom) is widespread throughout the Indo-Pacific region but is uncommon and infrequently encountered. This large, aggressive species is surprisingly well armed, with a large dorsal spine that can be locked into place, a stout ventral spine formed by the pelvic fin, rows of spines along the base of its tail, and strong jaws.*

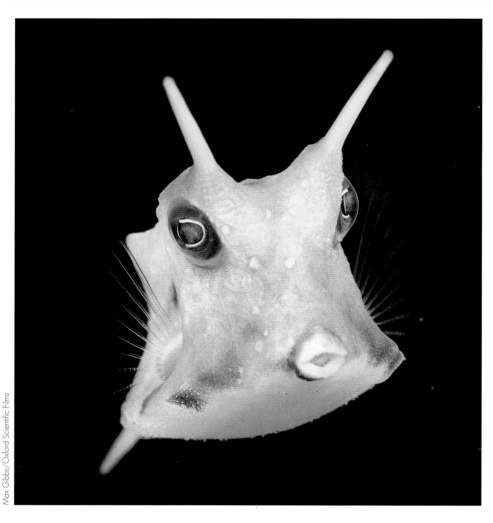

▲ *Sometimes reaching a length of 50 centimeters (20 inches), the long horn cowfish* Lactoria cornuta *of Australia's Great Barrier Reef is very common in deeper and larger coral pools. The caudal fin grows throughout life until it may equal the length of the body.*

composed of two minute spines, is not necessarily found on all individuals of the species. The pelvis, however, is a long shaft that can be rotated downward to flare a huge hanging piece of expansible skin bearing an eye-like spot, black surrounded by white on a yellowish-tan body. This probably serves the same function as the dewlap in balistids and monacanthids. When the apparent size of the body is suddenly increased, a potential predator may have second thoughts about trying to eat it, or at least pause in its pursuit and give its prey a better chance to escape. The pursefish is found in the deep waters of the Indo-west Pacific and is only rarely collected.

Pufferfishes

Pufferfishes (family Tetraodontidae) are the group of tetraodontoids with the most species. They are characterized by beak-like jaws made up of four pieces (one to each side of the midline in both the upper and lower jaws), no pelvis, and no dorsal fin spines. They are found in shallow temperate and tropical seas worldwide, in habitats ranging from coral and rocky reefs to seagrass beds and estuaries, with some species also found in waters greater than 100 meters (about 330 feet) in depth. A few species are found in fresh water, such as *Tetraodon* in the rivers and lakes of tropical Africa, Asia, and New Guinea, and *Colomesus* in the Amazonian drainages of South America. Most pufferfishes are small to medium in size, but some species of *Takifugu* and *Lagocephalus* attain lengths of 70 to 100 centimeters (28 to 39 inches).

Pufferfishes, too, change their body shape when under threat. They pump water (or air) into a distensible sac in the stomach, greatly increasing the size of their bodies. Since their bodies are usually covered by modified scales with spines that protrude from the skin when the body is inflated, enlarged pufferfishes become prickly objects that would be difficult for most predators to swallow. While they often have complex color patterns, most pufferfishes are relatively drab.

Some pufferfishes are highly poisonous, especially those in the Indo-Pacific. The poison, tetrodotoxin, is concentrated in the liver, ovaries, and gut, but it is also found in the other organs, in the skin, and even in the blood. It is a powerful neurotoxin, much stronger than cyanide, and is responsible for the death of a number of people each year in the Orient. Although parts of the viscera of certain species are poisonous, many species have highly palatable flesh and in eastern Asia, especially in Japan and Korea, pufferfishes are commercially important. Connoisseurs enjoy eating the flesh and testes of many species of *Takifugu*, which are highly esteemed and served in specially licensed restaurants. However, some species of *Lagocephalus* and *Takifugu* are

immune to predation from all but the largest of fishes. However, the small juvenile stages of boxfishes live in the open seas and are preyed upon heavily by billfishes and tunas. At least some boxfishes further protect themselves by secreting a poisonous substance (ostracitoxin) from their skin into the surrounding water.

SUBORDER TETRAODONTOIDEI

Members of the remaining families of tetraodontiforms are placed in the suborder Tetraodontoidei and have the teeth incorporated within the jaw bones to form a parrot-like beak.

Pursefishes

Triodon macropterus is the only species of pursefish (family Triodontidae) that is alive today. However, like most other groups of tetraodontiforms, they have a fossil record that extends back to the beginning of the Eocene epoch (about 55 million years ago). The beak in *Triodon* differs from that in other tetraodontoids in having three pieces; two in the upper jaw (one to each side of the midline) and a single piece fused across the middle in the lower jaw. *Triodon* also differs from the other tetraodontoids in having both a pelvis and a rudimentary spiny dorsal fin. This small structure,

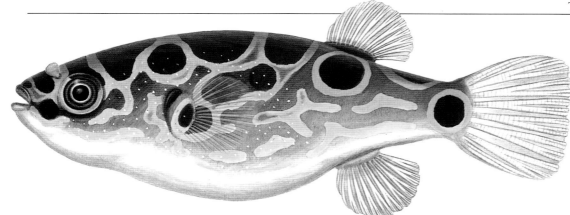

◄ *Unlike most members of its order, the figure-eight puffer* Tetraodon biocellatus *of Southeast Asia thrives in fresh water. Like other puffers it inflates its body when threatened, confusing predators and making it very difficult to handle.*

particularly dangerous, not only because the flesh is toxic but also because they look like other species of those genera that are delectable. Captain James Cook almost died in 1774 at New Caledonia during the second of his three great voyages of Pacific exploration when he indulged in just a small taste of the liver and ovaries of a specimen of *Lagocephalus sceleratus*. This same specimen became the basis for the original description of the species in 1789 in the thirteenth edition of Linnaeus's *Systema Naturae*. It has only recently been discovered that a pufferfish can also secrete its toxin into the surrounding water when frightened by a potential predator, halting any further advance of the predator.

Many pufferfishes are group spawners. Huge schools of some species of *Takifugu* arrive to spawn on Japanese beaches from May to July. *Takifugu niphobles* is the most striking example. One to five days after each full and new moon during the reproductive season, thousands of individuals rush up on to pebbly beaches on rising tides to spawn.

Porcupinefishes

Porcupinefishes (family Diodontidae) are similar to pufferfishes in many respects. However, the spines on the body are much larger and stronger, and the tooth plates are fused into a single piece across the middle in both the upper and lower jaws. They are slow swimmers of medium to large size and are mostly found in shallow tropical and temperate coastal waters throughout the world. A few species of *Diodon* inhabit the open seas and sometimes occur in aggregations of thousands of young individuals.

Porcupinefishes, like pufferfishes, can inflate their bodies. When the adults enlarge themselves into spiny spheres there are very few predators large enough to swallow them. However, just like some of the juvenile boxfishes that live on the open seas, juvenile porcupinefishes are an important source of food for large oceanic fishes like tunas, dolphins, and marlins.

Ocean sunfishes

Ocean sunfishes (family Molidae), which are like porcupinefishes in having a single tooth plate in each jaw, are easily recognized by their unique body shape. The rear end of the drab body is abbreviated; there is no caudal fin and the dorsal and anal fins are placed at the rear to take its place. Two of the three species of ocean sunfishes reach an enormous size of up to 3 meters (almost 10 feet) in length and more than 2 tonnes in weight. They are found in temperate and tropical seas, usually far offshore in oceanic waters. Even though ocean sunfishes can become very large, they feed mostly on soft-bodied invertebrates like jellyfishes, salps, and comb jellies, supplemented by small crustaceans and young fishes. Little is known about the biology of these large and highly modified fishes. They are often sighted slowly moving along at the ocean surface nibbling at drifts of jellyfishes, or lying quietly on their sides. A ripe female may produce 300 million tiny buoyant eggs.

KEIICHI MATSUURA & JAMES C. TYLER

▼ *Like pufferfish, the extraordinary porcupinefish can pump its body up with water when alarmed to present to any potential predator a daunting spiny sphere. Several species are common in rock pools and coral reefs in tropical and warm temperate regions around the world.*

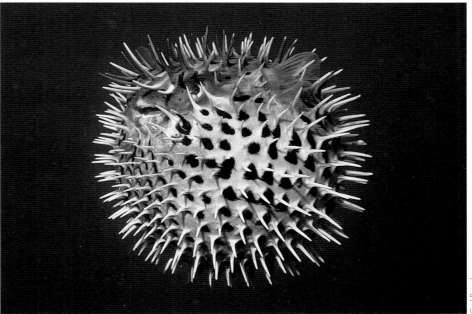

Carl Roessler

FURTHER READING

• Introducing Fishes (p. 14)

Bond, C.E., 1996. *Biology of Fishes* (2nd edn). Saunders College Publishing, Philadelphia. 750 pp.

Helfman, G.S., B.B. Collette & D.E. Facey, 1997. *The Diversity of Fishes*. Blackwell Science, Malden, USA. xii + 528 pp.

Jobling, M., 1995. *Environmental Biology of Fishes*. Chapman & Hall, London. 455 pp.

Marshall, N.B., 1971. *Explorations in the Life of Fishes*. Harvard University Press, Cambridge, USA. 204 pp.

Moyle, P.B. & J.J. Cech, Jr., 1996. *Fishes: An Introduction to Ichthyology* (3rd edn). Prentice Hall, Saddle River, USA. 590 pp.

• Classifying Fishes (p. 20)

Eschmeyer, W.N. (ed.), 1990. *The Genera of Recent Fishes*. California Academy of Sciences, San Francisco. vi + 697 pp.

Eschmeyer, W.N. (ed.), In Press, 1998. *Catalog of Fishes*. California Academy of Sciences, San Francisco. 3 vols.

Kocher, T.D. & C.A. Stepien (eds), 1997. *Molecular Systematics of Fishes*. Academic Press, San Diego. 314 pp.

Nelson, J.S., 1994. *Fishes of the World* (3rd edn). John Wiley & Sons, New York. xvii + 600 pp.

Stiassny, M.L.J., L.R. Parenti & G.D. Johnson (eds), 1996. *Interrelationships of Fishes*. Academic Press, San Diego. 496 pp.

• Fishes Through the Ages (p. 27)

Benton, M.J. (ed.), 1993. *The Fossil Record 2*. Chapman & Hall, London. 845 pp.

Long, J.A., 1995. *The Rise of Fishes, 500 Million Years of Evolution*. Johns Hopkins University Press, Baltimore. 223 pp.

Maisey, J.G., 1996. *Discovering Fossil Fishes*. Henry Holt & Co., New York. 223 pp.

• Habitats & Adaptations (p. 32)

Gon, O. & P.C. Heemstra (eds), 1990. *Fishes of the Southern Ocean*. J.L.B. Smith Institute of Ichthyology, Grahamstown, South Africa. xviii + 462 pp.

Marshall, N.B., 1979. *Developments in Deep-Sea Biology*. Blandford Press, Poole (England). 566 pp.

Matthews, W.J. & D.C. Heins (eds), 1987. *Community and Evolutionary Ecology of North American Stream Fishes*. University of Oklahoma Press, Norman, USA. 310 pp.

Moser, H.G., W.J. Richards, D.M. Cohen, M.P. Fahay, A.W. Kendall & S.L. Richardson (eds), 1984. *Ontogeny and Systematics of Fishes*. American Society of Ichthyologists and Herpetologists, Lawrence, USA. ix + 760 pp.

Sale, P.F. (ed.), 1991. *The Ecology of Fishes on Coral Reefs*. Academic Press, San Diego. xviii + 754 pp.

• Fish Behavior (p. 42)

Keenleyside, M. H. A. 1979. *Diversity and Adaptation in Fish Behavior*. Springer-Verlag, New York. 208 pp.

Krebs, J.R. & N.B. Davies, 1987. *An Introduction to Behavioural Ecology*. Blackwell Scientific Publications, London. 389 pp.

Parrish, J. K. & P. Turchin, 1997. 'Individual decisions, traffic rules, and emergent pattern in schooling fish', pp. 126-142 in J.K. Parrish & W.M. Hamner (eds), *Animal Groups in Three Dimensions*. Cambridge University Press, Cambridge.

Parrish, J.K., 1993. 'Comparison of the hunting behavior of four piscine predators attacking schooling prey', *Ethology*, 95: 233-246.

Pitcher, T.J., 1993. *Behaviour of Teleost Fishes* (2nd edn). Chapman & Hall, London. 715 pp.

• Endangered Species (p. 48)

Hudson, E. & G. Mace (eds), 1996. *Marine Fish and the IUCN Red List of Threatened Animals*. Report of the workshop held in collaboration with WWF and IUCN at the Zoological Society of London from April 29th to May 1st, 1996. IUCN, Gland, Switzerland. 26 pp.

IUCN, 1996. *1996 IUCN Red List of Threatened Animals*. IUCN, Gland, Switzerland. 368 pp. (Also on http://www.iucn.org/themes/ssc/redlist/redlist.htm)

Kottelat, M. & T. Whitten, 1996. *Freshwater Biodiversity in Asia, with Special Reference to Fish*. World Bank Technical Paper No. 343. 59 pp.

Rose, D.A., 1996. *An Overview of World Trade in Sharks and other Cartilaginous Fishes*. TRAFFIC International, Cambridge, UK. 106 pp.

Vincent, A.C.J., 1996. *The International Trade in Seahorses*. TRAFFIC International, Cambridge, UK. vii + 163 pp.

• Jawless Fishes (p. 56)

Hardisty, M.W. & I.C. Potter, 1971-1982. *The Biology of Lampreys*. Academic Press, London, Vol. 1 (1971), Vol. 2 (1972), Vol. 3 (1981), Vols 4 and 4B (1982).

Hardisty, M.W., 1979. *Biology of Cyclostomes*. Chapman & Hall, London. 428 pp.

Jørgensen, J., R. Weber, J. Lomboldt & H. Malte, 1998. *The Biology of Hagfishes*. Chapman & Hall, London. 278 pp.

Watson, J., (ed.), 1980. 'Proceedings of the Sea Lamprey International Symposium', *Canadian Journal of Fisheries and Aquatic Sciences*, 37 (11): 1585-2214.

• Sharks, Rays & Chimaeras (p. 60)

Last, P.R. & J.D. Stevens, 1994. *Sharks and Rays of Australia*. CSIRO Australia, Melbourne. 513 pp, 84 color plates.

Michael, S.W., 1993. *Reef Sharks and Rays of the World*. Sea Challengers, Monterey. 107 pp.

Springer, V.G. & J.P. Gold, 1989. *Sharks in Question*. Smithsonian Institution, Washington D.C. 187 pp.

Stevens, J.D., (ed.), 1987. *Sharks*. Golden Press, Sydney. 240 pp.

Taylor, L., (ed.), 1997. *The Nature Company Guide: Sharks and Rays*. Time-Life Books, USA. 288 pp.

• Lungfishes & Coelacanth (p. 70)

Bemis, W.E., W.W. Burggren & N.E. Kemp, 1987. *The Biology and Evolution of Lungfishes*. Alan R. Liss, New York. 383 pp.

Bruton, M.N., 1991. 'The ecology and conservation of the coelacanth *Latimeria chalumnae*', *Environmental Biology of Fishes*, 32: 313-339.

Fricke, H., 1988. 'Coelacanths, the fish that time forgot', *National Geographic Magazine*, 175: 824-838.

Thomson, K.S., 1991. *Living Fossil, the Story of the Coelacanth*. W.W. Norton & Co., New York. 252 pp.

• Bichirs & Their Allies (p. 75)

Birstein, V.J., J.R. Waldman & W.E. Bemis (eds), 1997. *Sturgeon Biodiversity and Conservation*. Kluwer Academic Publishers, Dordrecht, The Netherlands. 444 pp.

Lee, D.S. *et al.*, 1980. *Atlas of North American Freshwater Fishes*. North Carolina State Museum of Natural History, Raleigh, USA. 854 pp.

Schultze, H.P., & E. O. Wiley, 1984. 'The neotperygian *Amia* as a living fossil', pp 153-159 in N. Eldredge & S. M. Stanley (eds), *Living Fossils*. Springer Verlad, New York.

Wiley, E. O., and H.P. Schultze, 1984. 'Family Lepisosteidae (gars) as living fossils', pp 160-165 in N. Eldredge & S. M. Stanley (eds), *Living Fossils*. Springer Verlag, New York.

The Sturgeon Specialists Group: http://www.sturgeons.com

• Bonytongues & Their Allies (p. 80)

Braford, M.R., Jr., 1986. 'De gustibus non est disputandem: a spiral center for taste in the brain of a teleost fish, *Heterotis niloticus*', *Science*, 232: 489-491.

Greenwood, P. H., R.S. Miles & C. Patterson (eds), *Interrelationships of Fishes*. Academic Press, London. 536 pp.

Li, G.Q. & M.V.H. Wilson, 1996. 'Phylogeny of Osteoglossomorpha', pp. 163-174 in M.L.J. Stiassny, L.R. Parenti & G.D. Johnson (eds), *Interrelationships of Fishes*. Academic Press, San Diego.

Lissman, H.W., 1963. 'Electric location by fishes', *Scientific American*, 208: 50-59.

Roberts, T.R., 1975. 'Geographical distribution of African freshwater fishes', *Zoological Journal of the Linnean Society*, 57: 249-319.

• Eels & Their Allies (p. 85)

Böhlke, E.B. (ed.), 1989. 'Orders Anguilliformes and Saccopharyngiformes; Leptocephali', *Fishes of the western North Atlantic*. Part 9, vols. 1-2. Sears Foundation for Marine Research, New Haven. 1055 pp.

Forey, P.L., D.T.J. Littlewood, P. Ritchie & A. Meyer, 1996. 'Interrelationships of elopomorph fishes', pp. 175-191 in M.L.J. Stiassny, L.R. Parenti & G.D. Johnson (eds). *Interrelationships of Fishes*. Academic Press, San Diego.

McCosker, J.E., 1977. 'The osteology, classification, and relationships of the eel family Ophichthidae', *Proceedings of the California Academy of Science*, Ser. 4, 41: 1-123.

Tesch, F.W. & P.H. Greenwood, 1977. *The Eel, Biology and Management of Anguillid Eels* (2nd edn). Chapman & Hall, London. xiv+434 pp.

• Sardines & Their Allies (p. 91)

Lecointre, G., & G. Nelson, 1996. 'Clupeomorpha, sister-group of Ostariophysi', pp. 193-207 in M.L.J. Stiassny, L.R. Parenti & G.D. Johnson (eds), *Interrelationships of Fishes*. Academic Press, San Diego.

Siebert, D.J., 1997. 'Notes on the anatomy and relationships of *Sundasalanx* Roberts (Teleostei, Clupeidae), with descriptions of four new species from Borneo', *Bulletin of the Natural History Museum*, London (Zoology), 63(1): 13-26.

Whitehead, P.J.P., 1985. 'Clupeoid Fishes of the World (Suborder Clupeoidei)', *FAO Fisheries Synopsis*, No. 125, vol. 7, part 1. FAO, Rome. x+303 pp.

Whitehead, P.J.P., G.J. Nelson & T. Wongratana, 1988. 'Clupeoid Fishes of the World (Suborder Clupeoidei)', *FAO Fisheries Synopsis*, No. 125, vol. 7, part 2. FAO, Rome. 274 pp.

• Carps & Their Allies (p. 96)

Banister, K.E., 1973. 'A revision of the large *Barbus* (Pisces, Cyprinidae) of East and Central Africa', *Bulletin of the British Museum of Natural History (Zoology)*, 26(1): 1-148.

Banister, K.E. & G.J. Howes, 1985. 'Ostariophysi', pp. 70-83 in K.E. Banister & A.C. Campbell (eds), *Encyclopaedia of Underwater Life*. Unwin Animal Library, Vol IV. George Allen & Unwin, London.

Rainboth, W.J., 1996. *Fishes of the Cambodian Mekong*. FAO, Rome. 265 pp.

Winfield, J.J. & J.S. Nelson (eds), 1991. *Cyprinid Fishes: Systematics, Biology and Exploitation*. Chapman & Hall, London. 667 pp.

• Characins & Their Allies (p. 101)

Orti, G. & A. Meyer, 1997. 'The radiation of characiform fishes and the limits of resolution of mitochondrial DNA sequences', *Systematic Biology*, 46(1): 75-100.

Orti, G., 1997. 'Radiation of characiform fishes: evidence from mitochondrial and nuclear DNA sequences', pp. 219-314 in T.D. Kocher & C.A. Stepien (eds), *Molecular Systematics of Fishes*. Academic Press, San Diego.

Vari, R.P., In Press, 1998. 'Higher level phylogenetic concepts within Characiformes (Ostariophysi), an historical overview', in L.R. Malabarba, R.E. Reis & R.P. Vari (eds), *Phylogeny and Classification of Neotropical Fishes*. Porto Alegre, Brazil.

• Catfishes & Knifefishes (p. 106)

Burgess, W., 1989. *An Atlas of Freshwater and Marine Catfishes*. TFH Publications, Neptune City (New Jersey). 784 pp.

Ferraris, C., 1991. *Catfish in the Aquarium*. Tetra Press, Morris Plains, New Jersey. 199 pp.

Hieronimus, H., 1989. *Welse, Biologie und Haltung in der aquaristischen Praxis*. Verlag Eugen Ulmer, Stuttgart. 256 pp.

Moller, P., 1995. *Electric Fishes, History and Behavior*. Chapman & Hall, London. 584 pp.

• Salmons & Their Allies (p. 113)

McDowall, R.M., 1990. *New Zealand Freshwater Fishes: a Natural History and Guide*. Heinemann Reed, Auckland. 553 pp.

Pusey, B.J., 1990. 'Seasonality, aestivation and the life history of the salamanderfish, *Lepidogalaxias salamandroides* (Pisces: Lepidogalaxiidae)', *Environmental Biology of Fishes*, 29: 15-26.

Scott, W.B. & E.J. Crossman, 1973. *Freshwater Fishes of Canada*. Bulletin of the Fisheries Research Board of Canada, No. 184. 966 pp.

Smith, M.M. & P.C. Heemstra (eds), 1986. *Smiths' Sea Fishes*. Macmillan South Africa, Johannesburg. 1047 pp.

• Dragonfishes & Their Allies (p. 119)

Fink, W.L., 1985. 'A phylogenetic analysis of the family Stomiidae (Teleostei, Stomiiformes)', *Miscellaneous Publications of the Museum of Zoology, University of Michigan*, No. 171. 127 pp.

Harold, A.S. & S.H. Weitzman, 1997. 'Interrelationships of stomiid fishes', pp. 333-353 in M.L.J. Stiassny, L.R. Parenti & G.D. Johnson (eds), *The Interrelationships of Fishes*. Academic Press, San Diego.

Miya M. & M. Nishida, 1996. 'Molecular phylogenetic perspective on the evolution of the deep-sea fish genus *Cyclothone* (Stomiiformes: Gonostomatidae)', *Ichthyological Research*, 43(4): 375-398.

Sutton T.T. & T.L. Hopkins, 1966. 'Species composition, abundance, and vertical distribution of the stomiid (Pisces: Stomiiformes) fish assemblage of the Gulf of Mexico', Bulletin of Marine Science, 59(3): 530-542.

• Lizardfishes & Their Allies (p. 123)

Anderson, W.D. *et al.* (eds), 1966. 'Order Iniomi, Order Lyomeri', *Fishes of the western North Atlantic*. Part 5. Sears Foundation for Marine Research, New Haven. xv + 647 pp.

Baldwin, C. & G.D. Johnson, 1996. 'Interrelationships of Aulopiformes', pp. 355-404 in M.L.J. Stiassny, L.R. Parenti & G.D. Johnson (eds), *Interrelationships of Fishes*. Academic Press, New York.

Johnson, R.K., 1982. 'Fishes of the families Evermannellidae and Scopelarchidae: systematics, morphology, interrelationships, and zoogeography', Fieldiana Zoology, (n.s.) 12: 1-252.

• Lanternfishes (p. 127)

Gjøsaeter, J. & K. Kawaguchi, 1980. 'A Review of the World Resources of Mesopelagic Fish', *FAO Fisheries Technical Paper*, No. 193, 151 pp.

Hulley, P.A., 1992. 'Upper-slope distributions of oceanic lanternfishes (family: Myctophidae)', *Marine Biology*, 114: 365-383.

Karnella, C., 1987. 'Family Myctophidae, lanternfishes', pp. 51-168 in R.H. Gibbs, Jr. & W.H. Krueger (eds), *Biology of Midwater Fishes of the Bermuda Ocean Acre*. Smithsonian Contributions to Zoology, No. 452.

Stiassny, M.L.J., 1996. 'Basal ctenosquamate relationships and interrelationships of the myctophiform (scopelomorph) fishes', pp 405-426 in M.L.J. Stiassny, L.R. Parenti, & G.D. Johnson (eds), *Interrelationships of Fishes*. Academic Press, San Diego.

• Troutperches & Their Allies (p. 129)

Boltz, J.M. & J.R. Stauffer, 1993. 'Systematics of *Aphredoderus sayanus* (Teleostei: Aphredoderidae)', *Copeia*, 1993(1): 81-98.

Lee, D.S., *et al.*, 1980. *Atlas of North American Freshwater Fishes*. North Carolina State Museum of Natural History, Raleigh, USA. 854 pp.

Mayden, R.L. (ed.), 1992. *Systematics, Historical Ecology and North American Freshwater Fishes*. Stanford University Press, Stanford. xxi + 969 pp.

Rosen, D.E., 1985. 'An essay on euteleostean classification', *American Museum of Natural History Novitates*, No. 2827, 57 pp.

• Codfishes & Their Allies (p. 130)

Cohen, D.M., (ed.), 1989. 'Papers on the systematics of gadiform fishes', *Natural History Museum of Los Angeles County Science Series*, No. 32, 262 pp.

Cohen, D.M., T. Inada, T. Iwamoto & N. Scialabba, 1990. 'Gadiform fishes of the world', *FAO Fisheries Synopsis*, No. 125, vol. 10. FAO, Rome. x + 442 pp.
Markle, D.F., 1982. 'Identification of larval and juvenile Canadian Atlantic gadoids with comments on systematics of gadid subfamilies', *Canadian Journal of Zoology*, 60: 3420-3438.
Marshall, N.B. & D.M. Cohen, 1973. 'Order Anacanthini (Gadiformes)', pp. 479-495 in D.M. Cohen (ed.), *Fishes of the western North Atlantic*, No. 1, part 6. Memoirs of the Sears Foundation for Marine Research, New Haven.

• Cuskeels & Their Allies (p. 133)

Cohen, D.M. & J.G. Nielsen, 1978. 'Guide to the identification of genera of the fish order Ophidiiformes with a tentative classification of the order', *NOAA Technical Report, National Marine Fisheries Service Circular*, 417, 72 pp.
Markle, D.F. & J.E. Olney, 1990. 'Systematics of the pearlfishes (Pisces: Carapidae)', *Bulletin of Marine Science*, 47(2): 269-410.

• Toadfishes (p. 135)

Bachand, R.G., 1981. 'The oyster toadfish: a voice in Long Island Sound', *Sea Frontiers*, 27(3): 179-183.
Collette, B.B. & J.L. Russo, 1981. 'A revision of the scaly toadfishes, genus *Batrachoides*, with descriptions of two new species from the Eastern Pacific', *Bulletin of Marine Science*, 31: 197-233.
Gudger, E.W., 1910. 'Habits and life history of the toadfish *(Opsanus tau)*', *Bulletin of the United States Bureau of Fisheries*, 28: 1071-1109.
Hutchins, J.B., 1976. 'A revision of the Australian frogfishes (Batrachoididae)', *Records of the Western Australian Museum*, 4: 3-43.

• Anglerfishes (p. 137)

Bertelsen, E., 1951. 'The ceratioid fishes, ontogeny, taxonomy, distribution, and biology', *Dana Report*, 39, 281 pp.
Pietsch, T.W., 1976. 'Dimorphism, parasitism and sex: reproductive strategies among deep-sea anglerfishes', *Copeia*, 1976: 781-793.
Pietsch, T.W. & D.B. Grobecker, 1987. *Frogfishes of theWorld: Systematics, Zoogeography, and Behavioral Ecology*. Stanford University Press, Stanford. xxii + 420 pp.
Pietsch, T.W. & D.B. Grobecker, 1990. 'Frogfishes: masters of aggressive mimicry, these voracious carnivores can gulp prey faster than any other vertebrate predator', *Scientific American*, 262(6): 96-103, June 1990.

• Clingfishes & Their Allies (p. 142)

Briggs, J.C., 1955. 'A monograph of the clingfishes (order Xenoptergyii)', *Stanford Ichthological Bulletin*, No. 6, 224 pp.
Fricke, R., 1983. *Revision of the Indo-Pacific Genera and Species of the Dragonet Family Callionymidae (Teleostei)*. Cramer, Braunschweig, Germany. 774 pp.
Hutchins, J.B., 1991. 'Southern Australia's enigmatic clingfishes', *Australian Natural History*, 23(8): 626-633.
Shiogaki, M. & Dotsu, Y., 1983. 'Two new genera and two new species of clingfishes from Japan, with comments on head sensory canals of the clingfishes', *Japanese Journal of Ichthyology*, 30: 111-121.

• Flyingfishes & Their Allies (p. 144)

Breder, C.M. Jr., 1938. 'A contribution to the life histories of Atlantic Ocean flying fishes', *Bulletin of the Bingham Oceanographic Collection*, 6(5): 1-126.
Collette, B.B., 1974. 'The garfishes (Hemiramphidae) of Australia and New Zealand', *Records of the Australian Museum*, 29(2): 11-105.
Collette, B.B., *et al.* 1984. 'Beloniformes: development and relationships', pp. 334-354 in H.G. Moser *et al.* (eds), *Ontogeny and Systematics of Fishes*. American Society of Ichthyologists and Herpetologists, Lawrence, USA.
Parin, N.V., 1961.'The bases of classification of the flyingfishes (Exocoetidae) of the Pacific and Indian Oceans', *Trudy Institut Okeanologii*, 43: 92-183. (In Russian; English translation U.S. Bureau of Commercial Fisheries Ichthyology Laboratory Translation No. 67).

• Killifishes & Ricefishes (p. 148)

Huber, J.H., 1992. *Review of Rivulus, Ecobiogeography, Relationships*. Société Française d'Icthyologie, Paris. 572 p.
Parenti, L.R., 1981. 'A phylogenetic and biogeographic analysis of cyprinodontiform fishes (Teleostei, Atherinomorpha)', *Bulletin of the American Museum of Natural History*, 168(4): 335-557.
Scheel, J.J., 1968. *Rivulins of the Old World*. T.F.H. Publications, Neptune City, USA. 480 pp.
Wischnath, L., 1993. *Atlas of Livebearers of the World*. T.F.H. Publications, Neptune City, USA. 336 pp.

• Silversides & Their Allies (p. 153)

Allen, G.R., 1995. *Rainbowfishes*. Tetra Press, Germany. 180 pp.
Parenti, L.P., 1989. 'A phylogenetic revision of the phallostethid fishes (Atherinomorpha, Phallostethidae)', *Proceedings of the California Academy of Sciences*, 46: 243-277.
Parenti, L.P., 1993. 'Relationships of atherinomorph fishes', *Bulletin of Marine Science*, 52(1): 170-196.
Stiassny, M.L.J., 1990. 'Notes on the anatomy and relationships of the bedotiid fishes of Madagascar, with a taxonomic revision of the genus *Rheocles* (Atherinomorpha: Bedotiidae)', *American Museum Novitates*, 2979: 1-33.

• Oarfishes & Their Allies (p. 157)

Charter, S.R. & H.G. Moser, 1996. 'Lampridiformes', pp. 659-677 in H.G. Moser (ed.), *The early stages of fishes in the California Current region*. California Cooperative Oceanic Fisheries Investigations Atlas Number 33.
Olney, J.E., 1984. 'Lampridiformes: development and relationships', pp. 368-379 in H.G. Moser *et al.* (eds), *Ontogeny and Systematics of Fishes*. American Society of Ichthyology and Herpetology, Lawrence, USA.
Olney, J.E., G.D. Johnson & C.C. Baldwin, 1993. 'Phylogeny of lampridiform fishes', *Bulletin of Marine Science*, 52(1): 137-169.
Smith, M.M., 1986. 'Ateleopodidae', pp. 404-406 in M.M. Smith & P.C. Heemstra (eds), *Smiths' Sea Fishes*. Macmillan, Johannesburg.

• Squirrelfishes & Their Allies (p. 160)

Kotlyar, A.N., 1996. *Beryciform Fishes of the World*. VNIRO Publishing, Moscow. 368 pp. In Russian.
McCosker, J.E., 1977. 'Flashlight fishes', *Scientific American*, March 1977, 266(3): 106-114.
Merrett, N.R. & R.L. Haedrich, 1997. *Deep-sea Demersal Fish and Fisheries*. Chapman & Hall, London. xii + 282 pp.
Paxton, J.R., 1990. 'Whalefishes: little fish with big mouths', *Australian Natural History*, 23(5): 378-385.
Randall, J.E. & D.W. Greenfield, 1996. 'Revision of the Indo-Pacific holocentrid fishes of the genus *Myripristis*, with the description of three new species', *Indo-Pacific Fishes*, no. 25, 61 pp.

• Dories & Their Allies (p. 165)

Cuvier, G., 1829. *Histoire naturelle des Poissons*. Vol. 4. Strasbourg, Paris. 518 pp.
Last, P.R., E.O.G. Scott & F.H. Talbot, 1983. 'Order Zeiformes, Dories and allies', pp. 283-292 in *Fishes of Tasmania*. Tasmanian Fisheries Development Authority, Hobart.
Tighe, K.A. & M.J. Keene, 1984. 'Zeiformes: Development and relationships', pp. 393-398 in H.G. Moser *et al.* (eds), *Ontogeny and Systematics of Fishes*. American Society of Ichthyologists and Herpetologists, Lawrence, USA.
Wheeler, A., 1969. 'Dory and boar-fish, Zeomorphi', pp. 305-308 in *The Fishes of the British Isles and North-West Europe*. Macmillan, London.

• Pipefishes & Their Allies (p. 168)

Bell, M.A. & S.A. Foster (eds), 1994. *The Evolutionary Biology of the Threespine Stickleback*. Oxford University Press, Oxford. xii+571 pp.
Dawson, C.E., 1985. *Indo-Pacific Pipefishes (Red Sea to the Americas)*. Gulf Coast Research Laboratory, Ocean Springs, Mississippi. 230 pp.
Orr, J.W. & R.A. Fritzsche, 1993. 'Revision of the ghost pipefishes, family Solenostomidae (Teleostei: Syngnathoidei)', *Copeia*, 1993: 168-182.
Palsson, W. & T.W. Pietsch, 1989. 'Revision of the acanthopterygian fish family Pegasidae (order Gasterosteiformes)', *Indo-Pacific Fishes*, No. 18, 38 pp.
Vincent, A.C.J., 1996. *The International Trade in Seahorses*. TRAFFIC International, Cambridge. vii+163 pp.

• Swampeels & Their Allies (p. 173)

Graham, J.B., 1997. *Air-breathing Fishes. Evolution, Diversity, and Adaptation*. Academic Press, San Diego. 299 pp.
Lo Nostro, F.L. & G.A. Guerrero. 1996. 'Presence of primary and secondary males in a population of the protogynous *Synbranchus marmoratus*', *Journal of Fish Biology*, 49: 788-800.
Travers, R.A., 1984. 'A review of the Mastacembeloidei, a suborder of synbranchiform families. Part I: Anatomical descriptions; Part II: Phylogenetic analysis', *Bulletin of the British Museum (Natural History) Zoology series*, 46(1): 1-133; 47(2): 83-150.

• Scorpionfishes & Their Allies (p. 175)

Eschmeyer, W.N. & E.S. Herald, 1983. *A Field Guide to Pacific Coast Fishes of North America*. Houghton Mifflin, Boston. 336 pp.
Halstead, B.W, 1978. *Poisonous and Venomous Marine Animals of the World* (2nd edn). Darwin Press, New Jersey. 1043 pp.
Masuda, H., *et al.* (eds), 1988. *The Fishes of the Japanese Archipelago* (2nd edn). Tokai University Press, Tokyo. xix +456 pp. + 378 plates.
Smith, M.M. & P.C. Heemstra (eds), 1986. *Smiths' Sea Fishes*. Macmillan South Africa, Johannesburg. 1047 pp.

• Perches & Their Allies (p. 181)

Johnson, G.D., 1984. 'Percoidei: development and relationships', pp.464-498 in H.G. Moser *et al.* (eds), *Ontogeny and Systematics of Fishes*. American Society of Ichthyologists and Herpetologists, Lawrence, USA.
Johnson, G.D. & W.D. Anderson, Jr. (eds), 1993. 'Proceedings of the Symposium on Phylogeny of Percomorpha', *Bulletin of Marine Science*, No. 52(1). 626 pp.
Wheeler, A., 1975. *Fishes of the World, an Illustrated Dictionary*. Ferndale Editions, London. xiv + 366 pp.

• Groupers, Seabasses & Their Allies (p. 195)

Heemstra, P.C. &J.E. Randall, 1993. 'Groupers of the world', *FAO Fisheries Synopsis*, No. 125, vol. 16, vii+414 pp.
Randall, J.E. & P.C. Heemstra, 1991. 'Revision of Indo-Pacific groupers (Perciformes: Serranidae: Epinephelinae), with descriptions of five new species', *Indo-Pacific Fishes*, No. 20. 332 pp.
Randall, J.E. & L. Taylor, 1988. 'Review of the Indo-Pacific fishes of the serranid genus *Liopropoma*, with descriptions of seven new species', *Indo-Pacific Fishes*, No. 16. 47 pp.
Randall, J.E. & C.C. Baldwin, 1997. 'Revision of the serranid fishes of the subtribe Pseudogrammina, with descriptions of five new species', *Indo-Pacific Fishes*, No. 26. 56 pp.

• Cichlids (p. 200)

Fryer, G. & T.D. Iles, 1972. *The Cichlid Fishes of the Great Lakes of Africa. Their biology and evolution*. Oliver and Boyd, Edinburgh. 641pp.
Greenwood, P.H., 1974. *Cichlid Fishes of Lake Victoria, East Africa: the Biology and Evolution of a Species Flock*. Trustees of the British Museum (Natural History), London. 134pp.
Keenleyside, M.H.A., (ed.), 1991. *Cichlid Fishes: Behaviour, Ecology and Evolution*. Chapman Hall, London. 378pp.
Meyer, A., 1993. 'Phylogenetic relationships and evolutionary processes in East African cichlids', *Trends in Ecology and Evolution*, 8: 279-284.
Stiassny, M.L.J., 1993. 'What is a cichlid?', *Tropical Fish Hobbyist Magazine*, November, 1993: 141-146.

• Damselfishes (p. 205)

Allen, G.R., 1975. *Damselfishes of the South Seas*. T.F.H. Publications, Inc., Neptune City, New Jersey. 240 pp.
Allen, G.R., 1991. Damselfishes of the World. Mergus Verlag, Melle, Germany. 271 pp.
Fautin, D.G. & G.R. Allen, 1992. *Field Guide to Anemonefishes and their Host Sea Anemones*. Western Australian Museum, Perth. 160 pp.
Robertson, D.R. & B. Lassig, 1980. 'Spatial distribution patterns and coexistence of a group of territorial damselfishes from the Great Barrier Reef', *Bulletin of Marine Science*, 30: 177-203.
Wellington, G.M. & B.C. Victor, 1989. 'Planktonic larval duration of one hundred species of Pacific and Atlantic damselfish', *Marine Biology*, 101: 557-567.

• Wrasses & Parrotfishes (p. 209)

Bellwood, D.R. & J.H. Choat, 1990. 'A functional analysis of grazing in parrotfishes (family Scaridae): the ecological implications', *Environmental Biology of Fishes*, 28:189-214.
Choat, J.H., 1991. 'The biology of herbivorous fishes on coral reefs', pp. 120-155 in P.F. Sale (ed.), *The Ecology of Fishes on Coral Reefs*. Academic Press, San Diego.
Choat, J.H. & D.R. Bellwood, 1991. 'Reef fishes; their history and evolution', pp. 39-66 in P.F. Sale (ed.), *The Ecology of Fishes on Coral Reefs*. Academic Press, San Diego.
Kuiter, R.H., 1993. *Coastal Fishes of South-eastern Australia*. Univ. Hawaii Press, Honolulu. xxxi+437 pp.
Randall, J.E., G.R. Allen & R.C. Steene, 1990. *Fishes of the Great Barrier Reef and Coral Sea*. Crawford House Press, Bathurst, Australia. 507 pp.

• Blennies (p. 214)

Allen, G.R. & D.R. Robertson, 1994. *Fishes of the Tropical Eastern Pacific*. Crawford House Press, Bathurst, Australia. 332 pp.
Fricke, R., 1997. *Tripterygiid Fishes of the Western and Central Pacific*. Koeltz Scientific Books, Koenigstein, Germany. ix + 607 pp.
Nelson, J.S., 1994. *Fishes of the World* (3rd edn). John Wiley & Sons, New York. xvii + 600 pp.
Springer, V.G., 1993. 'Definition of the suborder Blennioidei and its included families (Pisces: Perciformes)', *Bulletin of Marine Science*, 52(1): 472-495.

• Gobies (p. 218)

Debelius, H., 1986. *Colorful Little Reef Fishes*. Meinders & Elftermann, Osnabrück, Germany. 160 pp.
Gomon, M.F., C.J.M. Glover & R.H. Kuiter (eds), 1994. *The Fishes of Australia's South Coast*. State Print, Adelaide. 992 pp.
Randall, J.E., G.R. Allen & R.C. Steene, 1990. *Fishes of the Great Barrier Reef and Coral Sea*. Crawford House Press, Bathurst, Australia. 507 pp.
Thresher, R.E., 1984. *Reproduction in Reef Fishes*. TFH Publications, Hong Kong. 399 pp.

• Flatfishes (p. 223)

Helfman, G.S., B.B. Collette, & D.E. Facey, 1997. *The Diversity of Fishes*. Blackwell Science, Malden, USA. xii + 528 pp.
Moser, H.G., *et al.* (eds), 1984. *Ontogeny and Systematics of Fishes*. American Society of Ichthyologists and Herpetologists, Lawrence, USA. ix + 760 pp.
Policansky, D., 1982. 'Asymmetry of flounders', *Scientific American*, 246: 116-122.

• Triggerfishes & Their Allies (p. 227)

Matsuura, K., 1980. 'A revision of Japanese balistoid fishes. I. family Balistidae', *Bulletin of the National Science Museum*, series A, 6(1): 27-69.
Matsuura, K. & J.C. Tyler, 1997. 'Tetraodontiform fishes, mostly from deep waters, of New Caledonia', *Mémoirs du Muséum national d'Histoire naturelle*, 174: 173-208.
Tyler, J.C., 1980. *Osteology, Phylogeny, and Higher Classification of the Fishes of the Order Plectognathi (Tetraodontiformes)*. NOAA Technical Report, (US) National Marine Fisheries Service Circular, No. 434, Washington, D.C. xi+422 pp.

INDEX

ACKNOWLEDGMENTS

The editors and publishers would like to thank the following people for their assistance and support: Sylvie Abecassis; Jean-Paul Ferraro, Auscape International; Kate Etherington; Janet Marando; Kylie Mulquin; Margaret Olds; Selena Quintrell; Jo Rudd; Tracy Tucker; Natalie Vellis; and Jill Wayment.

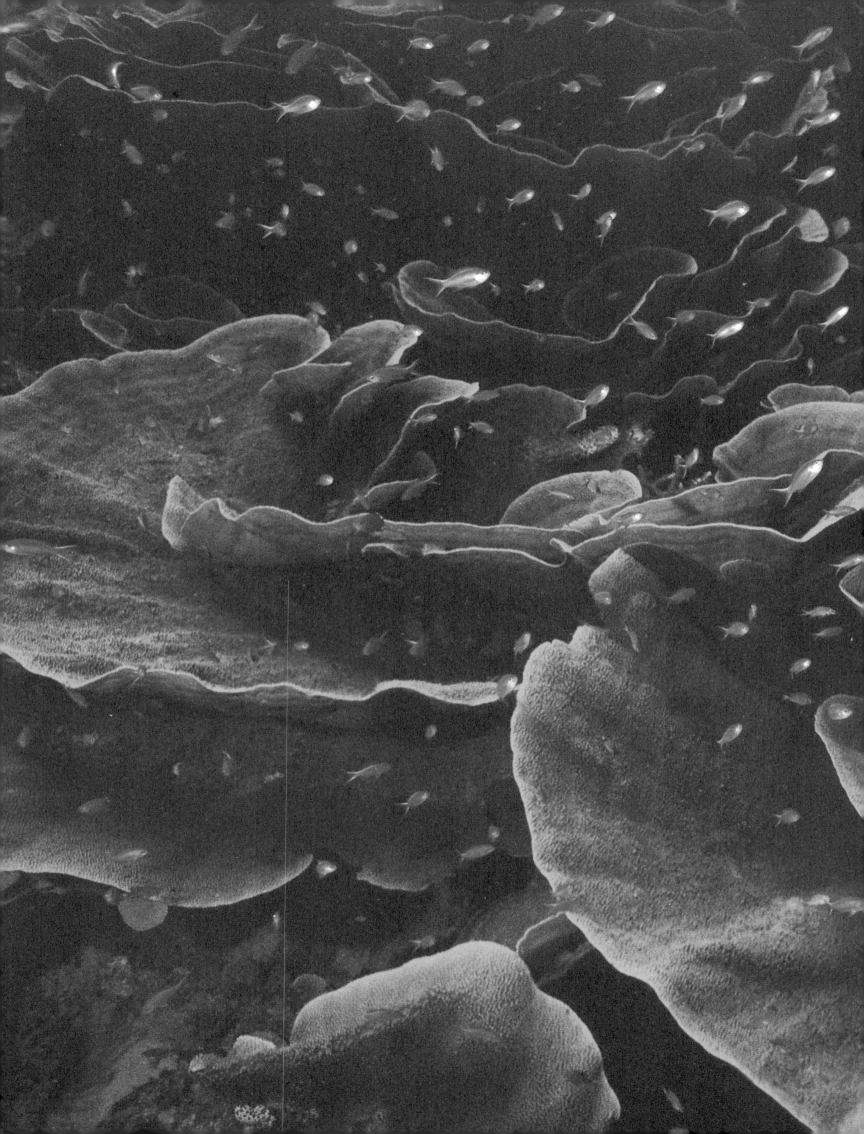